Cambridge Studies in Biological and Evolutionary Anthropology 52

Health Change in the Asia-Pacific Region

The Asia-Pacific region has seen great social, environmental and economic change across the past century, with great acceleration of change in the last 20 years or so, leading to dramatic changes in the health profiles of all populations represented in South East Asia, East Asia, Pacific Islands and the islands of Melanesia. This volume will consider recent evidence concerning prehistoric migration, and colonial, regional and global processes in the production of health change in the Asia-Pacific region. Notably, it will examine ways in which a health pattern dominated by undernutrition and infection has been displaced in many ways, and is being displaced elsewhere, by over-nutrition and the degenerative diseases associated with it. This book will present a cohesive view of the ways in which exchange relationships, economic modernization, migration and transnational linkages interact with changing rural subsistence ecologies to influence health patterns in this region.

RYUTARO OHTSUKA is President of the National Institute for Environmental Studies, Japan. Until 2005, he was Professor of Human Ecology in the University of Tokyo and conducted research among various populations in the Asia-Pacific region.

STANLEY J. ULIJASZEK is Professor of Human Ecology in the Institute of Social and Cultural Anthropology, University of Oxford. His research is focused on the biocultural ecology of human populations, past and present, and nutritional anthropology.

Cambridge Studies in Biological and Evolutionary Anthropology

Series editors

HUMAN ECOLOGY
C. G. Nicholas Mascie-Taylor, University of Cambridge
Michael A. Little, State University of New York, Binghamton
GENETICS
Kenneth M. Weiss, Pennsylvania State University
HUMAN EVOLUTION
Robert A. Foley, University of Cambridge
Nina G. Jablonski, California Academy of Science
PRIMATOLOGY
Karen B. Strier, University of Wisconsin, Madison

Health Change in the Asia-Pacific Region

Biocultural and Epidemiological Approaches

EDITED BY

RYUTARO OHTSUKA

National Institute for Environmental Studies, Tsukuba City, Japan

STANLEY J. ULIJASZEK

Institute of Social and Cultural Anthropology, University of Oxford, UK

CAMBRIDGE
UNIVERSITY PRESS

CAMBRIDGE UNIVERSITY PRESS
Cambridge, New York, Melbourne, Madrid, Cape Town, Singapore, São Paulo

Cambridge University Press
The Edinburgh Building, Cambridge CB2 2RU, UK

Published in the United States of America by Cambridge University Press,
New York

www.cambridge.org
Information on this title: www.cambridge.org/9780521837927

First published 2007

Printed in the United Kingdom at the University Press, Cambridge

A catalogue record for this publication is available from the British Library

Library of Congress Cataloguing in Publication data

Health change in the Asia-Pacific region : biocultural and epidemiological
approaches / edited by Ryutaro Ohtsuka and Stanley J. Ulijaszek.
    p.   ; cm. – (Cambridge studies in biological and evolutionary anthropology ; 52)
    Includes bibliographical references.
    ISBN-13: 978 0 521 83792 7 (hardback)
1. Health transition – Southeast Asia. 2. Health transition – East Asia. 3. Health
transition – Islands of the Pacific. 4. Nutrition – Southeast Asia. 5. Nutrition – East
Asia. 6. Nutrition – Islands of the Pacific. I. Ohtsuka, Ryutaro. II. Ulijaszek,
Stanley J. III. Series.
    [DNLM: 1. Health Transition – Asia, Southeastern. 2. Health Transition – Far
East. 3. Health Transition – Pacific Islands. 4. Anthropology, Cultural – Asia,
Southeastern. 5. Anthropology, Cultural – Far East. 6. Anthropology, Cultural –
Pacific Islands. 7. Nutritional Status – Asia, Southeastern. 8. Nutritional Status –
Far East. 9. Nutritional Status – Pacific Islands. 10. Socioeconomic Factors – Asia,
Southeastern. 11. Socioeconomic Factors – Far East. 12. Socioeconomic Factors –
Pacific Islands. WA 105 H4342 2007] I. Title. II. Series.

RA650.7.S68H43 2007
362.10959–dc22   2006036793

ISBN-13 978 0 521 83792 7 hardback

Contents

Contributors

Robert Attenborough
The School of Archaeology and Anthropology, A. D. Hope Building, The
Australian National University, Canberra, ACT 0200, Australia

Tony Blakely
Department of Public Health, Wellington School of Medicine and Health Sciences,
University of Otago, 23A Mein Street, Newtown, Wellington, New Zealand

Youngtae Cho
School of Public Health Seoul, National University Seoul, Korea

T. Elizabeth Durden
Department of Sociology and Anthropology, Bucknell University, Lewisburg, PA
17837

W. Parker Frisbie
Population Research Center and Department of Sociology, College of Liberal
Arts, University of Texas at Austin, 1 University Station A1700, Austin,
TX 78712-0543, USA

Robert A. Hummer
The University of Texas at Austin, Population Research Center, 1 University
Station G1800, Austin, TX 78712-0543, USA

Tsukasa Inaoka
Department of Environmental Sociology, Saga University, 1 Honjo-machi, Saga
City, Japan

Ember D. Keighley
International Health Institute, Brown University, Box G-B495, Providence, RI
02912, USA

Gary T. C. Ko
Hong Kong Alice Ho Miu Ling Nethersole Hospital, 11 Chuen On Road, Tai Po,
New Territory, Hong Kong

Uto'ofili A. Maga
Department of Health, Pago Pago, American Samoa 96799

Geoffrey C. Marks
Australian Centre for International and Tropical Health and Nutrition, University of Queensland, Level 4, Public Health Building, Herston Road, Herston, Brisbane, Queensland 4006, Australia

Yasuhiro Matsumura
National Institute of Health and Nutrition, Toyama 1-23-1 Shinjuku-ku, Tokyo 162-8636, Japan

Charles McCuddin
Tafuna Family Health Center, Department of Health, Pago Pago, American Samoa 9679

Stephen T. McGarvey
International Health Institute, Brown University, Box G-B497, Providence RI 02912, USA

Ryutaro Ohtsuka
National Institute for Environmental Studies, 16-2 Onogawa, Tsukuba City, Ibaraki 305-8506, Japan

Stephen Oppenheimer
Green College, Woodstock Road, Oxford OX2 6HG, UK

Christine Quested
Nutrition Centre, Ministry of Health, Government of Samoa, Apia, Samoa

Kazuhiro Suda
Hokkai-Gakuen University, Asahi-machi 4-1-40 Toyohira-ku, Sapporo 062-8605, Japan

Stanley J. Ulijaszek
Institute of Social and Cultural Anthropology, University of Oxford, 51 Banbury Road Oxford OX2 6PF, UK

Satupaitea Viali
Oceania University of Medicine Samoa, Apia Campus and Tupua Tamasese Meaole Hospital, Ministry of Health, Government of Samoa, Apia, Samoa

Alistair Woodward
School of Population Health, University of Auckland, Private Bag 92019, Auckland, New Zealand

Taro Yamauchi
Department of Human Ecology, School of International Health, Graduate School of Medicine, University of Tokyo, 7-3-1, Hongo, Bunkyo-ku, Tokyo 113-0033, Japan

Acknowledgements

The meeting which formed a basis for this volume took place in Tokyo, as part of the International Union of Anthropological and Ethnographic Sciences Congress, held in 2002. We thank the Japan Society for the Promotion of Science for helping to support this symposium financially. We also thank Robin Hide, Tony McMichael, Vicki Luker, Don Gardner and Donald Denoon for their comments and suggestions at various stages of this project.

1   *Health change in the Asia-Pacific region: disparate end-points?*

STANLEY J. ULIJASZEK AND RYUTARO OHTSUKA

Introduction

The Asia-Pacific region encompasses South East Asia, East Asia, Pacific Islands and the islands of Melanesia. In the present day, strong economic forces link it to the Pacific Rim nations of the United States, Australia and New Zealand. While epidemiologists have studied some of the relationships across geographical and population units within this region, there is thus far no formal consideration of health impacts of linkages within and across these units in historical and evolutionary contexts. While migrations across the region are known and common, these have both evolutionary and colonial histories. The nature and extent to which knowledge of population movements, past and recent, can impact on present-day human biology in this region has not been synthesized, despite having been considered separately by various authors and research groups. This volume considers recent evidence concerning prehistoric migration, and colonial, regional and global processes in the production of health in the Asia-Pacific region. Using their own research findings and/or by synthesizing those of others, the contributors to this volume describe health change in various populations in relation to their biological, cultural and/or socioeconomic attributes at various scales of time.

This region, consisting of the southeastern frontier of the Eurasian continent and the vast South Pacific, was the geographical locale of the first crossing of wide seas and oceans by human groups. A consequence of this was the adaptation of such migrant groups to a variety of novel environments. Many Pacific population maritime range expansions are likely to have taken place throughout the late Pleistocene, earlier than anywhere else in the world. Between the Pleistocene and the Holocene, various migrations, subsistence introductions and human biological changes took place. However, these took place at rates much slower than the rates of those introduced by European contact and the subsequent subjection of local populations to colonial regimes. Great variation in

Health Change in the Asia-Pacific Region: Biocultural and Epidemiological Approaches, ed. Ryutaro Ohtsuka and Stanley J. Ulijaszek. Published by Cambridge University Press.
© Cambridge University Press 2007.

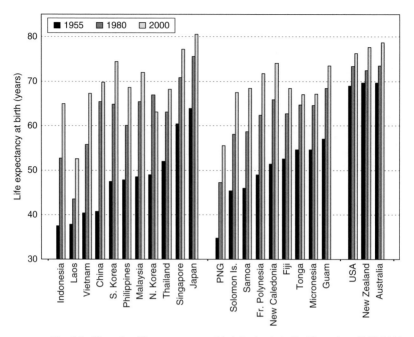

Fig. 1.1. Changing life expectancy at birth in the Asia-Pacific region, 1955–2000 (data from United Nations University 2006).

exposure to, and development of, new political and economic structures across the region has led to varying health profiles across and within populations. Throughout the past century, this region has seen great social, cultural, economic and environmental changes, triggering hasty health transitions in many populations, with this trend accelerating in the past few decades. The forces driving this change now include the following: the emergence of Asian economies as significant global influences; political change; globalization of trade; the penetration of the world food system to all parts of the region; adoption of Westernized foods and dietary habits; widespread dissemination of primary health care; increased adoption of health-conscious behaviours; and increasing urbanization and migration of populations.

Young-child mortality has declined considerably in the Asia-Pacific region since the 1950s. One consequence of this has been increases in life expectancy at birth (LEB) across the region between 1955 and 1980, and further increases between 1980 and 2000 in all nations except North Korea (Fig. 1.1). Almost 30 years have been added to LEB in the last 130 to 150 years in New Zealand, Australia and the United States, most of this increase having taken place before 1980. Life expectancy at birth in Singapore and Japan has increased similarly across the twentieth century, while increases of similar proportion have taken place since the 1950s in Indonesia, Vietnam, China and South Korea. Increases in LEB of between 20 and 30 years across the period 1955 to 2000 have taken place in the Philippines, Malaysia, Papua New Guinea (PNG), the Solomon Islands, Samoa, French Polynesia and New Caledonia. In the 1950s, the United States, New Zealand and Australia had the highest LEB in the region. By the year 2000, they had been overtaken by Japan, whose LEB came to exceed 80 years. Furthermore, South Korea, Malaysia, Singapore, French Polynesia, New Caledonia and Guam had joined the United States, New Zealand and Australia with LEBs exceeding 70 years.

A potential brake on these dramatic increases in human longevity is the rise in the burden of non-infectious diseases across the region since the 1980s, as well as the persistence and emergence of infectious diseases in some nations, including PNG. Evidence for decline in LEB after generations of increase comes from various nations where there are significant increases in infectious disease mortality, as in Zimbabwe, South Africa, Lesotho, Swaziland, Namibia, Zambia and Botswana (HIV/AIDS and tuberculosis) and increased mortality associated with severe undernutrition (North Korea). It has been suggested that LEB in the United States may decline by up to five years across the next two or three decades unless the rising rates of obesity are somehow controlled (Olshansky *et al.* 2005). In the United States, obesity rates (as defined by body mass index (BMI) exceeding $30 \, kg/m^2$) currently stand at 26% and 32% of males and females respectively, with rates of increase of 0.4% per year in both sexes (Nishida and Mucavele 2005). If an obesity rate of more than 30% can be taken as a level beyond which serious reversals in LEB due to chronic disease mortality may occur, various Pacific Island nations, including Nauru, the Cook Islands, French Polynesia, Tonga, Samoa, and American Samoa, exceeded this value several decades ago and do not appear to be undergoing a reversal in LEB yet. Obesity levels are rising in most nations where records are available (Nishida and Mucavele 2005), and some Pacific Islander populations are the most obese in the world. Indeed, the populations of Nauru, the Cook Islands, French Polynesia, Tonga, Samoa and American Samoa may be close to a possible ceiling on obesity prevalence and its associated chronic disease mortality. Given that it is unlikely that any nation outside of the Pacific has reached a ceiling in obesity prevalence, careful observation and understanding of obesity and chronic disease patterns in this region is therefore of much more than local interest.

While most Asian and Pacific Island nations experienced colonization by European nations in the nineteenth and twentieth centuries, the latter gained independence later. They also experienced slower economic growth, and, in Polynesia in particular, higher levels of out-migration to industrialized nations led to the formation of significant transnational communities by the end of the twentieth century. The most characteristic health outcomes of this rapid change among Pacific Islander populations are the extraordinarily high levels of obesity, non-insulin dependent (type 2) diabetes and cardiovascular diseases in urban populations, in contrast to rural ones, and the continuing high prevalence of malaria and malnutrition in rural populations in Melanesia. In this volume, various authors examine ways in which a health pattern dominated by undernutrition and infection has been displaced in many places by obesity and the degenerative diseases associated with it. The potential impacts of emerging and resurgent infectious diseases on the trend of increasing LEBs are not ignored, since they have the potential to reverse all gains in LEB at some stage in the future. The influence of infant and young-child mortality on LEB is much greater than that of mortality in later life; furthermore, the factors influencing mortality in earlier life are much stronger agents for natural selection than those influencing mortality in later life.

Forces driving increases in levels of obesity and chronic disease include modernization and the geographical and economic relationships between Pacific Island nations and the industrialized and industrializing nations that surround the Pacific. These relationships are also explored by various authors in this volume for the Samoas, Tonga, the Cook Islands, the Solomon Islands and PNG, and in respect of emergent transnational communities that link Tonga with the United States, New Zealand and Australia; the Cook Islands with New Zealand and Australia; American Samoa with the United States; and Samoa with New Zealand.

Physical and human geography in prehistory and its implications for present-day human biology

The Asia-Pacific region is characterized by many islands bounded by Pacific Rim nations, which include China, the United States, Australia and New Zealand. The largest islands, such as Sumatra, Java, Borneo (Kalimantan), Sulawesi (Celebes) and New Guinea, lie in the west of the region, in contrast with the eastern range of this area, where scattered small islands are separated by long distances. Geomorphologically, several deep ocean troughs run in the south–north direction between the island of Bali in the west and

New Guinea in the east, these two being approximately 1,500 km apart. This zone, with its many islands, including Sulawesi, Flores, Timor, Maluku (Mollucas), Halmahera and Serum, is called Wallacea (Dickerson 1928), and makes a zoogeographic boundary between the Palaeotropical region (the Oriental subregion) and the Australian region. Among several zoogeographic lines here, the westernmost is Wallace's (or Huxley's) Line, while the easternmost is Lydekker's Line, close to Weber's Line (Simpson 1977; Hayami 1987). Tropical rainforest predominates from island South East Asia to Melanesia, with monsoon forest in several islands of Wallacea and in some portions of New Guinea, and mountainous vegetation occurring sporadically on several of the large islands. Zoogeographically, the Oriental subregion differs from the Australian region, in that placental mammals dominate in the former and marsupial mammals dominate in the latter. However, flora scarcely differs between the two regions. Another important biogeographic difference among the islands in Oceania is their size and land formation. Large continental islands such as New Guinea, New Caledonia and New Zealand contrast with medium-sized volcanic islands and small coral reefs which abound in Polynesia and Micronesia. Terrestrial flora and fauna are richer on the continental than on the volcanic islands, and both more so than on the coral reefs. In Wallacea (present-day East Indonesia), island Melanesia, Polynesia and Micronesia, however, marine food resources are abundant and were much more abundant in the past.

Wallacea had long been a barrier to human migration from the west to the east. Palaeoanthropological evidence suggests that the island of Java was inhabited by *Homo erectus* more than a half million (perhaps 750,000) years ago (Jacob *et al.* 1978), whereas Oceanian islands located east of Wallacea were not inhabited until much more recently. According to archaeological studies in Oceania, the earliest dates for two sites of human habitation, determined by thermoluminescence in northern Australia, range between 60,000 and 50,000 years before the present (Roberts *et al.* 1990, 1994). Furthermore, the oldest artefact, a stone tool discovered in the Huon Peninsula of the northeastern tip of New Guinea, has been dated to between 60,000 and 40,000 years ago (Groube 1986). Lower global temperature around 50,000 years ago was associated with a sea level lower by between 100 and 150 m than that at present. Land formations on both sides of Wallacea also differed markedly from the present day. Its western and eastern sides were, respectively, Sunda Land, comprising the Asian continent and islands of South East Asia, including Borneo, Java and Bali, and Sahul Land (Australasia), comprising New Guinea, Australia and Tasmania.

The first settlers of Oceania, who were hunter-gatherers, are likely to have crossed Wallacea by 60,000 to 50,000 years ago, using water craft such as logs or weed-bundled rafts (White and O'Connell 1982; Denoon 1997). Several tens of millennia later, when the sea level had risen to the present level, another human group, who had a markedly different material culture, crossed Wallacea. This group's habitation is evidenced by their unique Lapita red-slipped pottery with its intricate geometric patterns, and settlement remains which were discovered in the Bismarck and Solomon Archipelagos. The oldest sites in the Mussau Islands of the Bismarck Archipelago have been dated to between 3,550 and 3,500 years ago (Kirch 2000). Long-distance maritime movement of obsidian probably started from around 20,000 years ago, its trade being carried out across increasing distance by 8,000 years ago. Horticulture and arboriculture began in Near Oceania by 9,000 to 10,000 years ago, earlier than in island South East Asia to the west of the Wallace's Line. The Neolithic subsistence base of Oceanic populations was derived from South East Asia and New Guinea, never having been rice-based.

The bulk of the contemporary inhabitants of the South Malay Peninsula, Taiwan, island South East Asia and Oceania speak Austronesian languages. Historical linguistic analysis has shown that nine of ten Austronesian language subgroups were spoken by indigenous (non-Han speaking) Taiwanese, with the implication that all Austronesian languages outside Taiwan may have diversified from the same proto-Austronesian language in or near Taiwan, and then spread to the wider area (Blust 1999). The Austronesian language sphere abuts the territories of speakers of three language families, one to the east of Wallacea and the other two in mainland Asia (Capell 1969; Wurm 1982; Wurm and Hattori 1983; Bellwood 1985; Bellwood *et al.* 1995). The first is the Non-Austronesian (Papuan) complex of language families in New Guinea and its surrounding islands. The second is the Austro-Asiatic family in mainland South East Asia, while the third is the Thai family in the central and northern parts of mainland South East Asia, extending to South China.

South East Asia and the Pacific region combine elements of ancient and recent colonizations and of admixture and entry into chains of uninhabited islands with extreme founder effects, alongside the powerful and interacting selective effects of nutrition and infectious diseases. The next two chapters, by Stephen Oppenheimer and Ryutaro Ohtsuka, respectively, paint broad pictures of adaptation and health among various human populations in the Asia-Pacific region, from peopling to the present. Oppenheimer (in Chapter 2) presents new insights into population genetic traits, paying attention to their selective interactions with nutrition and infectious diseases. Human iron deficiency is likely to have emerged as a major culturally induced change in the South West Pacific at an earlier time than in the Near East, because of the earlier transition to agriculture in New Guinea. High rates of alpha thalassaemia are shown not only to be a major cause of anaemia in coastal New Guinea populations, but, as with iron deficiency, to be protective against malaria. This disorder was probably selected for by malaria in lowland areas of Near Occania. In the early 1980s, detailed DNA mapping of the alpha globin gene identified three deletional mutations that caused the disorder, and which were indigenous to, and geographically distributed across, this region. One of them was a good candidate marker for the population expansion giving rise to the Polynesian dispersal, while the relative distribution of the other two suggested that the Polynesian dispersal had bypassed the New Guinea mainland. Detailed analysis of mitochondrial and Y-chromosome DNA in the past decade has been consistent with this view. This has led to a now dominant view of the peopling of this region, which involves several Pacific expansions across the Holocene, from admixed communities in Wallacea.

Populations that successfully colonized the Pacific Islands may have been adapted to periodic food shortages through biological selection of individuals with more efficient metabolism; while this would have favoured them in the past, the emergence of plentiful diet in the second half of the twentieth century has probably penalized this adaptation. Some present-day Pacific Island nations, such as those of Tonga, Nauru, the Cook Islands, American Samoa, Samoa and French Polynesia, have among them the highest rates of obesity in the world, as well as very high rates of cardiovascular diseases and type 2 diabetes. Ohtsuka (in Chapter 3) examines migration histories of Pacific populations from a biocultural perspective, and identifies some of the changing environmental circumstances which may have led to changes in adaptation and health of present-day Melanesian populations in PNG.

Modernization and health change

While patterns of health change in the Asia-Pacific region are outcomes of powerful economic and political forces across the twentieth century, they are also contingent upon cultural and ecological processes in history and prehistory. Factors influencing health in the prehistoric and pre-colonial past include patterns of migration, transitions in subsistence ecologies, and economic change associated with changing exchange patterns across the region. Social factors influencing human population size, distribution and health during the colonial period in many countries of the region include different models of colonial administration and different patterns of economic modernization. Western health workers reached the Pacific from the 1880s onwards, after devastating epidemics that seemed to threaten the survival of whole populations. Depopulation was the focus of government anxiety, missionary alarm and scholarly concern up to, and including, the 1950s (Ulijaszek 2006). Before the divorce of anthropology and psychology from medicine, multi-disciplinary analysis was the usual way of trying to understand such crises. Once depopulation fears faded in the decades after the Second World War (Ulijaszek 2006), and segregation, quarantine and anti-malarial medication were seen to keep Europeans alive, the Pacific became an arena for the development of new public health responses. The new availability of penicillin and sulfa drugs prompted quasi-military health campaigns in the Pacific against specific diseases, until the advent of primary health care in the 1970s. Since colonial administrations accepted a mandate to improve population health, and usually had adequate resources for this task, overt arguments over health policies and programmes surfaced mainly after political independence in many nations of the Asia-Pacific region. Resources became more limited, and the cost-effectiveness of fulfilling colonial mandates for health and other areas came to be questioned increasingly. Global health campaigns, as promoted by the World Health Organization and other agencies, often came into opposition with local realities. Some populations have seen far-reaching changes associated with entry into the cash economy in the context of broader economic modernization, including changes in diet, morbidity and mortality, while others have seen less change. Patterns of nutritional health varied and continue to vary enormously, from high levels of undernutrition in some societies, to a dominance of overnutrition in others.

The nations of South East Asia have varied enormously in their economic profiles across the second half of the twentieth century, poor nutrition remaining a significant contributor to morbidity and mortality in most nations, but overnutrition increasingly becoming a contributor to poor health in places where increased economic prosperity has emerged. In Chapter 4, Geoffrey Marks describes how urbanization, and social and political changes in South East Asia have led to more complex patterns of nutritional health. Where poor nutrition remains prevalent, protein-energy malnutrition and deficiencies of vitamin A, iodine and iron are the most common manifestations of this. Overnutrition has led to increased rates of cardiovascular disease, type 2 diabetes and cancer. Marks argues that while changes in nutritional health are clear to see, there are also changes in community expectations, governance and other factors that affect how agencies might respond to improved population health. The traditional divide between urban and rural populations is now less important as a classifier of nutritional health in South East Asia, and the nutrition agenda in all countries now involves both undernutrition and overnutrition among most age groups, and not just among young children and women of reproductive age.

Overnutrition has become an increasingly important contributor to chronic disease morbidity and mortality in one nation of South East Asia – Hong Kong. Recent studies of body fat percentage and BMI in the Hong Kong Chinese population have identified lower BMI cut-offs as being more salient for the identification of overweight and obesity than in the case of European populations; in contrast, BMI cut-off values for Pacific Islander populations are higher than those recommended for Europeans. In Chapter 5, Gary Ko shows that Hong Kong Chinese people, and perhaps Chinese elsewhere, may be prone to obesity-associated morbidities at lower BMI levels than Europeans.

The large body size and muscular build of Pacific Islanders was noted by Europeans from the time of Captain Cook's voyages (Pollock 1995). Photographs taken in the 1800s also indicate that body fatness was common among the higher classes of Islander societies (Baker 1984). However, there is little evidence of significant body fatness more generally across all levels of Pacific Islander societies prior to the Second World War. The emergence of fatness and obesity generally among Pacific Islander populations began during the second half of the twentieth century, this being largely attributed to health impacts of economic modernization. Most importantly, dietary change and changes in patterns of physical activity associated with levels of education, occupational status, and rural residence have been invoked as being central to the emergence of obesity in this region (Evans and Prior 1969; Bindon and Baker 1985; McGarvey 1991).

Historically, traditional diets of the populations of the Pacific Islands and Melanesia have been very low in fat, and high in complex carbohydrates, dietary fibre, and foods of plant origin (Shintani and Hughes 1994). Dietary change in the Pacific region has been documented, showing a higher contribution of fat and protein to total energy intake among urban communities than among those practising traditional subsistence (Ringrose and Zimmet 1979; Hanna *et al.* 1986; Hezel 1992). In Chapter 6, Yamauchi describes changes in diet and physical activity among Highlanders in PNG and among coastal Solomon Islanders. Although the increased consumption of energy-dense store-bought foods explains some of the difference in body fatness between urban and rural New

Guinea Highlanders, reduced physical activity is as important. In particular, the decline in gender inequality in the division of labour in the urban populations, which caused the greater decline in physical activity of women, may go some way to explaining their greater body fatness relative to males. In contrast, there was no difference in body fatness between the rural and urban Solomon Islander populations. The traditional diet in the Solomon Islands, based on fish, root crops, and coconut, is nutritionally good, and it is perhaps unsurprising that the nutritional status of rural villagers is similar to that of the more modernized villagers. Unlike the groups seen in PNG, Yamauchi finds no clear difference in physical activity between traditional and more modernized Solomon Islander women, although traditional village men were more physically active than their more modernized counterparts. Yamauchi concludes that the less modernized Solomon Islander population is at an earlier stage of transition from subsistence to cash economy. Thus the influence of modernization on nutritional health is only partly manifest, but constitutes a potential health risk for the future.

While modernization and urbanization took place quickly in the Pacific after the Second World War, with far-reaching effects on human biology, the effects of these changes continue to penetrate many isolated Pacific islands to the present day. In Chapter 7, Tsukasa Inaoka and colleagues describe changing lifestyles, and associations between obesity and metabolism-related factors among the Tongan population in Tonga. The majority of Tongans live in urban areas, although rural populations are not very isolated when compared with populations in many other countries in the region, including PNG, the Solomon Islands and Vanuatu. In addition, about 100,000 Tongan citizens are out-migrants to various developed countries, including the United States, Australia and New Zealand. Many Tongans move between their own country and that of overseas residence, and flows of information between Tonga and the larger developed nations to which they migrate are great, as are remittances from out-migrants to Tongan relatives. For these reasons, it is unlikely that the rate of lifestyle modernization and the concomitant rise in obesity and non-infectious disease mortality will slow down there in the near future.

In Chapter 8, Ember Keighley and colleagues present data on physical and dietary characteristics of modernizing Samoans on independent Samoa, American Samoa and Hawaii respectively. In both Samoas, the prevalence of obesity increased between 1976 and 2003, to one of the highest levels on earth. Levels of overweight and obesity among adults in American Samoa in the 1970s matched the levels in Samoa some 30 years later. Furthermore, levels of overweight in children and adolescents from

Hawaii in the 1970s presaged overweight in youth living in American Samoa in 2002. If current trends continue, Keighley and colleagues speculate, they may also be indicative of future levels of overweight among youth living in Samoa. Prevalence of type 2 diabetes is over one-third higher for all age-sex groups in American Samoa relative to Samoa. Furthermore, rates of type 2 diabetes are increasing rapidly in both Samoas. Exposure to environments predisposing to obesity and diabetes is high and ubiquitous at all levels of Samoan society.

Migration, transnationalism and nutritional health

While European contact with Pacific Islander populations largely took place over 200 years ago, significant migrations of Pacific Islanders to colonizing nations began only in the 1920s, when Samoans and Tongans migrated to Hawaii in large numbers. The greatest influx of Samoans to Hawaii came in the 1950s with the end of US naval administration in American Samoa, but migration has continued at a steady rate to the present day. Of Pacific Islanders living abroad, most are in the United States (about a quarter of a million subjects in the year 2000) (United States 2000 Census 2002). Pacific Islanders have a more recent demographic history in New Zealand. In 1945, there were just over 2,000 people of Pacific origin there. While there was a period of high immigration in the early 1970s, this inflow slowed in the late 1970s, as social, economic and labour market conditions in New Zealand became less favourable. In the early 1980s, the flow reversed, when return migration to the Pacific and chain migration to Australia combined exceeded immigration to New Zealand. Immigration increased again by the end of the decade, and by the year 2001, the number of Pacific Islanders in New Zealand was about 6% of the total resident population, at 232,000 (Statistics New Zealand 2002). In France, Pacific Islander immigration from French colonies has taken place since the beginning of the twentieth century. France is estimated to have had a Pacific Islander population (mostly from French Polynesia) of around 233,000 at the turn of the twenty-first century (Ulijaszek 2005). In Australia, migration of Pacific Islanders was already in place prior to 1950, but increased greatly in the early 1980s. In 1971, the Pacific Islander population of Australia was 23,000 strong; by 1996, there were 96,000 of them, mostly living in the cities of Sydney and Melbourne.

Increasingly, Pacific Islander migrants have been able to maintain closer links with their communities in their nation of origin than were ever possible prior to cheap communication and transport. This has both reduced the economic risk associated with migration and has favoured the returning of remittances to home communities. Most transnational populations possess urban character. In the early 1970s, first the Niuean and then the Cook Island populations in New Zealand exceeded those of their home island populations (Hau'ofa 1994). In the Cook Islands, the population has shown little increase between 1950 and 2000 (National Statistical Office 2001). However, Cook Islander migrants in New Zealand and Australia came to outnumber island residents by about two to one by the year 2001 (Ulijaszek 2005). Other Pacific Island nations where migrants outnumber, or are likely to outnumber island residents, include Tonga, Wallis and Fortuna, Tokelau and Niue. While there are no strong links between the proportion of migrants from Pacific Island nations and major indicators of economic performance and prosperity in those nations, links between migrants and relatives on their island nations of origin operate to generate tastes for imported foods and in providing remittances which can be used to buy foods and other goods by those living on the home islands. Such migrations have had profound impacts on health of Pacific Islanders both at home and abroad. Most important has been the emergence of obesity and fatness, cardiovascular disease and type 2 diabetes as significant public health problems in Pacific Island nations.

Among migrants from the Asia-Pacific region to more industrialized and increasingly post-industrial nations, there have been dramatic changes in health patterns, sometimes regressing to the greater health pattern of the host nation, but often differing from it in important ways. Parker Frisbie and colleagues (in Chapter 9) examine the extent of heterogeneity in health outcomes among various Asian and Pacific Islander populations in the United States. While most of the larger Asian American subgroups, including those of Chinese, Philippino, Korean and Japanese origin, exhibit fairly healthy profiles, indigenous Hawaiians and Samoans do not. Nor do some of the smaller Asian American subgroups, such as Laotians, Cambodians and Hmong.

Cook Islander transnationalism predominantly spans the Cook Islands, New Zealand and Australia. Stanley Ulijaszek (in Chapter 10) describes changes in the extent of obesity and fatness since the 1950s among adult Cook Islanders living on Rarotonga, the most economically developed of the Cook Islands. The influences of diet, physical activity, modernization and transnationalism on blood pressure, body fatness (as assessed by BMI) and fasting blood glucose of adult Cook Islanders on Rarotonga in the 1990s are examined, and comparisons made with the same measures carried out among Cook Islanders living in Melbourne, Australia. Cook Islanders on Rarotonga have undergone increases in BMI since the 1950s, although rising blood pressure, while already high by the 1960s, appears to have been kept in check by good compliance to anti-hypertension medication. The importance of transnational connections for elevated blood pressure is demonstrated, as is the importance of physical activity for BMI of males but not females. Diet does not emerge as a factor in the prevalence of obesity, hypertension and diabetes in the Cook Islands, perhaps because transnational factors play stronger roles in all aspects of health-related behaviour and not dietary patterns alone. The Cook Islander population of Australia is younger than that resident in Rarotonga, and shares physical and dietary characteristics of the Rarotongan population of similar age.

In Chapter 11, Alistair Woodward and Tony Blakely examine inequalities in health and wealth across Pacific Island nations, Australia and New Zealand, and their impacts on mortality. They find that the greatest differences in mortality between indigenous and non-indigenous populations are found within the latter two nations. While social and economic deprivation contributes to the gap in life expectancy between Maori and non-Maori, and between Australian Aboriginal and non-Aboriginal people, these differences cannot be attributed wholly to economic factors. Within economically deprived categories, mortality rates for indigenous peoples are higher than those for non-indigenous people. Woodward and Blakely draw a parallel between industrialization in Europe in the 1800s and changes taking place in the Asia-Pacific region now. As the former brought new health problems, particularly for those with few resources, while empires were built and fortunes accrued by others, so modernization in the latter has led to new health problems, but of a different kind from the former. The social consequences of rapid social and economic change in the Pacific are mixed, having brought benefits to some, but having placed others into economic hardship and poor health.

Health transition and biocultural adaptation

The dominant narrative of the majority of the chapters preceding the final one is that of decline in infectious disease mortality and increase in obesity and the chronic disease mortality associated with it. Robert Attenborough (in Chapter 12) shows that this is not a universal narrative in the Asia-Pacific region. By focusing on health change in PNG, he emphasizes that the immediate health outlook there is not one dominated by the diseases of modernization, as it makes its transition from tradition to modernity. While adaptation to long-term stressors such as malaria has been demonstrated, Attenborough points to breakdown in adaptation during rapid change in environment or way of life, and with this, the emergence of new health problems, including circulatory disease, diabetes and asthma. While the diseases associated with modernization are becoming epidemiologically important in some communities, non-infectious disease is also a matter for real and rising concern (Temu 1991; Tefuarani *et al.* 2002). In PNG as elsewhere, reality is more complex than many of the models of health change, and Attenborough notes that it would be dangerously misleading to ascribe any type of automatism to the health transition process. It is by no means clear in PNG that time and economic development will be sufficient to ensure that lifestyle-related diseases will eventually overtake infections, or even that gains so far in the control of infection and increases in LEB are secure for the future. Malnutrition, diarrhoea and measles continue to cause uncontrolled problems; tuberculosis is increasing; and most saliently, the emergence of HIV/AIDS threatens a large increase in the infectious death toll, while malaria is developing in increasingly dangerous ways.

The final chapter begs a broader question: how likely is it that epidemiological and nutritional transitions will run their course in predictable ways in this region? The biological and physical environment of the Asia-Pacific region continues to change at an unprecedented rate, and these changes are likely to continuously have effects on health. Transition models would predict continued increases in obesity levels and the chronic diseases associated with it. Various nations, including Singapore, Malaysia, Japan and South Korea have undergone health transition, but have attained low obesity prevalences relative to most European and North American nations. In the Pacific Island nations, health transition has taken place in French Polynesia and continues to take place in Samoa, Tonga and the Cook Islands; in these nations, obesity rates exceed enormously even the very highest rates of non-Pacific Island nations. It is thus possible to postulate two health transition models for the Asia-Pacific region: one that involves rapidly increasing rates of obesity to exceptionally high levels (as in the Pacific Island nations), and another that involves increases in obesity only to comparatively low levels (as in the modernized nations of East Asia).

The present patterns of obesity in the Pacific Island nations are outcomes of conjoining forces across the past 50 years or so. These include (1) continuing economic development with comparatively few serious setbacks; (2) food security that has increased for most of the region; (3) the penetration of the world food system, even to some of the remotest islands of the Pacific; (4) declining prices for energy-dense foods; (5) progressive mechanization of labour-intensive tasks; (6) urbanization and sedentarization of work life; and (7) increasing mechanization of transport. It is unlikely that there will be a reversal in obesity prevalence trends in the Pacific Island nations at any stage in the near future. A ceiling on the potential for obesity prevalence in the populations of the Pacific Islands is unlikely to have been reached anywhere apart from perhaps Tonga, where 79% of the adult population has BMI greater than $30 \, kg/m^2$. Some hope for declines in obesity prevalence rates comes from studies of appropriate body size in modernizing societies. Obesity is also an outcome of cultural and symbolic over-valuation of food in the context of plenty and such over-valuation declines as subsequent generations are born into good times. A number of communities and societies in which obesity has risen across recent decades, and who previously were shown to desire and/or accept larger bodies and obesity, now prefer thinner bodies. This has been observed among Pacific Islanders (Craig *et al.* 1996; Brewis *et al.* 1998; Becker *et al.* 2005) and Korean children (Lee *et al.* 2004), as well as groups elsewhere in the world (Story *et al.* 1995; Rinderknecht and Smith 2002; Katz *et al.* 2004).

While infant mortality rates have declined universally in the Asia-Pacific region between 1978 and 1998, a number of infectious diseases, including tuberculosis and malaria, have spread geographically, increasingly in more virulent and drug-resistant forms. HIV/AIDS has emerged as a new infectious disease which, without drug therapy, is almost universally fatal. Countries in the region with continuing high infectious disease mortalities include Indonesia, Thailand, Laos, Cambodia and Vietnam. Although HIV/AIDS and tuberculosis are predicted to account for the overwhelming majority of deaths from infectious diseases in Africa by 2020, this is much less likely to be so for the Asia-Pacific region. Adult HIV prevalence rates vary from well below 0.1% in Japan and South Korea, to 0.4% in Malaysia, 0.6% in PNG and 1.5% in Thailand (UNAIDS 2005). These rates are well below the rate necessary to cause a decline in LEB. Figure 1.2 shows adult HIV prevalence rates in 36 African nations according to whether or not they experienced a decline in LEB across the years 1990 to 2000. In the vast majority of African nations where adult HIV prevalence rates exceed 5%, LEB is declining, due to high AIDS mortality. Nearly all African nations with rates below this level have stable or increasing LEB. Even if HIV/AIDS proceeded unabated and untreated, it would take a decade or more for the two nations with the highest current adult HIV prevalence rate in the region (Thailand and PNG) to reach 5% adult HIV prevalence, beyond which LEB could undergo decline.

What goes around, comes around. And to complete the circle it is necessary to return to genetic adaptation. A range of genetic susceptibilities and

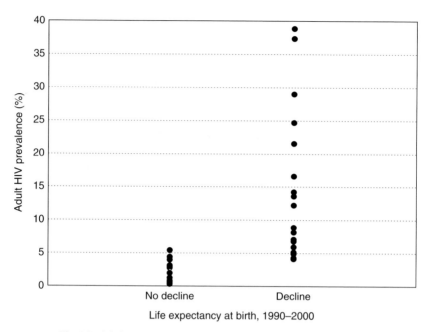

Fig. 1.2. Adult HIV prevalence in 36 African nations according to whether or not life expectancy at birth underwent decline between 1990 and 2000 (data from UNAIDS 2005).

resistances to infectious diseases have been demonstrated among various populations across the world (Hill 1999).

Human leucocyte antigen (HLA) allele variants have been associated with resistance to various infectious diseases such as malaria, AIDS, and hepatitis B and C (Segal and Hill 2003). HLA allele variants have also been associated with susceptibility to tuberculosis, AIDS, typhoid fever and leprosy (Segal and Hill 2003). It is probable that genetic adaptation to malaria will continue in the Asia-Pacific region, even as malarial parasites develop resistance to the medication used for treating it. Genetic suscept-ibility to tuberculosis has been demonstrated in the region, specifically in Cambodia (Goldfeld 2004), as well as elsewhere in the world (Fernando and Britton 2006). Human host genetic variants of CCR2 and CCR5 genes have been shown to influence susceptibility to, and progression of, HIV infection in various populations (Blanpain *et al.* 2002; Anastassopoulou and Kostrikis 2003; Julg and Goebel 2005; Shrestha *et al.* 2006), including one in South East Asia (Capoulade-Metay *et al.* 2004). It is likely that further evidence for genetic variation in susceptibility to HIV-1 will emerge in the Asia-Pacific region, as elsewhere in the world. Concurrent with emerging knowledge of genetic variation in susceptibility to HIV, the virus continues to develop resistance to medication used to manage the process of the infection.

To conclude, it seems both unlikely that an epidemiological end-point will occur in the Asia-Pacific region at any time in the near future, and that there will be only one epidemiological end-point. While liberal democracies and free markets may be significant drivers of the nutrition and health transition in the region, it is not yet clear whether they will continue to be so into the future with no significant challenge, either environmental or ideological. In the meantime, infectious disease continues to present new challenges, if not yet a serious threat to the health transition in this region.

References

Anastassopoulou, C. G. and Kostrikis, L. G. (2003). The impact of human allelic variation on HIV-1 disease. *Current HIV Research* **1**, 185–203.

Baker, P. T. (1984). Genetics and the degenerative diseases of South Pacific Islanders. In *Migration and Mobility*, ed. A. Boyce. London: Taylor and Francis Ltd, pp. 209–40.

Becker, A. E., Gilman, S. E. and Burwell, R. A. (2005). Changes in prevalence of overweight and in body image among Fijian women between 1989 and 1998. *Obesity Research* **13**, 110–17.

Bellwood, P. (1985). *Prehistory of Indo-Malaysian Archipelago*. Sydney: Academic Press.

Bellwood, P., Fox, J. J. and Tryon, D. (1995). The Austronesians in history: common origins and diverse transformations. In *The Austronesians: Historical and Comparative Perspectives*, ed. P. Bellwood, J. J. Fox and D. Tryon. Canberra: The Australian National University, pp. 1–16.

Bindon, J. R. and Baker, P. T. (1985). Modernization, migration and obesity among Samoan adults. *Annals of Human Biology* **12**, 67–76.

Blanpain, C., Libert, F., Vassart, G. and Parmentier, M. (2002). CCR5 and HIV infection. *Receptors and Channels* **8**, 19–31.

Blust, R. (1999). Subgrouping, circulatory and extinction: some issues in Austronesian comparative linguistics. *Symposium Series of the Institution of Linguistics of Academia Sinica* **1**, 31–94.

Brewis, A. A., McGarvey, S. T., Jones, J. and Swinburn, B. A. (1998). Perceptions of body size in Pacific Islanders. *International Journal of Obesity and Related Metabolic Disorders* **22**, 185–9.

Capell, A. (1969). *A Survey of New Guinea Languages*. Sydney: Sydney University Press.

Capoulade-Metay, C., Ma, L. Y., Truong, L. X., *et al.* (2004). New CCR5 variants associated with reduced HIV coreceptor function in Southeast Asia. *AIDS* **18**, 2243–52.

Craig, P., Swinburn, B., Matenga-Smith, T., Matangi, H. and Vaughan, F. (1996). Do Polynesians still believe that big is beautiful? Comparison of body size perceptions and preferences of Cook Islands, Maori and Australians. *New Zealand Medical Journal* **14**, 200–3.

Denoon, D. (1997). Production in 'Pacific Eden? Myths and realities of primitive affluence'. In *The Cambridge History of the Pacific Islanders*, ed. D. Denoon. Cambridge: Cambridge University Press, pp. 83–90.

Dickerson, R. E. (1928). *Distribution of Life in the Philippines*. Philippines Bureau of Sciences Monograph No. 21, Manila.

Evans, J. G. and Prior, I. A. M. (1969). Indices of obesity derived from height and weight in two Polynesian populations. *British Journal of Preventive and Social Medicine* **23**, 56–9.

Fernando, S. L. and Britton, W. J. (April 2006). Genetic susceptibility to myco-bacterial disease in humans. [Review] [164 refs] [Journal Article. Review] *Immunology & Cell Biology* **84**(2), 125–37.

Goldfeld, A. E. (2004). Genetic susceptibility to pulmonary tuberculosis in Cambodia. *Tuberculosis* **84**(1–2), 76–81.

Groube, L. M. (1986). Waisted axes of Asia, Melanesia, and Australia. In *Archaeology at ANZAAS Canberra*, ed. G. K. Ward. Canberra: The Australian National University Press, pp. 168–77.

Hanna, J. M., Pelletier, D. L. and Brown, V. J. (1986). The diet and nutrition of contemporary Samoans. In *The Changing Samoans: Behavior and Health in Transition*, ed. P. T. Baker, J. M. Hanna and T. S. Baker. New York: Oxford University Press, pp. 275–96.

Hau'ofa, E. (1994). Our sea of islands. *The Contemporary Pacific* **6**, 147–62.

Hayami, I. (1987) Geohistorical background of Wallace's Line and Jurassic marine biogeography. In *Historical Biogeography and Plate Tectonic Evolution of Japan and East Asia*, ed. A. Taira and M. Tashiro. Tokyo: Terra Publisher, pp. 111–33.

Hezel, F. X. S. J. (1992). Expensive taste for modernity: Caroline and Marshall Islands. In *Social Change in the Pacific Islands*, ed. A. B. Robillard. London: Kegan Paul International, pp. 203–19.

Hill, A. V. S. (1999). Genetics and genomics of infectious disease susceptibility. *British Medical Bulletin* **55**, 401–13.

Jacob, T., Soejono, R. P., Freeman, L. G. and Brown, R. H. (1978). Stone tools from Mid-Pleistocene sediments in Java. *Science* **202**, 885–7.

Julg, B. and Goebel, F. D. (2005). Susceptibility to HIV/AI|D|S: an individual characteristic we can measure? *Infection* **33**, 160–2.

Katz, M. L., Gorden-Larsen, P., Bentley, M. E., *et al.* (2004). 'Does skinny mean healthy?' Perceived ideal, current, and healthy body sizes among African-American girls and their female caregiver. *Ethnicity and Disease* **14**, 533–41.

Kirch, P. V. (2000). *On the Road of the Wind: An Archaeological History of the Pacific Islands before European Contact*. Berkeley: University of California Press.

Lee, K., Sohn, H., Lee, S. and Lee, J. (2004). Weight and BMI over 6 years in Korean Children: relationships to body image and weight loss efforts. *Obesity Research* **13**, 1959–66.

McGarvey, S. T. (1991). Obesity in Samoans and a perspective on its etiology in Polynesians. *American Journal of Clinical Nutrition* **53**, 1586S–94S.

National Statistical Office (2001). *Cook Islands Annual Statistical Bulletin.* Rarotonga: National Statistical Office.

Nishida, C. and Mucavele, P. (2005). Monitoring the rapidly emerging public health problem of overweight and obesity: the WHO global database on body mass index. *United Nations System Standing Committee on Nutrition* **29**, 5–12.

Olshansky, S. J., Passaro, D. J., Hershow, R. C., *et al.* (2005). A potential decline in life expectancy in the United States in the 21st century. *New England Journal of Medicine* **352**, 1138–45.

Pollock, N. J. (1995). Social fattening patterns in the Pacific – the positive side of obesity. A Nauru case study. In *Social Aspects of Obesity*, ed. I. de Garine and N. J. Pollock. Amsterdam: Gordon and Breach, pp. 87–109.

Rinderknecht, K. and Smith, C. (2002). Body-image perceptions among urban Native American youth. *Obesity Research* **10**, 315–27.

Ringrose, H. and Zimmet, P. (1979). Nutrient intakes in an urbanized Micronesian population with a high diabetes prevalence. *American Journal of Clinical Nutrition* **32**, 1334–41.

Roberts, R. G., Jones, R. and Smith, M. A. (1990). Thermoluminescence dating of a 50,000-year old human occupation site in northern Australia. *Nature* **345**, 153–6.

Roberts, R. G., Jones, R., Spooner, N. A., *et al.* (1994). The human colonization of Australia: Optical dates of 53,000 and 60,000 years bracket human arrival at Deaf Adder Gorge, Northern Territory. *Quaternary Science Review* **13**, 575–86.

Segal, S. and Hill, A. V. S. (2003). Genetic susceptibility to infectious disease. *Trends in Microbiology* **11**, 445–8.

Shintani, T. T. and Hughes, C. K. (1994). Traditional diets of the Pacific and coronary heart disease. *Journal of Cardiovascular Risk* **1**, 16–20.

Shrestha, S., Strathdee, S. A., Galai, N., *et al.* (2006). Behavioral risk exposure and host genetics of susceptibility to HIV-1 infection. *Journal of Infectious Diseases* **193**, 16–26.

Simpson, G. G. (1977). Too many lines: the limits of the Oriental and Australian zoogeographical regions. *Proceedings of the American Philosophical Society* **121**, 63–6.

Statistics New Zealand (2002). 2001 Census of population and dwellings. http://www.stats.govt.nz (accessed September 2005).

Story, M., French, S. A., Resnick, M. D. and Blum, R. W. (1995). Ethnic/racial and socioeconomic differences in dieting behaviors and body image perceptions in adolescents. *International Journal of Eating Disorders* **18**, 173–9.

Tefuarani, N., Sleigh, A. and Hawker, R. (2002). Congenital heart diseases: a future burden for Papua New Guinea. *Papua New Guinea Medical Journal* **45**, 175–7.

Temu, P. I. (1991). Adult medicine and the 'new killer diseases' in Papua New Guinea: an urgent need for prevention. *Papua New Guinea Medical Journal* **34**, 1–5.

Ulijaszek, S. J. (2005). Modernisation, migration, and nutritional health of Pacific Island populations. *Environmental Sciences* **12**, 167–76.

(ed.) (2006). *Population, Reproduction and Fertility in Melanesia*. Oxford: Berghahn Books.

UNAIDS (2005). Country data. www.unaids.org (accessed March 2006).

United Nations University (2006). Globalis – an interactive world map. Life expectancy at birth. www.gvu.unu.edu (accessed March 2006).

United States 2000 Census (2002). http://www.tetrad.com/pcensus/usa/census2000data.html (accessed September 2005).

White, J. P. and O'Connell, J. F. (1982). *A Prehistory of Australian, New Guinea and Sahul*. Sydney: Academic Press.

Wurm, S. A. (1982). *Papuan Languages of Oceania*. Tubingen: Gunter Narr Verlag.

Wurm, S. A. and Hattori, S. (eds.) (1983). *Language Atlas of the Pacific Area*, Part 2. Canberra: Australian Academy of Humanities.

2   Interactions of nutrition, genetics and infectious disease in the Pacific: implications for prehistoric migrations

STEPHEN OPPENHEIMER

Introduction

Like some huge natural experiment in population genetics and evolution, South East Asia and the Pacific region combine elements of ancient and recent colonizations, of admixture and of entry into chains of uninhabited islands with extreme founder effects alongside the powerful and interacting evolutionary selective effects of nutrition and infectious disease. New Guinea was one of the first places in the world to achieve its own Neolithic revolution. This may have signalled the onset of specific micronutrient deficiencies, in particular of iron. Neolithic sedentary behaviour may have also increased transmission of malaria. Malaria is a major lethal disease in South East Asia and lowland Near Oceania although not in Far Oceania and, since the Late Pleistocene, has exerted strong selective effects promoting genetic disorders of globin-chain and red cell production. Overlaying this selective mechanism is an exquisite three-way interaction between micronutrient availability (in particular iron), infectious disease (in particular malarial susceptibility) and genetic protection (in particular alpha-globin gene deletions). The Holocene change from mobile hunting and gathering to arboriculture, horticulture and sedentary living may have acted to increase the level of iron deficiency in early childhood both as a result of nutritional inadequacy and as a secondary result of malaria and genetic causes of anaemia in the iron stores of newborn infants. Over-enthusiastic attempts to correct such deficiency have the unfortunate effect of increasing both malarial and non-malarial infectious morbidity.

    Early clinical research on such interactions had the serendipitous result of unearthing evidence of ancient migrations and genetic relationships in the region. Unique genotypes for red cell disorders are geographically

Health Change in the Asia-Pacific Region: Biocultural and Epidemiological Approaches, ed. Ryutaro Ohtsuka and Stanley J. Ulijaszek. Published by Cambridge University Press.

highly specific in South East Asia and the Pacific. Precise mapping of the distribution of these disorders in the 1980s indicated close relationships between Austronesian-speakers of Melanesia and Far Oceania but not so much with Austronesian speakers in South East Asia much to the west of the biogeographic boundary described by Wallace. Although not fully appreciated at the time, this key finding undermined the conventional view of Polynesians as recent descendants of Taiwanese rice farmers and argued for a more local origin of the argonauts of the Pacific east of the Wallace's Line. More recent research using mitochondrial DNA and Y chromosome markers has suggested an Asian/Melanesian admixed colony in Wallacea during the early Holocene as the source of many typical Polynesian lineages.

Late Pleistocene explorers in the Pacific

South East Asia and the South West Pacific region were geographic back-drops for two unique, epic, widely separated periods in modern human evolution and the prehistoric colonization of the planet. In their archae-ology, physical anthropology, genetics, modern-day languages, culture and human disease, these areas still hold static after-images of those events. They were unique events since, with the possible exception of the presence of *Homo erectus* in Flores from around 800,000 years ago and the much more recent finding of *Homo floresiensis* on the same island (Morwood *et al.* 1998, 2004), no other human species had previously crossed the Wallace's Line, let alone reached the more distant islands of the Pacific.

The first period saw the primary spread of anatomically modern humans out of Africa probably as a single expansion (Oppenheimer 2003; Macaulay *et al.* 2005), subsequently replacing previous human types and giving rise to all of today's non-Africans. Based on genetic evidence, this group most likely migrated across the mouth of the Red Sea initially following the coastline of the Indian Ocean (Oppenheimer 2003). A contrary view is that the ancestors of Europeans and most Asians left Africa via the Sahara desert rather later than the ancestors of the Australians (Underhill *et al.* 2001). This two-exit view does not signifi-cantly alter the paradigm of the Australian southern coastal route, apart from different out-of-Africa date based on the Y chromosome. The choice of a single southern route is also supported by the finding of the earliest dates for Asian colonization in South East Asia and Australia. Australia has a controversial luminescence date of first evidence of occupation in the region, of 62,000 years ago. This fits the window of opportunity provided by a low sea level for crossing the sea gap at 65,000 years ago (Oppenheimer 2003). Although presently the earliest evidence, based on thermoluminescence data, points to human presence in New Guinea over 40,000 years ago (Groube *et al.* 1986), genetic dating there suggests a first colonization at least as old as Australia if not older (Redd and Stoneking 1999). Even before the advent of detailed mitochondrial DNA gene mapping, it was generally thought that these initial colonies, which were established during the Late Pleistocene, provided the bulk of the gene pool that still exists in Near Oceania (also known as Melanesia), with the highlands of New Guinea showing less admixture with later migrations than lowland and coastal regions (Serjeantson and Hill 1989: 286–94).

Up until about 10,000 years ago (the start of the Holocene), low sea levels meant that land-bridges extended the Asian mainland southeast to include most of the Sunda shelf (Fig. 2.1), allowing free migration as far as Bali and Borneo (Oppenheimer 1998). Beyond this point, however, the sea has, for millions of years, slowed down further migration towards the Pacific. This is the practical basis of the biogeographic boundary known as Wallace's Line, and is the most likely reason why New Guinea and Australia had until recently, to a great extent, preserved their Pleistocene cultural and genetic isolation (Oppenheimer 1998). The boundary of Wallace's Line can still be seen in genetic, linguistic, cultural and morphological features as one passes across Wallacea from Borneo to New Guinea (Oppenheimer 2004a). Perhaps the best evidence of this isolation is demonstrated by the fact that South East Asian intrusive human lineages only appear in Melanesia during the early Holocene (Oppenheimer 2004a), as do new foodstuffs such as betel nut (Swadling 1997). What is perhaps surprising is a similar apparent cultural and genetic boundary still seen between Australia and New Guinea, which, although separated by the Coral Sea today, were part of the same continental landmass (the Sahul) up until a few thousand years ago.

There is a tendency to regard Pleistocene peoples of Near Oceania as lacking in maritime and other skills. As evidenced by their initial sea crossings and later efforts, this was clearly not the case. The Bismarck Archipelago was colonized about 35,000 years ago and Buka Island in the North Solomons about 29,000 years ago (Fig. 2.2) (Allen and Gosden 1996). The latter showed evidence of working of the tuber taro on a grindstone and required an open sea crossing of 180 km. About 20,000 years ago, obsidian was being moved across large distances (over 350 km) over coastal routes between the two main Bismarck Islands (Allen and Gosden 1996). Manus, in the Admiralty Islands, was also colonized about

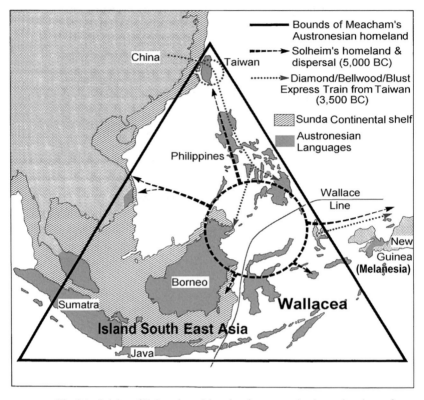

Fig. 2.1. Origins of Polynesians. Map showing two main alternative views of Austronesian origins, on-shore and off-shore. The oldest view represented by Meacham (1985: solid triangle), Terrell (1986) and Solheim (1994, 1996): interrupted solid black line and circle) argues an Island South East Asian homeland (>5,000 BC). The 'out of Taiwan' view of a recent rapid migration from China via Taiwan (3,000–4,000 BC), spreading to replace the older populations of Indonesia after 4,000 years, is shown as a dotted line.

20,000 years ago, across an open-sea distance of 200 km (Anderson 2000). Evidence of maritime mobility and marine exploitation increased between 10,000 and 4,000 years ago (Allen and Gosden 1996). During the period 9,000 to 6,000 years ago, East New Britain began to show increased occupation and increased use of obsidian (Allen and Gosden 1996). Around 8,000 years ago, obsidian started being moved across even greater coastal and inter-island distances, such as from Talasea on the north coast of New Britain to the east coast of New Ireland. At the same time, new animal bones such as shark's teeth appeared at the same sites and shell-tool use increased and became more diverse (White *et al.* 1991). About 6,000

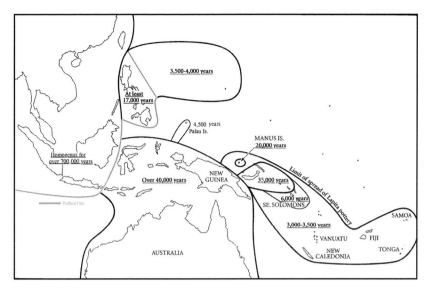

Fig. 2.2. Dates of human colonization of the southwest Pacific and Australia up to the Lapita horizon.

years ago, the southern Solomons were occupied (Roe 1992), and the Palau islands were occupied perhaps about 4,500 years ago (Welch 2002). Evidence of the start of root-crop horticulture in the highlands of New Guinea about 10,000 years ago has recently been confirmed (Denham *et al.* 2003). Whether this extremely early agricultural event was independent or part of a regional South West Pacific trend remains to be seen. However, notable (but still controversial to some) Asian domestic intrusions in the Bismarck Archipelago and mainland New Guinea in the early Holocene (5,500 to 8,000 years ago) are systematic tree cropping (Swadling 1997; Latinis 2000), dogs (Bulmer 2001) and pigs (Allen and Gosden 1996).

The long-standing controversy concerning dates of the first pigs in Melanesia has been admirably reviewed by Hide (2003), with a touch of wry humour on the arguments used either way. The older dates are fiercely contested by those archaeologists who feel that this would not fit their favoured linguistic model, which depends on pigs having been introduced by Austronesian speakers from Taiwan as part of an agricultural package. The linguistic dating of human language splits is a contentious issue, but dating pigs linguistically may be pushing the paradigm too far. Allen (2000: 159) argues that it is 'more than likely than not that pigs were in New Guinea and parts of the Bismarcks before the mid-Holocene'. Recent genetic work on pig mitochondrial DNA has shown that

> The New Guinea pig haplotypes ... cluster with pigs from Hawaii,
> Vanuatu, and Halmahera in a monophyletic group ... the 'pacific
> clade') ... and are well separated from any other individuals, domestic
> or wild ... This evidence is consistent with a Lapita dispersal from
> Near to Remote Oceania, but the lack of any genetic affinity
> between this group and Taiwanese wild boar offers no support for
> the 'Out-of-Taiwan' model of human and pig dispersal into Near
> Oceania.
>
> (Larson *et al.* 2005)

Or, one might add, 'for only recent introduction of pigs to the Pacific'.

'Two peoples, two periods?'

All this contrasts with the present dominant paradigm of the Express Train from Taiwan (ETT) to Polynesia (Diamond and Bellwood 2003), where the last phase of human world colonization, the conquest of the Pacific, is seen as a single rice-growing Neolithic expansion and exodus from Taiwan in East Asia, carrying Austronesian languages and Taiwanese genes to South East Asia, from thence immediately to the Pacific 3,500 years ago (Fig. 2.1). The archaeological label for this hypothetical single expansion as it went through Near Oceania and out as far as Samoa in Western Polynesia was the so-called Lapita pottery, initially thought to be the first ceramic horizon in the region (Bellwood 1997).

The implied view of 'Two Peoples, Two Periods' (the first – non-sailing negrito hunter-gatherers in the Late Pleistocene; and the second – skilled maritime Neolithic Mongoloids spreading Lapita in the Late Holocene) has been castigated not only as simplistic 'racial migrationism' but also as wrong (Terrell *et al.* 2001). That it is an oversimplification is apparent in the first instance from the multiple other post-glacial, pre-Lapita maritime expansions between 20,000 and 6,000 years ago. The early appearance and spread of domestic mammals with these expansions, as well as hook-line fishing (Szabó and O'Connor 2004), horticulture and tree cropping, characteristic of the wider Pacific colonization, with these expansions, undermines the 'Neolithic' newness of the Lapita-associated expansion. Nor was Lapita the first ceramic horizon of the region (Swadling 1997). Furthermore, the entire subsistence package of Pacific domesticates appears to have originated either in Island South East Asia or Melanesia (Oppenheimer and Richards 2003), while rice, the staple of the supposed expanding Taiwanese farming migrants, never made it across the Wallace's Line until 1,000 years ago, and was not associated with either of the two supposed key ceramic horizons of the ETT – red-slipped pottery in South East Asia and Lapita in Oceania (Paz 2003).

The core geographic structure of the ETT hypothesis depends on a particular interpretation of the Austronesian linguistic tree first proposed in the 1970s (Shutler and Marck 1975). There is little dispute that Austronesian languages were intrusive to the Pacific sometime within the past 10,000 years. Nor is there disagreement with the congruence of trajectories of reconstructed proto-languages, genes and archaeological records of initial colonization of Polynesia – in other words, the unitary nature of the last phase, the conquest of the deep Pacific (Far Oceania). However, there are dissenting voices over the reconstructed prehistory of Holocene expansions in Near Oceania arguing, at the least, for an appreciation of a more complex picture there, and with a deeper time-scale (Meacham 1985; Solheim 1994, 1996; Terrell *et al.* 2001; Oppenheimer and Richards 2001, reviewed in Oppenheimer 1998, 2004a). Perhaps the strongest dissent with the ETT relates to the assertion that there was a congruence of genes, languages and archaeology right back to a mid-Holocene homeland in Taiwan. The assertion, based on a controversial Austronesian linguistic reconstruction (Pawley 2003), that Taiwan is the homeland, rather than the geographic default homeland in the Philippines or elsewhere in Island South East Asia, depends not on strict rules of comparative linguistics, but on arguments of diversity and the application of a 'least moves parsimony' argument predicated on an ultimate Chinese mainland origin (Blust 1985). An additional problem in the dating of the nodes of the linguistic story is not depending on linguistic evidence, but on a presumed congruence with archaeological ceramic horizons. This is a major concern because of the risk of creative circularity in such arguments of congruence, when only archaeological dates are used for validation (reviewed in Oppenheimer 2004b).

Ideally, any reconstruction of demic migration prehistory should take primary note of genetic rather than cultural or linguistic evidence, because the former is a closer proxy for people than the latter. In the early 1980s, New Guinea entered the arena of analysis of unique, regionally defined and ancient genetic migration markers at an earlier stage than virtually any other region in the world. But the scientific entry to this work came serendipitously through another discipline – the study of the interaction of iron nutrition and anaemia with susceptibility to infections, in particular malaria. Before looking at that, it is worth describing the status and prevalence of iron nutrition and malaria in Near Oceania, at very recent initial contact with the West, in the context of the long Neolithic isolation of the South West Pacific region.

Iron deficiency and malaria: twin Holocene plagues in Neolithic New Guinea?

Iron deficiency may be relatively recent in human evolutionary history, as hunter-gatherers turned to agriculture (Fleming 1982), but is now the most common micronutrient deficiency in the world (Baker and De Maeyer 1979; Oppenheimer and Hendrickse 1983). New Guinea, which had its own Neolithic revolution 10,000 years ago, is no exception to this; one early study in coastal New Guinea recorded some of the lowest levels of liver iron in the world (Disler *et al.* 1973).

Malaria was endemic in Near Oceania at first contact, with a finely tuned latitudinal cline ranging from holo-endemicity along the north coast of New Guinea to absence in New Caledonia, and an altitudinal cline with hypo-endemicity due to temperature-dependent low transmission above an altitude of 1800 m in mainland New Guinea (Flint *et al.* 1986). Eastwards from Near Oceania in the Pacific, malaria decreases sharply in endemicity and is absent in Polynesia. Malaria remains at more or less the same endemicity in Near Oceania as found at first contact and is represented by all four human-specific species. It is likely to have been present from the Pleistocene, judging by the observation that all four malaria-protective mutations in the alpha-globin gene locus reside on characteristic Pacific flanking DNA haplotypes (Flint *et al.* 1986; Hill *et al.* 1989). Because of limitations in the flight-range of mosquitoes, maintenance of transmission is considerably enhanced in sedentary or village-based communities, so malaria would have increased with the onset of sedentism and the Neolithic at the beginning of the Holocene.

The interaction of iron status with malaria

The prevalence of iron deficiency and respiratory infections in New Guinea, combined with previous studies indicating that respiratory infections could be reduced by iron supplementation, suggested the benefit of iron supplementation to health. After it was established that iron deficiency was prevalent among infants in the study population, a prospective, randomized, double-blind, placebo-controlled trial of iron supplementation in infancy was carried out in Madang in north coastal New Guinea (Oppenheimer *et al.* 1984a,b, 1986a,b). Malarial endemicity varied from meso- to holo-endemic in the area. To avoid the known risks of iron therapy in the early neonatal period (Barry and Reeve 1977), a single dose of iron dextran (150 mg elemental iron) was administered at 2 months of age to the treatment group ($n = 236$); control infants ($n = 250$) received an injection of sterile pyrogen-free saline. Infants were re-examined fully 1 week after the injection (with no ill-effects detected) and then again at 6 and 12 months of age. At the field follow-up, blood was also taken for a range of haematological indices, including thick and thin film examination for malaria. In addition, all admissions to hospital were carefully documented. On analysis, no significant differences in malaria rates were seen at the 1-week follow-up visit, but at both the 6- and 12-month visits, malaria slide positivity and spleen rates were higher in the iron treatment group. The increase in malaria slide positivity rate associated with iron was 64% for each visit while spleen rates were 30% to 40% higher (Oppenheimer *et al.* 1986a). Twenty five percent of the iron treatment group had malaria-associated admissions to hospital in the first year of life compared with 17% of the placebo group ($p < 0.05$) (Oppenheimer *et al.* 1986b). However, no effect of iron was seen on parasite densities.

An additional finding was that infants with higher birth haemoglobins (and thus higher total body iron) were significantly more likely to have had malaria at follow-up and were also more likely to have been admitted to hospital with malaria. The latter effect was synergistic with that obtained with iron treatment (Oppenheimer *et al.* 1986b). Thus, infants that both received iron and had a higher birth haemoglobin also had a 40% greater risk of admission with evidence of malaria in the first year of life.

Although increased malaria rates both in hospital admissions and in the field were the most clearly demonstrable results of this study, clinical attacks of malaria were a relatively less common primary reason for admission from the cohort than were acute lower respiratory infections (16% against 63%). All 12 deaths in the study were also primarily due to pneumonia. Rates of admission for pneumonia and time spent in hospital with pneumonia were higher in the group receiving iron dextran supplement. In addition, infants with higher birth haemoglobins were significantly more likely to be admitted to hospital with pneumonia (Oppenheimer *et al.* 1986b). Since this is in conflict with all results of longitudinal studies in infants previously reported from non-malarious areas, it is worth examining the possibility that malaria might have had an effect on susceptibility to pneumonia. Circumstantial evidence is available for this: 89% of pneumonia admissions had evidence of malaria (blood slide positive and/or significant splenomegaly), a much higher rate than that observed in the same cohort in the field. If only cases of pneumonia without evidence of malaria are examined, then there were slightly fewer admissions with pneumonia from the iron dextran group than from the placebo group ($7/231$ vs $10/250$) (Oppenheimer *et al.* 1986b). This was the first prospective long-term intervention study that analysed the interaction of iron and malaria. The results obtained indicate a subtle balance of competition for iron between host and malarial parasite that may have acted more in favour of the host since the start of the Neolithic, with its associated increase in prevalence of iron deficiency; but there were two additional twists.

Alpha-thalassaemia protects against malaria and interacts with the iron effect

A major determinant of iron deficiency developing in the infants under study was the presence of lower than normal birth haemoglobin with consequently lower total body iron stores (Oppenheimer *et al.* 1986a,b). This low birth haemoglobin could not be explained in regression analysis by maternal iron status (Oppenheimer *et al.* 1986c). Suspecting that there could be other causes of low haemoglobin at birth, a parallel study of possible genetic causes had been in place from the start of the cohort recruitment. Initial starch gel electrophoresis on newborns' cord blood revealed that 80.6% of the study cohort had an abnormal band of haemoglobin 'Hb Bartholomew' – evidence of alpha thalassaemia, a known genetic cause of anaemia, resulting from deletion of one or more of the genes coding for haemoglobin (Oppenheimer *et al.* 1984c, 1987). Those newborns whose parents were from the New Guinea highlands did not show this trait, suggesting that the high rate of alpha thalassaemia in the lowland infants was the result of differential malarial selection. This was the highest rate ever detected for any abnormal human genetic trait in the world and was the first indication that there *was* an evolutionary link between alpha thalassaemia and protection against malaria (Oppenheimer *et al.* 1984c).

What was more, there was analytic evidence within the cohort study not only of a protective effect of alpha thalassaemia against the anaemic effects of malaria, but of further interactions, in that the presence of alpha thalassaemia prevented the deleterious effects of iron intervention on malarial experience. This implies a complex differential, ecological interaction between genes, host, nutrition and parasite, affecting and involving the bulk of the lowland New Guinea population. This is the only study to have looked at the three-way interactions between malaria, iron and thalassaemia.

Alpha-thalassaemia gene deletions act as migration markers in the Pacific

Following preliminary communication of these results in the *Lancet* (Oppenheimer *et al.* 1984c, 1987), detailed mapping of alpha-globin gene loci was carried out on DNA from infants from Madang, the Bismarck Archepelago, the neighbouring highlands and then the surrounding regions of the Solomon Islands, Vanuatu and New Caledonia (Flint *et al.* 1986; Oppenheimer 2004a). Three unique alpha-globin gene deletions were found in these regions, which appeared to have potential as markers for the study of migration (Oppenheimer *et al.* 1984c; Flint *et al.* 1986). One of these, the '$\alpha^{3.7III}$ type', has a 3.7-kilobase deletion inactivating one of the two genes that encode for the alpha-globin part of the haemoglobin molecule. The $\alpha^{3.7III}$ type constitutes 77% of alpha-globin gene deletions found in Austronesian speakers of the Bismarck Archipelago. It is also the main type found throughout the rest of island Melanesia, although at lower rates than elsewhere in Melanesia. It is the only type found in Polynesia and constitutes 68% of alpha-globin gene deletions in Micronesia. The $\alpha^{3.7III}$ type is rare in the mainland of New Guinea. This distribution makes the $\alpha^{3.7III}$ type a good candidate marker for the Lapita-associated expansion and also indirectly, by simultaneous geographic association today, for the putative expansion of the Oceanic branch of the Austronesian language family. This was one of the first examples in the literature of the use of highly resolved direct DNA mapping to trace specific migrations (Hill *et al.* 1985). A similar but genetically distinct 3.7-kilobase deletion, type $\alpha^{3.7I}$, is characteristic of the southern half of the New Guinea mainland. It is found only at low rates elsewhere in Melanesia and Micronesia and is absent in Polynesia. The third type, $\alpha^{4.2}$, has a 4.2-kilobase deletion in the other two alpha-globin genes. In Melanesia, $\alpha^{4.2}$ is the main type in, and characteristic of, the northern half of the New Guinea mainland, especially on the north coast where it is found either as a heterozygote or homozygote in up to 90% of the population, and in 100% of deletion alleles in the Admiralty Islands. It also occurs throughout Austronesian speakers (in 25% to 50% of island Melanesian deletion alleles) but is a less common type than the $\alpha^{3.7III}$ deletion. The $\alpha^{4.2}$ deletion is notably not found at all in Polynesia but is present in 26% of Micronesian deletion alleles.

In all of these three alpha-thalassaemia deletions, analysis of the flanking DNA sequences indicates that they are local mutations and unlikely to be recent South East Asian intrusions (Flint *et al.* 1986; Hill *et al.* 1989). In other words, the $\alpha^{3.7III}$ deletion may have travelled with Austronesian speakers out to eastern Polynesia, but it arose on (or acquired by gene conversion) a local DNA framework, somewhere either in Wallacea or along the voyaging corridor around or off the north coast of New Guinea. Not only are they local, but these deletions may also be quite ancient. The unique $\alpha^{3.7III}$ deletion has been around northern island Melanesia long enough to acquire a further mutation that produces a variant haemoglobin molecule called Hb J Tongariki, which is found in a small number of people on Kar Kar Island, off the north coast of New Guinea (Old *et al.* 1980; Hill *et al.* 1989: 246–85).

These latter observations suggest that those Polynesian ancestors carrying the $\alpha^{3.7III}$ deletion, whether they originated in Wallacea or were local to the Bismarcks, must have stopped in Kar Kar Island off the north coast of New Guinea at least long enough to intermarry locally. There is, however, a problem with that supposed genetic interaction. If the Polynesians' ancestors stopped along the north coast of New Guinea long enough to leave the $\alpha^{3.7III}$ deletion it seems strange that they failed to pick up the $\alpha^{4.2}$ deletion at the same time because that mutation is present in over 80% of the people now living on the north coast mainland. The $\alpha^{4.2}$ deletion was carried only to the Solomon Islands and Vanuatu and not further into Polynesia (Hill *et al.* 1985). The only possibility that could explain this selective genetic divergence in northern Melanesia, apart from small canoes and extreme founder effects, is that the contact area where the pre-Polynesians took on (or originated) the $\alpha^{3.7III}$ deletion was exclusively offshore from the New Guinea mainland. By offshore we might include the Bismarck Archipelago (the Western Islands, Manus Island, New Ireland and New Britain) or the nearer volcanic islands off the New Guinea north coast (Kairiru, Mushu, the Schoutens, Manam, Kar Kar and Long Island). But this interpretation would suppose that these sailors bypassed the mainland north coast of New Guinea more or less entirely, in their race out to the Pacific 3,500 years ago. In which case they may have originated in offshore islands such as the Bismarcks (or Wallacea, since the $\alpha^{3.7III}$ deletion is not found in Western Indonesia (Hill *et al.* 1989: 246–85)). Such an interpretation fits the common archaeological model that identifies the proto-Polynesians with Lapita pottery because, with one exception, there are no Lapita pottery sites anywhere on the New Guinea mainland.

Asians in the Pacific? A 1980s view of autosomal loci

The finding of the $\alpha^{3.7III}$ deletion with its clear links from Near Oceania to Polynesia and Micronesia, combined with the unifying simplicity of the rice-driven Taiwan-to-Polynesia hypothesis, stimulated a trawl of genetic evidence in the late 1980s to demonstrate specific genetic substrates from origin-to-terminus of the so-called Express Train expansion, named as such by Diamond (1988). What was remarkable from this huge trawl was how few specific genotypes were shared across the Wallace's Line and that none were shared between Taiwan and Polynesia. Only three genetic marker types were singled out as being shared between South East Asia and the Pacific, but not with the New Guinea highlands (Serjeantson and Hill 1989: 286–8). Unfortunately, even these few lacked sufficient specificity. The first, a triplicated zeta-globin gene, has a low distribution rate across the world, including Africa, while the other two markers are common to all East Asian populations. Thus, while they are indicative of Asian intrusion to the Pacific they can give no measure of dates, and as non-specific haplogroup markers, they cannot be called 'shared haplotypes' or route-markers. Another of these three haplogroup markers, the 9-base pair deletion in mitochondrial DNA, is estimated to be over 50,000 years in age (Redd *et al.* 1995). The third group of shared markers, 'alpha-globin gene haplotypes' (defined by restriction fragment length polymorphisms (RFLP) of the flanking DNA of the alpha-globin gene locus) (Hill *et al.* 1989; Oppenheimer 2004a) gave a clear suggestion of Asian intrusions to the Pacific.

The alpha-globin gene RFLP haplotype distributions (Oppenheimer 2004a) are instructive in that although there is significant overlap, there is a clear tendency for one set of haplotypes to predominate in Central Australia and the New Guinea highlands, and less so in lowland Melanesia, Polynesia, Micronesia and North West Australia. These characteristic Pacific haplotypes could be seen as proxies for the original Pleistocene populations in contrast to the haplotypes characteristic of East Asia which predominate in Micronesia and Polynesia. This differential picture suggests Asian intrusions across the Wallace's Line into lowland Pacific regions and supports the perspective of a New Guinea highlands population that did not admix significantly with these intrusive Asian lineages. Despite the tendency in the late 1980s to talk about 'Mongoloid' admixture (e.g., O'Shaughnessy *et al.* 1990), these labels were misleading and tend to imply a racial stereotype. None of the markers actually determine appearance. It is perhaps more appropriate to label the intrusions as 'Asian', since they are equally characteristic of the Asian mainland and Island South East Asia. In which case Asian intrusions to Micronesia were estimated as 81%, Polynesia 66% and Fiji 42%. It is crucial to realize that these results do not give us any direct dates for these intrusions and thus these non-specific Asian haplotypes cannot, by themselves, be regarded as markers of a recent expansion least of all from Taiwan (Oppenheimer 2004a).

The Wallace's Line and specific alpha-globin gene deletions

When additional highly discriminating alpha-globin markers on the same locus (the specific and unique alpha-globin gene deletions referred to above) were used in combination with the RFLP alpha haplotypes, the picture of a recent massive gene flow from South East Asia to the Pacific appeared even less secure. Haplotypes comprising a unique combination of deletion type and alpha-globin gene RFLP haplotype were clearly divided by Wallace's Line. None of the common combined alpha-gene deletion haplotypes seen in South East Asia are found in the Pacific and vice versa. All the Pacific deletions including the Oceanic '$\alpha^{3.7III}$-deletion' found in Micronesians and Polynesians are characteristic to the Pacific and reside on typical Pacific alpha-globin gene haplotypes (Oppenheimer 2004a). This implies a relative barrier between the Greater Sundas of Island South East Asia and the Pacific, at least as far as the malarial protective markers are concerned. Only North West Australia shows any such evidence of specific gene flow across the Wallace's Line from South East Asia with other intrusions coming apparently from southern New Guinea ($\alpha^{3.7I}$ deletion) (Oppenheimer 2004a).

An important monograph was written in the 1980s summarizing the then current knowledge on numerous genetic markers of the genetic trail into the Pacific, including specifically the globin gene work. The editing authors rejected a sole 'Indigenous Melanesian Origin' for Polynesians. However, they stated in the concluding summary chapter that 'The genetic data have not located a precise "homeland" for the pre-Polynesians, but evidence clearly indicates that they are mainly derived from a South East Asian population prior to Mongoloid expansion' (Serjeantson and Hill 1989: 286). This was in reality a rejection of the out-of-Taiwan view based on current genetic knowledge, but this genetic statement remained unacknowledged until the later work on uniparental loci in the late 1990s confirmed the issue.

Pacific intrusions of Asian mitochondrial DNA haplogroups

The picture of intrusive Asian lineages shown by the alpha-globin haplotypes was echoed in the 1980s with new mitochondrial DNA locus results (Table 2.1). The nine-base-pair deletion which defines the Asian B haplogroup was found throughout lowland regions of the Pacific. A less common Asian intruder, the F haplogroup, was found at low rates in Polynesia and Eastern Indonesia, both of which share several specific haplotypes

Table 2.1. *'Intrusion' of known Asian mitochondrial DNA lineages (mainly B and F haplogroups) beyond the Wallace's Line into Oceania and the proportions of the derivative type the 'Polynesian motif'*

	Eastern Indonesia (%)	PNG Coast (%)	Vanuatu (%)	Micronesia (%)	Samoa (%)	Cook Islands (%)
Admixture of intrusive 'Asian' lineages	31	48	21	84	93	79
Proportion of the above due to the oceanic 'Polynesian motif'	24	79	37	51	85	85

Table 2.2. *Finding exact matches for Polynesian mitochondrial DNA haplotypes in putative homeland regions: New Guinea (Highlands: NH; Coast: NC); Wallacea (W); South East Asia (S); Taiwan (T); and China (C)*

Intrusive Asian haplotypes	HgpB 189, 217	HgpB 189, 217, 261	HgpB-PM 189, 217, 261, 247	HgpF 172, 304	HgpF 172, 304, 311
Matches found	NC, W, S, T, C	NC, W, S, T, C	NC, W	NC, W, S, C	NC, W, S, C
Indigenous SW Pacific haplotypes	M11: 129, 144148, 223241, 265311, 343	R10: 266, 357			
Matches found	NH, NC, W	NH, NC, W			

(Table 2.2), but the F haplogroup is found nowhere else in the Pacific, suggesting a bypass of New Guinea (Oppenheimer 2004a).

The high admixture rates of B haplogroup in the deep Pacific are very similar to the previous alpha-globin gene haplotype results and suggest Asian intrusion. But, again there is a problem of dating based on this high-order haplogroup marker; as mentioned before, the nine-base-pair deletion in mitochondrial DNA is over 50,000 years in age (Redd *et al.* 1995), so cannot be used to date the intrusion. The most important observation in Table 2.1, however, is that the majority of so-called 'intrusive' mitochondrial lineages found in the Pacific belong to a unique Oceanic derivative of the Asian B haplogroup known as the Polynesian motif (PM) (row 2 of Table 2.1).

The distribution of haplogroup B and its derivatives is vast, including most of East Asia and the Americas. All derivatives carry the Asian nine-base-pair deletion but one common East Asian sub-cluster B4a (Kivisild *et al.* 2002) defined initially by mutations in hypervariable segment 1 (HVS 1) at base numbers 16217 and 16261 has mutated further at 16247 to form the main Oceanic variant, PM. Phylogeographic study indicates that the PM variant originates east of the Wallace's Line probably in Wallacea or northern Melanesia, where its diversity is greatest (Richards *et al.* 1998; Oppenheimer 2004a). So haplogroup B distribution suggests Asian intrusions to the Pacific but with further modification after arrival. This again does not support rapid expansion after crossing the Wallace's Line from the Southern Philippines as suggested by the ETT model, rather a period of reflection and development in Wallacea or Melanesia. The difference between the newer highly resolved mitochondrial DNA markers and the older classical and other autosomal markers is that with mitochondrial DNA, there is the opportunity to measure diversity of the derived lineages of a cluster in order to estimate the age of the new mutation.

Examination of diversity in derivatives of the PM shows there was a succession of founding events as the PM cluster spread out to the Pacific. When the divergence times are calculated, the colonization of Central Polynesia (Samoa) and of Eastern Polynesia (Cook Islands) correspond fairly well with known archaeological dates respectively at 3,500 and 1,000 years ago (Richards *et al.* 1998; Oppenheimer 2004a). Further west, there is a different story. Wallacea has the oldest estimate at 17,000 years, but this latter figure, being based on small numbers, has a large standard error. More relevant for intrusions to Melanesia, two independent estimates based on two *much larger* data sets from the New Guinea mainland come out at around 10,000 years (Oppenheimer 2004a). While it is possible that the figure for the New Guinea mainland is over-estimated as a result of diversity carried over from Wallacea, this simply strengthens the evidence for the antiquity of the PM at the southwestern Pacific edge. Indeed, it could be that Melanesia is site of origin of the PM. In any case, this suggests a much older Asian intrusion across the Wallace's Line than that imposed by the ETT.

Not only is the characteristic Polynesian haplogroup B type (the PM) absent from Taiwan and from most of South East Asia, but study of the remaining small percentage of intrusive Asian mitochondrial DNA lineages found in Polynesia fails to reveal any specific haplotype matches with Taiwan apart from the root type precursor to the PM, which in any case has universal distribution throughout China and South East Asia

(Melton *et al.* 1995; Oppenheimer 2004a). By contrast, all mitochondrial DNA haplotypes found in Polynesia, including those indigenous to the Pacific, are also present in Wallacea and northern Melanesia (Table 2.2), consistent with that being the homeland of the expansion giving rise to Polynesians (Oppenheimer 2004a).

Y Chromosome: intrusive Asian lineages in the Pacific

Leaving the bi-parental autosomes and maternally transmitted mitochondrial DNA, we come to the Y chromosome, which is only passed down by males. Dealing first with the two typically Asian haplogroups that may have intruded to the Pacific, there is a slightly different picture in that, at first appearance, the degree of intrusion is surprisingly less, with very low rates in the New Guinea mainland and northern Melanesia and rather higher rates in the deep Pacific. The Y chromosome type M119 is very uncommon in Oceania, but evenly scattered, while M122 achieves frequencies of 30% and 60% respectively in the two locations of Tahiti and Tonga (Oppenheimer 2004a). Analysis of short tandem repeat (STR) haplotypes indicates that the Tahiti figure results from historic Chinese immigration, but the Tonga picture looks like a prehistoric founder event, although no more likely from Taiwan than anywhere else in East Asia (Oppenheimer 2004a).

This surprising picture of apparently low Asian Y chromosome type M119 intrusion to the Pacific has been interpreted as reflecting a predominantly Melanesian origin of Polynesian Y chromosomes; but a slow boat from South East Asia fits the picture better, as one can see looking at the other Y haplogroups. The Y haplotype that dominates Polynesia is known as M38, a type that is extremely uncommon in the putative Pleistocene New Guinea highlands populations. Ultimately it belongs to a widespread ancient Asian 'haplogroup 10' (Hg10) defined by an RPS4Y marker. Being one of the earliest introductions to Asia and Australia, Hg10 is mainly found in those regions as locally mutated derivatives. The ancestral form has been found only in Wallacea, a region where Hg10 has also acquired a new Oceanic mutation M38. M38 is the only Hg10 type found in the rest of the Pacific (Oppenheimer 2004a). Arguably, M38 could be the male analogue of the Polynesian motif in that it is ultimately Asian, may have originated around the beginning of Holocene in Wallacea or lowland Melanesia, is absent from the New Guinea highlands and has achieved high founder frequencies in Polynesia. Age estimates for the gene cluster holding the M38 mutation are about 11,500 years for the western Pacific, with an expansion signal around 5,000 years ago. When the analysis was restricted to Polynesians, a strong expansion signal dated to about 2,200 years ago (Kayser *et al.* 2000).

The Polynesian New Guinea bypass supports a separate early Holocene mainland Austronesian colony

The characteristic or indigenous mitochondrial DNA and Y chromosome markers in New Guinea seem to echo the distribution of the indigenous alpha-thalassaemia mutations mentioned earlier ($\alpha^{3.71}$ and $\alpha^{4.2}$ deletions), in that they do not feature among Polynesians. A total of six common markers found in both Austronesian and non-Austronesian speakers of north coastal New Guinea fail to appear in Polynesians (Oppenheimer 2004a). This seems to be consistent with the view that the expansion giving rise to Polynesians bypassed the New Guinea mainland. However, if one accepts that the ancestors of the Polynesians did not call in on mainland New Guinea, it then follows that the very high rates of intrusive Asian lineages found in lowland New Guinea (48% Asian intrusion, with 79% of that being PM (Table 2.1) (Oppenheimer 2004a)) must derive from another independent expansion. The dates of that expansion based on the imprecise genetic clock cited above appear to lie somewhere between 10,000 and 5,000 years ago. There is some archaeological and linguistic evidence to suggest this may have been an earlier Austronesian-speaking expansion (Oppenheimer 2004a).

Conclusions

By virtue of its shallow continental shelf, Island South East Asia has been the recipient of mass Asian expansions for as long as humans have been in Asia. Beyond Borneo and Java on the other hand, Wallacea, New Guinea, Near Oceania and Australia have been partially protected from mass migration by the permeable marine filter of Wallace's Line. The morphological and genetic transition zone of Wallacea, the other side of the line, may reflect this filter effect rather than complete physical and cultural isolation. In spite of the problems of open-sea voyages, archaeological evidence of human migrations and multiple maritime colonizations of the South West Pacific region go back to 40,000 to 60,000 years.

A long-held view of "Two Peoples, Two cultures, Two Periods", equating the Pleistocene with primitiveness and the Late Holocene with sophistication, which found its ultimate expression in the rice-driven 'Express Train from Taiwan to Polynesia' 3,500 years ago has been argued to be too simplistic. There have been many Pacific expansions, not two, and there is evidence that maritime range expansions in the Pacific continued throughout the Late Pleistocene, earlier than anywhere else in the world. Long-distance maritime movement of obsidian (presumed trading) started from 20,000 years ago. Horticulture and arboriculture started in the early Holocene in Near Oceania, earlier apparently than in Island South East Asia to the west of the Wallace's Line; and the Neolithic subsistence base of Oceanic populations was derived from South East Asia and New Guinea, never having been rice-based.

An early indigenous Neolithic in New Guinea may have meant that iron deficiency started as a major culturally induced change in the South West Pacific even earlier than it did in the Near East. It was the attempt to study the effect of correcting severe iron deficiency in infancy in north coast New Guinea that serendipitously offered the first opportunity to find a genetic substrate for these supposed migrations. This gave rise to one of the earliest examples of the use of unique mutational variants to trace migrations.

As a by-product of a trial of iron supplementation in Madang, an extraordinarily high rate of alpha thalassaemia was discovered. Not only was this a major cause of anaemia in the population, but it was shown that, like iron deficiency, it was protective against malaria with a complex interaction with iron status. Alpha thalassaemia appeared to have been selected for by malaria in lowland areas of Near Oceania. In the early 1980s, detailed DNA mapping of the alpha-globin gene in these areas revealed three different and unique gene deletions that caused the disorder. These deletional mutations were indigenous and geographically distributed. One, the '$\alpha^{3.7III}$ deletion', appeared to be a good candidate marker for the expansion giving rise to the Polynesian dispersal. The relative distribution of the other two appeared to suggest that the Polynesian dispersal had bypassed the New Guinea mainland. Further detailed analysis of uniparental DNA markers (mitochondrial DNA and the Y chromosome) in the last eight years has been consistent with this view and has allowed an alternative theory of several Pacific expansions from admixed Asian/Pacific communities in Wallacea, spanning the Holocene, carrying both Oceanic and typically Asian branch lineages, the latter represented by specific variants derived from further mutation while in Wallacea. While the genetic connections across the Wallace's Line imply a complex early Holocene or even Pleistocene history of secondary colonization from Asia, there is no genetic evidence for any direct demic connection between Taiwan and Polynesia as required by the Express Train hypothesis.

References

Allen, J. (2000). From beach to beach: the development of maritime economies in prehistoric Melanesia. In *East of Wallace's Line: Studies of Past and Present Maritime Cultures of the Indo-Pacific Region. Modern Quaternary Research in Southeast Asia*, ed. S. O'Connor and P. Veth, Vol. 16. Rotterdam/Brookfield, VT: A.A. Balkema, pp. 139–76.

Allen, J. and Gosden, C. (1996). Spheres of interaction and integration: modelling the culture history of the Bismarck Archipelago. In *Oceanic Culture History: Essays in Honour of Roger Green*, ed. J. M. Davidson, G. Irwin, B. F. Leach, A. Pawley and D. Brown. Dunedin: New Zealand Journal of Archaeology Special Publication, pp. 183–97.

Anderson, A. J. (2000). Slow boats from China: issues in the prehistory of Indo-Pacific seafaring. *Modern Quaternary Research in Southeast Asia* **16**, 13–50.

Baker, S. J. and De Maeyer, E. M. (1979). Nutritional anaemia: its understanding and control with special reference to the work of the World Health Organization. *American Journal of Clinical Nutrition* **32**, 368–417.

Barry, D. M. J. and Reeve, A. W. (1977). Increased incidence of gram-negative neonatal sepsis with intramuscular iron administration. *Pediatrics* **60**, 908–12.

Bellwood, P. (1997). *Prehistory of the Indo-Malaysian Archipelago*. Hawaii: University of Hawaii Press.

Blust, R. (1985). The Austronesian homeland, a linguistic perspective. *Asian Perspectives* **26**, 45–67.

Bulmer, S. (2001). Lapita dogs and singing dogs and the history of the dog in New Guinea. In *The Archaeology of Lapita Dispersal in Oceania, Papers from the Fourth Lapita Conference, June 2000*, Canberra, Australia, ed. G. R. Clark, A. J. Anderson and T. Vunidilo. *Terra Australis* **17**. Canberra, Australia: Pandanus Books, pp. 183–201.

Denham, T. P., Haberle, S., Lentfer, C., *et al.* (2003). Origins of agriculture at Kuk Swamp in the Highlands of New Guinea. *Science* **301**, 189–93.

Diamond, J. (1988). Express train to Polynesia. *Nature* **336**, 307–8.

Diamond, J. and Bellwood, P. (2003). Farmers and their languages: the first expansions. *Science* **300**, 597–603.

Disler, P., Charlton, R. W., Bothwell, T. H. and Karik, J. (1973). Hepatic iron storage concentrations in the inhabitants of Papua New Guinea. *South African Journal of Medical Science* **38**, 91–4.

Fleming, A. F. (1982). Iron deficiency in the tropics. *Clinical Haematology* **11**, 365–88.

Flint, J., Hill, A. V. S., Bowden, D. K., *et al.* (1986). High frequencies of alpha thalassaemia are the result of natural selection by malaria. *Nature* **321**, 744–50.

Groube, L., Chappell, J., Muke, J. and Price, D. (1986). A 40,000 year-old human occupation site at Huon Peninsula, Papua New Guinea. *Nature* **324**, 453–5.

Hide, R. (2003). *Pig Husbandry in New Guinea. A Literature Review and Bibliography*. ACIAR Monograph No. 108, 12–13.

Hill, A. V. S., Bowden, D. K., Trent, R. J., *et al.* (1985). Melanesians and Polynesians share a unique alpha thalassaemia mutation. *American Journal of Human Genetics* **37**, 571–80.

Hill, A. V. S., O'Shaughnessy, D. F. and Clegg, J. B. (1989). Haemoglobin and globin gene variants in the Pacific. In *The Colonisation of the Pacific: A Genetic Trail*, ed. A. V. S. Hill and S. W. Serjeantson. Oxford: Clarendon Press, pp. 246–85.

Kayser, M., Brauer, S., Weiss, G., *et al.* (2000). Melanesian origin of Polynesian Y chromosomes. *Current Biology* **10**, 1237–46.

Kivisild, T., Tolk, H.-V., Parik, J., *et al.* (2002). The emerging limbs and twigs of the East Asian mtDNA tree. *Molecular Biology and Evolution* **19**, 1737–51.

Larson, G., Dobney, K., Albarella, A., *et al.* (2005). Worldwide phylogeography of wild boar reveals multiple centers of pig domestication. *Science* **307**, 1618–21.

Latinis, D. K. (2000). The development of subsistence system models for Island Southeast Asia and Near Oceania: the nature and role of arboriculture and arboreal-based economies. *World Archaeology* **32**, 41–67.

Macaulay, V., Hill, C. Achilli, A., *et al.* (2005). Single, rapid coastal settlement of Asia revealed by analysis of complete mitochondrial genomes. *Science* **308**, 1034–6.

Meacham, W. (1985). On the improbability of Austronesian origins in South China. *Asian Perspectives* **26**, 89–106.

Melton, T., Peterson, R., Redd, A. J., *et al.* (1995). Polynesian genetic affinities with Southeast Asian populations as identified by mtDNA analysis. *American Journal of Human Genetics* **57**, 403–14.

Morwood, M. J., O'Sullivan, P. B., Aziz, F. and Raza, A. (1998). Fission-track ages of stone tools and fossils on the east Indonesian island of Flores. *Nature* **392**, 173–6.

Morwood, M. J., Soejono, R. P., Roberts, R. G., *et al.* (2004). Archaeology and age of a new hominin from Flores in eastern Indonesia. *Nature* **431**, 1087–91.

Old, J. M., Clegg, J. B., Weatherall, D. J. and Booth, P. B. (1980). Hb J Tongariki is associated with alpha thalassaemia. *Nature* **273**, 319–20.

Oppenheimer, S. (1998). *Eden in the East: The Drowned Continent of Southeast Asia*. London: Weidenfeld and Nicolson.

(2003). *Out of Eden: The Peopling of the World*. London: Constable.

(2004a). Austronesian spread into Southeast Asia and Oceania: where from and when. In *Pacific Archaeology: Assessments and Prospects: Proceedings of the International Conference for the 50th Anniversary of the First Lapita excavation*, ed. C. Sand. Kone, Noumea: Département Archéologie, Caledonie, Service des Musées de Nouvelle Caledonie, pp. 54–70.

(2004b). The 'Express Train from Taiwan to Polynesia': on the congruence of proxy lines of evidence. *World Archaeology* **36**, 591–600.

Oppenheimer, S. and Hendrickse, R. (1983). The clinical effects of iron deficiency and iron supplementation. *Nutrition Abstracts and Reviews: Reviews in Clinical Nutrition* **53**, 585–98.

Oppenheimer, S. and Richards, M. (2003). Polynesians: devolved Taiwanese rice farmers or Wallacean maritime traders with fishing, foraging and horticultural skills. In *Examining the Farming/Language Dispersal Hypothesis*, ed. P. Bellwood and C. Renfrew. McDonald Institute Monographs. Cambridge: McDonald Institute for Archaeological Research, pp. 287–97.

Oppenheimer, S. J. and Richards, M. (2001). Polynesian origins: slow boat to Melanesia? *Nature* **410**, 166–7.

Oppenheimer, S. J., Macfarlane, S. B. J., Moody, J. B., *et al.* (1984a). Iron and infection in infancy. Report on field studies in Papua New Guinea. 1. Demographic description and pilot surveys. *Annals of Tropical Paediatrics* **4**, 135–43.

Oppenheimer, S. J., Hendrickse, R. G., Macfarlane, S. B. J., *et al.* (1984b). Iron and infection in infancy. Report of field studies in Papua New Guinea. 2. Protocol and description of study cohort. *Annals of Tropical Paediatrics* **4**, 145–53.

Oppenheimer, S. J., Higgs, D. R., Weatherall, D. J., Barker, J. and Spark, R. A. (1984c). Alpha thalassaemia in Papua New Guinea. *Lancet* **I**, 424–6.

Oppenheimer, S. J., Gibson, F. D., Macfarlane S. B. J., *et al.* (1986a). Iron supplementation increases prevalence and effects of malaria. Report on clinical studies in Papua New Guinea. *Transactions of the Royal Society of Tropical Medicine and Hygiene* **80**, 603–12.

Oppenheimer, S. J., Macfarlane, S. B. J., Moody, J. B., Bunari, O. and Hendrickse, R. G. (1986b). Effect of iron prophylaxis on morbidity due to infectious disease. Report on clinical studies in Papua New Guinea. *Transactions of the Royal Society of Tropical Medicine and Hygiene* **80**, 596–602.

Oppenheimer, S. J., Macfarlane, S. B. J., Moody, J. B. and Harrison, C. (1986c). Total dose iron infusion malaria and pregnancy in Papua New Guinea. *Transactions of the Royal Society of Tropical Medicine and Hygiene* **80**, 818–22.

Oppenheimer, S. J., Hill, A. V. S., Gibson, F. D., *et al.* (1987). The interaction of alpha thalassaemia with malaria. *Transactions of the Royal Society of Tropical Medicine and Hygiene* **81**, 322–6.

O'Shaughnessy, D. R., Hill, A. V. S., Bowden, D. K., Weatherall, D. J. and Clegg, J. B. (1990). Globin gene in Micronesia: origin and affinities of Pacific Island peoples. *American Journal of Human Genetics* **46**, 144–55.

Pawley, A. (2003). The Austronesian dispersal: languages, technologies and people. In *Examining the Farming/Language Dispersal Hypothesis*, ed. P. Bellwood and C. Renfrew. McDonald Institute Monographs. Cambridge: McDonald Institute for Archaeological Research, pp. 251–74.

Paz, V. (2003). Island Southeast Asia: spread or friction zone? In *Examining the Farming/Language Dispersal Hypothesis*, ed. P. Bellwood and C. Renfrew. McDonald Institute Monographs. Cambridge: McDonald Institute for Archaeological Research, pp. 275–86.

Redd, A. J. and Stoneking, M. (1999). Peopling of Sahul: mtDNA variation in Aboriginal Australian and Papua New Guinean populations. *American Journal of Human Genetics* **65**, 808–28.

Redd, A. J., Takezaki, N. J., Sherry, S. T., *et al.* (1995). Evolutionary history of the COII/tRNA(Lys) intergenic 9-base-pair deletion in human mitochondrial DNAs from the Pacific. *Molecular Biology Evolution* **12**, 604–15.

Richards, M., Oppenheimer S. J. and Sykes, B. (1998). MtDNA suggests Polynesian origins in eastern Indonesia. *American Journal of Human Genetics* **63**, 1234–6.

Roe, D. (1992). Investigations into the prehistory of the central Solomons: some old and some new data from Northwest Guadalcanal. In *Poterie Lapita et Peuplement*, ed. J.-C. Galipaud. Nouméa: ORSTOM.

Serjeantson, S. W. and Hill, A. V. S. (1989). The colonization of the Pacific, the genetic evidence. In *The Colonization of the Pacific, a Genetic Trail*, ed. A. V. S. Hill and S. W. Serjeantson. Oxford: Clarendon Press, pp. 286–94.

Shutler, R. and Marck, J. C. (1975). On the dispersal of the Austronesian horticulturalists. *Archaeology and Physical Anthropology in Oceania* **10**, 81–113.

Solheim II, W. (1994). Southeast Asia and Korea from the beginnings of food production to the first states. In *The History of Humanity*, ed. S. J. De Laet. London: Routledge, pp. 468–81.

   (1996). The Nusantao and north–south dispersals. In *The Chiang Mai Papers*, ed. P. Bellwood, Vol. 2. *Bulletin of the Indo-Pacific Prehistory Association* **15**, 101–9.

Swadling, P. (1997). Changing shorelines and cultural orientations in the Sepik-Ramu, Papua New Guinea: implications for Pacific prehistory. *World Archaeology* **29**, 1–14.

Szabó, K. and O'Connor, S. (2004). Migration and complexity in Holocene Island Southeast Asia. *World Archaeology* **36**, 621–8.

Terrell, J. E. (1986). *Prehistory in the Pacific Islands*. Cambridge: Cambridge University Press.

Terrell, J. E., Kelly, K. M. and Rainbird, P. (2001). Foregone conclusions? In search of 'Austronesians' and 'Papuans'. *Current Anthropology* **42**, 97–124.

Underhill, P. A., Passarino, G., Lin, A. A., *et al.* (2001). The phylogeography of Y chromosome binary haplotypes and the origins of modern human populations *Annals of Human Genetics* **65**, 43–62.

Welch, D. J. (2002). Archaeological and palaeoenvironmental evidence of early settlement in Palau. *Bulletin of the Indo-Pacific Prehistory Association* **22**, 161–73.

White, J. P., Flannery, T. F., O'Brien, R., Hancock, R. V. and Pavlish, L. (1991). The Balof Shelters, New Ireland. In *Report of the Lapita Homeland Project*, ed. J. Allen and C. Gosden. Occasional Paper in Prehistory 20. Canberra: Department of Prehistory, Research School of Pacific Studies, Australian National University, pp. 46–58.

3   Biocultural adaptation and population connectedness in the Asia-Pacific region

RYUTARO OHTSUKA

Introduction

This chapter focuses on physiological adaptations of speakers of Non-Austronesian (NAN) and Austronesian (AN) languages in space and time, with some reference to Aboriginal Australian populations. The language-based dichotomous classification of Oceanian populations into the NAN and AN language groups has been, in general, in agreement with biological traits observed in either (Kirk and Szathmary 1985; Hill and Serjeantson 1989; Attenborough and Alpers 1992; Yoshida *et al.* 1995; Ohashi *et al.* 2000, 2003). In relation to colonization histories, the descendants of first-wave migrants belong to the NAN language group. Their ancestors reached Sahul Land from Asia, subsequently some of them dispersing into present-day Australia, becoming Aboriginal Australians. Second-wave migrants and their descendants, represented by Lapita people, were and continue to be AN speakers. The Lapita people dispersed later from their homeland in island Melanesia in Near Oceania (Green 1991) into Remote Oceania, or islands in South East Melanesia as well as Polynesia and Micronesia. The extent of admixture of NAN and AN language groups in island Melanesia and later in wider Melanesia remains debated, however. As demonstrated by several studies (Serjeantson *et al.* 1983; Friedlaender 1987), genetic traits and languages do not always correlate, due to replacement of languages and/or pidginization or creolization of the languages themselves; for similar reasons, links between Aboriginal Australian languages and NAN languages have scarcely been found (Foley 1986).

Present-day AN speakers (except their outliers in Madagascar, who exceed 10 million in number) are roughly distributed as follows: 212 million in Indonesia, 23 million in Malaysia, 80 million in the Philippines, 300,000 in Taiwan, 1 million in Vietnam and 2 million in Oceania. NAN speakers number some four million, and exclusively inhabit Papua New Guinea

Health Change in the Asia-Pacific Region: Biocultural and Epidemiological Approaches, ed. Ryutaro Ohtsuka and Stanley J. Ulijaszek. Published by Cambridge University Press.

(PNG), some of the Solomon Islands in Oceania, and Irian Jaya (West New Guinea) and its vicinities in Indonesia.

Adaptation of Pleistocene migrants

The earliest migrants to Sahul Land came around 60,000 years ago; however, there is uncertainty about migrant population size and the number of migrations that took place. A comparison of within-population nucleotide diversity of the complete mitochondrial genomes of six populations (African, European, East Asian, South East Asian, NAN speaking and Native American) shows it is highest among Africans (2.11), followed by four populations with similar levels of diversity (1.37 for East Asians, 1.33 for South East Asians, 1.31 for NAN and 1.30 for Native Americans) (Horai *et al.* 1996). Similar results have been obtained from a different analysis using the same genetic traits for five populations: African, New Guinean (both NAN and AN language speakers), Asian, Aboriginal Australian and European (Ingman and Cyllensten 2003). It is therefore more plausible that multiple migrations took place rather than a single migration, even if with very heterogeneous groups.

Little is known of the ways of life of early inhabitants of Wallacea (present-day East Indonesia), or of the early settlers in Sahul. According to a palaeoecological review of Nusa Tenggara and Maluku in East Indonesia (Monk *et al.* 1997), the earliest signs of human occupation, in Serum and Morotai Islands, are at about 40,000 years ago, with no concrete evidence of the subsistence adaptations of the migrant population (Bellwood 1991; Ellen 1993). Ethnographic studies of hunter-gatherer societies can be used to shed some light on past subsistence practices. Among these, the ecological and nutritional surveys carried out among Aboriginal Australian populations in the coastal savanna of northern Arnhem Land in the late 1940s, when there was no external influence (McCarthy and McArthur 1960), and in the 1970s, when there was slight external influence (Jones 1980), are particularly important. In both cases, there was heavy dependence on animal foods. A variety of fish, crustaceans and shellfish in brackish and fresh waters and birds in swamps accounted for 70% to 95% of the total caloric intake, despite wild plant foods such as tubers and cycad seeds being plentiful. This adaptive strategy differs from the common pattern of tropical hunter-gatherers, who utilize many more foods of plant origin. This discrepancy has been explained by abundance of animal foods and smaller labour input required for obtaining and processing them, relative to plant foods.

The majority of early migrant groups to Sahul may have inhabited animal-rich environments. For instance, the Arafura Plain, which once existed around the present-day Arafura Sea, would have been populated, with subsistence adaptations changing in response to new environments encountered with population movement. On the Australian side, people moved southwards along the eastern coast and later along the western coast, and finally entered the dry land in the inner part, where their subsistence changed to diets dominated by plant foods. On the New Guinea side, sago palms of the *Metroxylon* species growing in the lowland plains would have been an important natural starch resource available to settlers (Swadling 1983); this palm does not grow in Australia. Botanical studies suggest that *Metroxylon* sago originated in New Guinea and Wallacea (Flach and Rumawas 1996), and it is very possible that the first wave of migrants were already acquainted with sago-utilizing techniques. Peoples dependent on palm sago effectively obtain adequate food energy from it, although sago flour contains negligible amounts of nutrients other than carbohydrate, so that such people are required to find animal and other plant foods to obtain nutritional balance (Townsend 1971; Ohtsuka 1983; Ulijaszek 2001).

The descendants of the first wave of migrants dispersed across the New Guinea island and into island Melanesia. The New Guinea Highlands was peopled up to an altitude of around 2,000 m above sea level more than 20,000 years ago (White and O'Connell 1982). In island Melanesia, archaeological studies have shown that New Ireland and New Britain, in the Bismarck Archipelago, were inhabited about 35,000 years ago (Allen and Gosden 1996). Findings from archaeological sites dated to around 20,000 years ago suggest that people depended heavily on animal foods, consuming a variety of terrestrial animals such as birds, lizards, snakes and rats, and relatively fewer fish and shellfish, as well as naturally growing *Colocasia* tubers (Spriggs 1997).

Adaptation of Austronesian migrants

Following the 'slow boat' scenario of the peopling of this region, AN-speaking peoples who inhabited island South East Asia migrated to New Guinea and then moved along its northern coast to reach island Melanesia. The timing of this migration is unclear because there is little archaeological evidence from western and northern New Guinea that can shed light on this issue.

A two-stage model is widely accepted as the way in which agriculture was developed in South East Asia (Gorman 1977). The first stage is viewed as having involved domestication of locally grown plant crops, such as taro, yam and banana, more than 10,000 years ago. Following this, rice cultivation, which began perhaps in South China, was introduced on the other side of the Wallace's Line. Rice cultivation along the Yangtze River took place around 10,000 years ago, becoming widespread across South China about 7,000 to 5,000 years ago, and then to mainland South East Asia, including northern Vietnam, Thailand and Peninsular Malaysia, about 3,500 to 3,000 years ago (M. Nishitani, personal communication). However, the colonization of island South East Asia by AN-speaking peoples is viewed to have progressed from about 5,000 years ago (Horridge 1995). Subsistence in island South East Asia is likely to have been based on the cultivation of tubers and bananas and exploitation of sago palms at the time when the AN-speaking migrants came to Oceania (Spencer 1963).

Information concerning the subsistence patterns of early inhabitants of East Indonesia remains poor. However, a food production economy was in place about 6,000 to 5,000 years ago. Pigs, which are likely to have been introduced from the west, were discovered among fossils in sediments in Timor and Sawu/Roti, dated to 5,500 years ago (Glover 1971; Ellen 1993). The oldest potteries, dated to between 5,000 and 4,500 years ago, were discovered in Timor (Bellwood 1991; Rowland 1992). Furthermore, bones of cuscuses, the marsupial species endogenous in the Australian zoogeographic region, were excavated from the same sediment in Timor, suggesting that a trade network between East Indonesia and New Guinea was already in place by then (Bellwood 1991).

Pigs are the most important domestic animals throughout Melanesia for consumption at rituals as ceremonial gifts, and to a much lesser extent as a daily source of protein. The earliest evidence for pig husbandry in the Pacific comes from fossils found in the Kria cave in the western end of Irian Jaya, dated to around 4,000 years ago using accelerator mass spectroscopy (Hide 2003). It is most likely that pigs were brought to New Guinea by the early AN-speaking settlers, carried to island Melanesia with their descendants, and then spread to mainland New Guinea, occasionally in feralized forms.

From the point of view of subsistence, the Lapita cultural complex was characterized by taro, yam and banana horticulture of South East Asian origin, and the husbandry of pigs, dogs and chickens. The Lapita people were also skilled foragers of marine resources, although it has been questioned whether these techniques were brought from their homeland or developed in island Melanesia. In case of the former, long habitation in East Indonesia would have favoured the development of such foraging techniques, although in immature form. In case of the latter,

NAN-language–speaking people would have already developed such techniques, while the Lapita people improved and incorporated them within their subsistence strategies. Either way, their subsistence repertoire was highly adaptive to their island environment, allowing population increase. This is likely to have led to shortages of agricultural land and natural resources in subsequent generations. Shortages of food and of the forest, which provided building materials and fuel, may have triggered migration into Remote Oceania from 3,200 years ago.

Change and diversification in New Guinea

At the time of the arrival of Europeans in Melanesia, a vast majority of populations were agriculturalists to varying degrees, in the extent and intensity of the use of different crops, and in their husbandry of animals, pigs in particular. This is in contrast to Aboriginal Australians, who did not use much in the way of domesticated plants and animals. All major food crops, except the naturally occurring *Metroxylon* sago palms, were cultivars of Asian origin.

The prehistory of agricultural development in Melanesia has been best investigated in relation to change of plant resource use by NAN-language–speakers in the New Guinea Highlands, based on extensive archaeological excavations in the Wahgi Valley. Six agricultural phases have been identified from evidence of development of the drain networks in this wetland environment and the archaeological dating of ashes and other remains (Golson 1977; Golson and Gardner 1990). Phase one is dated to between 9,000 and 6,000 years ago, in which pioneering agriculture began, although the plant species cultivated have not been identified. Phase two of 6,000 to 5,500 years ago and phase three of 4,000 to 2,500 years ago involved the cultivation of *Colocasia* taro, Fe'i banana (*Australimusa*; *Musa fei*) and some other indigenous plants. Phase four of 2,000 to 1,200 years ago and phase five of 400 to 250 years ago involved intensive taro cultivation. Phase six (250 to 100 years ago) involved the cultivation of sweet potato, which subsequently became the dominant crop.

Agricultural development in the Wahgi Valley, even during phases one to four, was a consequence of human responses to resource depletion in the Highlands, where the absence of the malarial parasites *Plasmodium vivax* and *P. malariae* and of anopheline mosquito vectors favoured human population increase (Groube 1989, 1993). Taro grown in phases four and five might have involved the use of cultivars introduced by AN-language–speakers. Sweet potato, which has been a dominant crop throughout the

New Guinea Highlands prior to colonial contact, was introduced about 350–300 years ago. This crop is of South American origin and was introduced to the Philippines by Spanish colonists, and thereafter carried to East Indonesia (perhaps, Maluku) and New Guinea by traditional trade routes.

Although the agricultural revolution in the Wahgi Valley was based on the control of water, many populations in New Guinea have depended on slash-and-burn cultivation, differing markedly from place to place in productivity mostly because of differences in the duration of fallow practised. Records of agriculture among contemporary New Guinea populations show that fallow periods can vary by a factor of ten. The short fallow of a few months before new planting is observed in sweet potato cultivation in the Highlands, whereas the opposite extreme of a fallow of 30 years or more for the cultivation of taro can be found in the highland fringe, giving a difference of more than 20-fold greater long-term land productivity (Kuchikura 1994). Short-fallow or semi-permanent cultivation of sweet potato in the Highlands is possible only because of geographical advantages of this location. These include high soil fertility, suitable rainfall patterns (1,500 to 2,500 mm per year, with seasonal variation) and temperatures low enough to deter the rapid growth of weeds (Allen 1983). Differences in agricultural productivity such as those seen across the New Guinea island strongly influence both human carrying capacity and population density (Brown and Podolefsky 1976).

Using fertility and mortality patterns of New Guinea Highland populations soon after European contact in the 1950s and the 1960s, van Baal (1961) and Stanhope (1970) estimated that prior to European contact, the Highlands probably underwent slow population expansion, with population overspill into the highland fringe and the lowlands, where population decreased due to mortality from malaria and other infectious diseases. Epidemiological studies (Riley 1983; Riley and Lehmann 1992) have supported the view that malaria was, and continues to be, the most significant determinant of population distribution and population density in New Guinea. The 'population sink' model of highland fringe depopulation is an extreme view of population change. However, diverse adaptations to varying agricultural productivities, and malarial and other infectious disease mortalities have contributed to varying population densities in recent years, from more than 60 persons per km^2 among sweet potato cultivators in the Highlands to less than 5 (and usually less than 2) persons per km^2 among sago-dependent populations in the swampy lowlands, with intermediate population densities in coastal lowlands and islands (Hanson *et al.* 2001).

From Lapita homeland to Remote Oceania

Successful subsistence adaptations of descendants of AN-language–speaking second-wave settlers in island Melanesia (Bismarck and the main Solomon Islands Archipelago) allowed rapid population increase, which, in turn, triggered their dispersal towards the uninhabited islands of Remote Oceania located further south and east. Vanuatu and New Caledonia had seen human habitation about 3,200 years ago, the same being true of Fiji, Tonga and Samoa about 3,000 years ago (Spriggs 1997). The settlers of Polynesia may have been direct descendants of the bearers of Lapita culture, with various genetic and linguistic studies suggesting that these people are likely to have been highly homogeneous. Furthermore, contemporary Polynesians are genetically more similar to AN-language–speaking Asians than to NAN-language–speaking Melanesians, even though AN-language speakers may have undergone some admixture with NAN-language speakers in the Lapita homeland of the Bismarck and Solomon Islands Archipelago.

The first arrival of the Lapita people in Micronesia, for example in the Mariana Islands, took place around 3,200 to 3,000 years ago, at a rate similar to the southeastward expansion of Lapita peoples. Linguistic and archaeological studies (Pawley and Ross 1993; Intoh 1997) suggest, however, that western Micronesia was colonized directly from South East Asia, presumably the Philippines.

The Lapita people who moved to the southeast experienced a sedentary period of about 500 years in West Polynesia, followed by a return to long-distance voyaging to the east. They reached the Society Islands about 2,500 years ago, the Marquesas Islands about 2,000 years ago, the Hawaii and Rapanui (Easter) Islands about 1,500 years ago, and finally New Zealand and its adjacent islands about 1,000 to 500 years ago (Spriggs 1997). The original Lapita culture changed considerably across this period, especially between 2,500 and 2,000 years ago. Pottery gradually changed in style and eventually dropped from their cultural repertoire. The terms 'Polynesian culture' and 'Polynesians' are commonly used to denote the culture and the people from this stage onwards, replacing the terms 'Lapita culture' and 'Lapita people', which denote earlier stages. Within the past 2,000 years, some Polynesians migrated westwards and northwards to become Outlier Polynesians in Melanesia and Micronesia.

These migrations took place in response to population increase in association with decreases in land availability and resources. For the same reasons, Polynesians occupied some 20 tiny islands, such as Henderson, Pitcairn and Fanning Islands, scattering themselves across Polynesia. Because archaeological traces of Polynesian habitation have been found in these islands in the absence of human habitation at the time of early European exploration, they have been called 'mystery islands' (Kirch 1988). Various archaeologists believe that Polynesians voyaged to South America (Finney 1994), the successful returnees bringing with them sweet potato of South American origin, perhaps around 1,000 years ago (Kirch 2000). Parallel to these efforts of colonization of new habitats, agricultural expansion and intensification progressed (Denoon 1997) and stratified chiefdoms were established (Kirch 2000) in Polynesia and to lesser extents in Micronesia, as observed ethnographically in the nineteenth and twentieth centuries.

The large-scale dispersal of Polynesians into island Pacific also involved physiological adaptation; they developed a large and muscular body physique to adapt to cold temperatures, to which they were exposed during sea voyages in particular, in contrast to the smaller body size of inland peoples on the large islands of Oceania (Houghton 1991). Alternatively, Katayama (1996) has suggested that the hypermorphic features of Polynesians were formed as the result of adaptation to marine-resource–rich environments in the Lapita homeland, this adaptation being strengthened when they populated previously uninhabited Polynesian islands also rich in such resources. From skeletal evidence, Katayama (1996) observed that the morphology of the Lapita people was similar to, but slightly smaller than, that of Polynesians. The hypotheses of both Houghton (1991) and Katayama (1996) are in good agreement with the known living and migration histories of AN-language–speaking populations of the Pacific.

Contemporary changes in human biology: some case studies from Papua New Guinea

Oceanian peoples changed their adaptive strategies across prehistory, differing not only between NAN- and AN-language–speaking groups but also within the former, who inhabited much more diversified environments and whose subsistence patterns have been much more heterogeneous. With European contact and colonization, many things changed, with significant impacts on health and survivorship. However, many Papua New Guinea (PNG) populations remained relatively isolated, experienced differing environmental stresses and underwent differing adaptations. In what follows, anthropometric and other biological data for several populations in PNG, which were collected by the author and his colleagues in the last several decades, are compared to identify some of the spatio-temporal differences in biocultural adaptation among them.

Papua New Guinea, an independent nation since 1975, comprises the eastern half of the New Guinea island and the Bismarck Archipelago. Human populations include both NAN- and AN-language–speaking groups living in diverse environments, including island, coastal, lowland, highland fringe and highland locations. The term 'Highlands', as used here, refers to the high-altitude area where sweet potato cultivation has been a dominant subsistence activity, population density has been high and cultural heritage has been similar (Feil 1987; Sillitoe 1988). The inhabitants of this country have survived in small groups that operate within well-defined boundaries and in relative isolation, compared to the island nations in Polynesia and Micronesia (Bayliss-Smith 1977).

Comparison of body composition among five populations

Adult body composition and protein intake are compared among populations of five different language groups in the 1980s and the early 1990s. These populations had scarcely admixed with other groups at least for the past several decades, and these are (1) Gidra (NAN-language speakers) of the southern coastal plain; (2) Samo/Kubo (NAN-language speakers) of the southern lowlands; (3) Mountain Ok (NAN-language speakers) in the highland fringe; (4) Huli (NAN-language speakers) in the Highlands; and (5) Balopa (AN-language speakers) in the Manus Islands (Fig. 3.1). The five groups differ markedly in their human–environment relations, representing the major traditional subsistence patterns of PNG (Table 3.1).

Figure 3.2 shows large variations in stature and body weight and, consequently, body mass index (BMI; body weight (kg) divided by stature (m) squared) among the five groups. In decreasing order of BMI, Balopa had the highest value, followed by the Huli, the Mountain Ok, then the Gidra and Samo/Kubo. Gidra and Samo/Kubo people, both of whom have traditionally depended heavily on sago exploitation and hunting, expended much more energy in physical work than the others, a likely reason for their lower BMI. Inter-population differences in protein intake depended on two factors. The first was the availability of animal sources of protein obtained by traditional means (land animals for the Gidra and marine animals for the Balopa). The second was the amount of food purchased with money, a consequence of modernization. The Balopa are the sole AN-speaking population among these groups. Their genetic traits are more similar to Polynesians and Asians than the NAN-language–speaking Gidra according to analysis of ABO blood group genes

Table 3.1. *Characteristics of five Papua New Guinea populations and their protein intake per day per adult male, broken down by food source*

Population	Habitat	Population density (/km²)	Major subsistence		Protein intake (g)			
			Plant food	Animal food	Local plant	Local animal	Purchased	Total
Gidra[a]	Lowlands	1	Sago, mixed cultivation	Productive hunting	24	36	6	66
Kubo[b]	Foothill	2	Sago, mixed cultivation	Less-productive hunting	24	6	0	30
Mountain Ok[a]	Highland fringe	2	Taro monoculture	Negligible	19	3	1	23
Huli[c]	Highlands	50–100	Sweet potato cultivation	Pig husbandry	39	10	7	56
Balopa[d]	Island	20	Mixed cultivation	Reef fishing	16	27	21	64

Note:
Only Balopa group has had experience of cash cropping for several decades.
Sources: [a] Ohtsuka (1994); [b] Suda (1997); [c] Umezaki *et al.* (1999); [d] Ataka and Ohtsuka (2000).

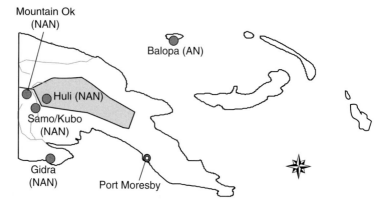

Fig. 3.1. The location of five Papua New Guinea populations. The shadowed area in the central part of the main island refers to the Highlands (NAN: Non-Austronesian, AN: Austronesian).

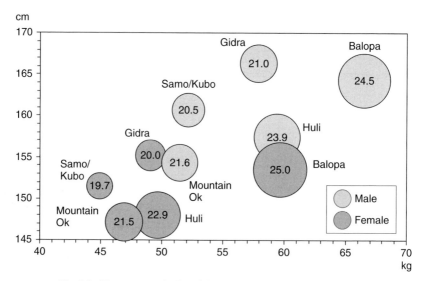

Fig. 3.2. Mean stature, body weight and body mass index (BMI) of adults of the five Papua New Guinea populations. The size of circle represents the BMI value mentioned in the circle.

(Ohashi *et al.* 2003). Not only do they have the largest BMI of the five groups, they also suffer elevated risk of degenerative disease (Inaoka *et al.* 1996), as do other AN-language–speaking populations such as Polynesians.

Secular change among the Gidra

Modernization began for the Gidra in the 1960s–1970s, and was accompanied by increased consumption of nutritious purchased foods, such as rice, wheat flour and tinned fish, as well as commercial salt (Ohtsuka and Suzuki 1990). The daily protein intake per adult male of an inland village increased from 47 g/day in 1971 to 66 g/day in 1981 and 89 g/day in 1989 (Ohtsuka 1993). Interestingly, no adult males and females had BMI exceeding 24.0 until 1981, but the proportion of 'obese' individuals (here, defined as BMI >24.0) in 1989 was 9.5% in males and 11.1% in females.

In another coastal Gidra village, where modernizing influences became manifest from the 1960s, proportions of obese males and females were 13.2% and 9.1% in 1981 and 26.4% and 13.3% in 1989, respectively. The proportion of hypertensive Gidra (as defined by systolic blood pressure ≥ 140 mmHg) varied more greatly than did BMI between villages. Fewer than 10% of either males or females in the less-modernized village were hypertensive in 1989. However, in the more-modernized village, 23.7% in males and 29.5% in females were hypertensive in 1981, increasing to 26.4% and 31.9% in males and females respectively in 1989. Risk of degenerative disease has increased for the Gidra, while they continue to suffer from infectious disease – malaria, in particular – and undernutrition, as represented by anaemia (Nakazawa *et al.* 1996).

Rural–urban comparisons

Comparisons of food and nutrient intakes and energy expenditure of the NAN-language–speaking Huli and the AN-language–speaking Balopa in their homelands, and migrants in Port Moresby, the capital of PNG, show clear modernization effects (Yamauchi *et al.* 2001; Umezaki *et al.* 2002). The mean BMIs of urban-dwelling Huli adult males and females (25.0 and 27.0, respectively) were higher than those of their counterparts in their homeland (24.0 and 22.7, respectively). This reflects increased intakes of protein and fat, and decreased energy expenditure, particularly among females.

Both urban-dwelling Huli and Balopa showed high cardiovascular disease risk, as represented by obesity, hypertension, high total cholesterol and high lipoprotein(a) (Natsuhara *et al.* 2000). The prevalence of hypercholesterolemia (defined as total serum cholesterol > 240 mg/dl) was almost 20% in both sexes of both groups. However, sex and between-group differences were found for other markers. Elevated serum

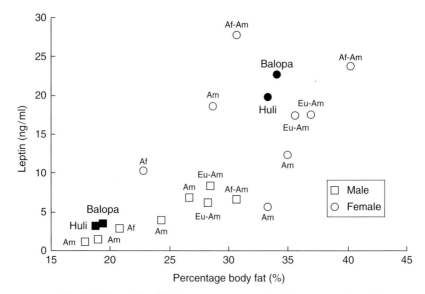

Fig. 3.3. The relationship between mean serum leptin concentration and mean body fat percentage for urban-dwelling Huli and Balopa, together with those of reported eight populations worldwide (Af: African, Am: Amerindian, Af-Am: African-American, Eu-Am: European-American).

lipoprotein(a) (>40 mg/dl) was about 30% for Huli and less than 5% for Balopa of both sexes, although other markers of chronic disease risk were higher in the Balopa. The prevalence of hypertension (systolic pressure ≥140 mmHg, or diastolic pressure ≥ 90 mmHg) was higher in the Balopa than the Huli for both sexes, being more than 10% in the former and less than 5% in the latter. The proportion of subjects who were obese (here, defined as BMI ≥30 kg/m^2) ranged from 25% to 36% in Balopa males and females and in Huli females, respectively, but was very low (only one case among 96 individuals) in Huli males. The low proportion of obese individuals among Huli males is attributable to high energy expenditure at work in labouring jobs in the informal sector (Yamauchi *et al.* 2001).

A comparison of serum leptin between the urban-dwelling Huli and Balopa showed negligible inter-group difference for each sex. Mean leptin concentration was about 3 ng/ml for males and 22–24 ng/ml for females, whose body fatness averaged, respectively, about 18% and 33%–34% of total body weight. The relationship between leptin level and body fat percentage was within the range of values for this relationship for eight populations elsewhere in the world (of which one is African, three are Amerindian, two are African-American, and two are European-American) (Fig. 3.3). Differences in serum leptin between the AN- and

NAN-language–speaking groups are largely determined by body composition, which may be related more to ecological conditions than to genetic differences (Tanaka *et al.* 2005).

Inter-population differences in risk of degenerative disease

Across the past, AN-language–speaking peoples, particularly Polynesians and Micronesians, have been reported as showing high prevalences of obesity and non-insulin–dependent diabetes mellitus, or type 2 diabetes (Beizer 1990; King 1992). The thrifty genotype hypothesis was put forward to explain this (Neel 1962); this supposes gene selection for physiological storability of dietary energy in the form of fat among ancestral groups, and increased pathogenic obesity and diabetes which is triggered by the 'thrifty' genes under modernization. However, there are many studies that point out alternative possibilities. For instance, high risk of type 2 diabetes among Polynesians may be caused intrinsically not by insulin-resistant traits but by their lifestyle-derived obesity (Simmons *et al.* 2001). It has also been suggested that if the thrifty genotype hypothesis is applicable to Oceanian populations, the targets might not be exclusively AN-speaking groups. King and his colleagues stated that NAN-language–speaking groups in PNG Highlands, who were almost free of genetic admixture with AN-language–speaking groups, were relatively, though not absolutely, resistant to the deleterious influence of acculturation on glucose tolerance (King *et al.* 1984). However, follow-up studies of NAN-language–speaking people in peri-urban Highlands villages indicated that they were in metabolic transition towards diabetes and other degenerative diseases (King *et al.* 1989). Also, strikingly high risks of type 2 diabetes have been observed in Aboriginal Australians in urban contexts (O'Dea 1991; Busfield *et al.* 2002).

Another inter-population difference, particularly between Asian and Oceanian populations and European populations, has been found in the risks of obesity and type 2 diabetes. For instance, the body fat percentage at BMIs of 25 and 30 respectively in female and male Australians of European origin was equivalent to the body fat percentage in Tongan women and men at BMIs of 28.8 and 35.1 and 27.5 and 35.8, respectively (Craig *et al.* 2001). In contrast, Aboriginal Australians with lower BMIs have an equivalent risk of type 2 diabetes, compared to Australians of European origin; the BMI-specific diabetes incidence rates are two to five times greater in the former than in the latter (Daniel *et al.* 1999, 2002). Similar population-specific characteristics were also found among Asians.

For example, BMI cut-off values for overweight and obesity of Singaporeans of Chinese, Malay and South Asian Indian origin are lower by two to three units than the cut-offs using European standards (Deurenberg-Yap *et al.* 2002).

Summary

All inhabitants of Oceania are descendants of the people who once lived in South East Asia, but their genotypic and phenotypic characteristics have undergone change and become diversified across prehistory and history. The current inhabitants of South East Asia may share genetic traits with those of Oceania to varying degrees from one group to another, but the majority are descendants of later migrants from Asia and thus may have scarcely shared direct ancestry with Oceanians. Recent modernization of lifestyle, and of dietary habits in particular, has progressed profoundly among many populations of Asia and Oceania, and with it, the risk of chronic and degenerative disease. Health research for these populations is needed to take into account their highly varied and ever-changing biocultural adaptations, given the diversity of human biological response to similar environmental changes associated with modernization.

Acknowledgements

I thank the Japanese Ministry of Education, Science and Culture and the Japan Society for the Promotion of Science for their support of various aspects of this work, and many colleagues who took part in these studies for their kindness and understanding.

References

Allen, B. (1983). Human geography of Papua New Guinea. *Journal of Human Evolution* **12**, 3–23.
Allen, J. and Gosden, C. (1996). Spheres of interaction and integration. In *Oceanic Culture History: Essays in Honour of Roger Green*, ed. J. Davidson, G. Irwin, F. Leach, A. Pawley and D. Brown. Dunedin: New Zealand Journal of Archaeology, pp. 183–97.

Ataka, Y. and Ohtsuka, R. (2000). Resource use of a fishing community on Baluan Island, Papua New Guinea: comparison with a neighboring horticultural-fishing community. *Man and Culture in Oceania* **16**, 123–34.

Attenborough, R. D. and Alpers, M. P. (eds.) (1992). *Human Biology in Papua New Guinea: The Small Cosmos.* Oxford: Clarendon Press.

Bayliss-Smith, T. P. (1977). Human ecology and island populations: the problem of change. In *Subsistence and Survival: Rural Ecology in the Pacific*, ed. T. P. Bayliss-Smith and R. Feachem. London: Academic Press, pp. 11–20.

Beizer, R. A. (1990). Prevalence of abnormal glucose tolerance in six Solomon Islands populations. *American Journal of Physical Anthropology* **81**, 471–82.

Bellwood, P. (1991). *Archaeological Survey and Excavation in the Halmahera Island Group, Maluku Utara, Indonesia.* Mimeo.

Brown, P. and Podolefsky, A. (1976). Population density, agricultural intensity, land tenure and group size in the New Guinea highlands. *Ethnology* **15**, 211–83.

Busfield, F., Duffy, D. L., Kesting, J. B., *et al.* (2002). A genomewide search for type 2 diabetes: susceptibility genes in indigenous Australians. *American Journal of Human Genetics* **70**, 349–57.

Craig, P., Halavatau, V., Comino, E. and Caterson, I. (2001). Differences in body composition between Tongans and Australians: time to rethink the healthy weight ranges? *International Journal of Obesity and Related Metabolic Disorders* **25**, 1806–14.

Daniel, M., Rowley, K. G., McDermott, R., Mylvaganam, A. and O'Dea, K. (1999). Diabetes incidence in an Australian Aboriginal population: an 8-year follow-up study. *Diabetes Care* **22**, 1993–8.

Daniel, M., Rowley, K. G., McDermott, R. and O'Dea, K. (2002). Diabetes and impaired glucose tolerance in Aboriginal Australians: prevalence and risk. *Diabetes Research in Clinical Practice* **57**, 23–33.

Denoon, D. (1997). Production in Pacific Eden? Myths and realities of primitive affluence. In *The Cambridge History of the Pacific Islanders*, ed. D. Denoon. Cambridge: Cambridge University Press, pp. 83–90.

Deurenberg-Yap, M., Chew, S. K. and Deurenberg, P. (2002). Elevated body fat percentage and cardiovascular risks at low body mass index levels among Singaporean Chinese, Malays and Indians. *Obesity Reviews* **3**, 209–15.

Ellen, R. F. (1993). *The Cultural Relations of Classification: An Analysis of Nuaulu Animal Categories from Central Serum.* Cambridge: Cambridge University Press.

Feil, D. K. (1987). *The Evolution of Highland Papua New Guinea Societies.* Cambridge: Cambridge University Press.

Finney, B. R. (1994). Polynesian voyagers to the New World. *Man and Culture in Oceania* **10**, 1–13.

Flach, M. and Rumawas, F. (eds.) (1996). *Plant Resources of South-East Asia No. 9: Plants Yielding Non-seed Carbohydrates.* Leiden: Backhuys Publishers.

Foley, W. (1986). *The Papuan Languages of New Guinea.* Cambridge: Cambridge University Press.

Friedlaender, J. S. (ed.) (1987). *The Solomon Islands Project: A Long-term Study of Health, Human Biology, and Culture Change.* Oxford: Clarendon Press.

Glover, I. C. (1971). Prehistoric research in Timor. In *Aboriginal Man and Environment in Australia*, ed. D. J. Mulvaney and J. Golson. Canberra: Australian National University Press, pp. 158–81.

Golson, J. (1977). No room at the top: agricultural intensification in the New Guinea Highlands. In *Sunda and Sahul: Prehistoric Studies in Southeast Asia*, Melanesia and Australia, ed. J. Allen, J. Golson and R. Jones. London: Academic Press, pp. 601–38.

Golson, J. and Gardner, D. (1990). Agriculture and sociopolitical organization in New Guinea Highlands prehistory. *Annual Reviews of Anthropology* **19**, 395–417.

Gorman, G. (1977). A priori models and Thai prehistory: a reconsideration of the beginnings of agriculture in Southeastern Asia. In *Origins of Agriculture*, ed. C. A. Reed. The Hague: Mouton, pp. 321–55.

Green, R. C. (1991). Near and Remote Oceania: disestablishing "Melanesia" in culture history. In *Man and a Half: Essays in Pacific Anthropology and Ethnobotany in Honour of Ralph Bulmer*, ed. A. Pawley. Auckland: Polynesian Society, pp. 491–502.

Groube, L. M. (1989). The taming of the rain forests: a model for late Pleistocene forest exploitation in New Guinea. In *Foraging and Farming: The Evolution of Plant Exploitation*, ed. D. R. Harris and D. C. Hillman. London: Unwin, pp. 292–304.

(1993). Contradictions and malaria in Melanesian and Australian prehistory. In *Sahul in Review: Pleistocene Archaeology in Australia, New Guinea and Island Melanesia*, ed. M. A. Smith, M. Spriggs and B. Frankhause. Canberra: Department of Prehistory, Australian National University, pp. 164–86.

Hanson, L. W., Allen, B. J., Bourke, R. M. and McCarthy, T. J. (2001). *Papua New Guinea Rural Development Handbook*. Canberra: The Australian National University.

Hide, R. (2003). *Pig Husbandry in New Guinea: A Literature Review and Bibliography*. Canberra: Australian Centre for International Agricultural Research.

Hill, A. V. S. and Serjeantson, S. W. (eds.) (1989). *The Colonization of the Pacific: A Genetic Trail*. Oxford: Clarendon Press.

Horai, S., Odani, S., Nakazawa, M. and Ohtsuka, R. (1996). The origin and dispersal of modern humans as viewed from mitochondrial DNA. *Gann Monograph of Cancer Research* **44**, 97–105.

Horridge, A. (1995). The Austronesian conquest of the sea – upwind. In *The Austronesians: Historical and Comparative Perspectives*, ed. P. Bellwood, J. J. Fox and D. Trion. Canberra: The Australian National University, pp. 134–51.

Houghton, P. (1991). Selective influences and morphological variation amongst Pacific Homo sapiens. *Journal of Human Evolution* **21**, 49–59.

Inaoka, T., Ohtsuka, R., Kawabe, T., Hongo, T. and Suzuki, T. (1996). Emergence of degenerative diseases in Papua New Guinea islanders. *Environmental Sciences* **4**, 79–93.

Ingman, M. and Cyllensten, U. (2003). Mitochondrial genome variation and evolutionary history of Australian and New Guinea aborigines. *Genome Research* **13**, 1600–6.

Intoh, M. (1997). Human dispersal into Micronesia. *Anthropological Science* **105**, 15–28.

Jones, R. (1980). Hunters in Australian coastal savanna. In *Human Ecology in Savanna Environments*, ed. D. R. Harris. New York: Academic Press, pp. 107–46.

Katayama, K. (1996). Polynesians the hypermorphic Asiatics: a scenario on prehistoric Mongoloid dispersals into Oceania. *Anthropological Science* **104**, 15–30.

King, H. (1992). The epidemiology of diabetes mellitus in Papua New Guinea and the Pacific: adverse consequences of natural selection in the face of sociocultural change. In *Human Biology in Papua New Guinea: The Small Cosmos*, ed. R. D. Attenborough and M. P. Alpers. Oxford: Clarendon Press, pp. 363–72.

King, H., Heywood, P., Zimmet, P., *et al.* (1984). Glucose tolerance in a highland population in Papua New Guinea. *Diabetes Research* **1**, 45–51.

King, H., Finch, C., Collins, A., *et al.* (1989). Glucose tolerance in Papua New Guinea: ethnic differences, association with environmental and behavioural factors and the possible emergence of glucose intolerance in a highland community. *Medical Journal of Australia* **21**, 204–10.

Kirch, P. V. (1988). Polynesia's mystery islands. *Archaeology* **41**, 26–31.

(2000). *On the Road of the Wind: An Archaeological History of the Pacific Islands before European Contact*. Berkeley: University of California Press.

Kirk, R. and Szathmary, E. (eds.) (1985). *Out of Asia: Peopling the Americas and the Pacific*. Canberra: The Journal of Pacific History.

Kuchikura, Y. (1994). A comparative study of subsistence patterns in Papua New Guinea. *Bulletin of the Faculty of General Education, Gifu University* **30**, 41–89.

McCarthy, F. D. and McArthur, M. (1960). The food quest and the time factor in aboriginal economic life. In *Records of the American-Australian Scientific Expedition to Arnhem Land*, Vol. 2: *Anthropology and Nutrition*, ed. C. P. Mountford. Melbourne: Melbourne University Press, pp. 145–94.

Monk, K. A., Fretes, Y. and Reksodiharjo-Lilley, G. (1997). *The Ecology of Indonesia Series*, Vol. 5: *The Ecology of Nusa Tenggara and Maluku*. Halifax: Dalhousie University.

Nakazawa, M., Ohtsuka, R., Kawabe, T., *et al.* (1996). Iron nutrition and anaemia in a malaria – endemic environment: Haematological investigation of the Gidra – speaking population in lowland Papua New Guinea. *British Journal of Nutrition* **76**, 333–46.

Natsuhara, K., Inaoka, T., Umezaki, M., *et al.* (2000). Cardiovascular risk factors of migrants in Port Moresby from the highlands and island villages, Papua New Guinea. *American Journal of Human Biology* **12**, 655–64.

Neel, J. V. (1962). Diabetes mellitus: a thrifty genotype rendered detrimental by 'progress'? *American Journal of Human Genetics* **14**, 353–62.

O'Dea, K. (1991). Westernisation, insulin resistance and diabetes in Australian aborigines. *Medical Journal of Australia* **155**, 258–64.

Ohashi, J., Yoshida, M., Ohtsuka, R., *et al.* (2000). Analysis of HLA-DRB1 polymorphism in the Gidra of Papua New Guinea. *Human Biology* **72**, 337–47.

Ohashi, J., Naka, I., Ohtsuka, R., *et al.* (2003). Molecular polymorophism of ABO blood group gene in Austronesian and non-Austronesian populations in Oceania. *Tissue Antigens* **63**, 355–61.

Ohtsuka, R. (1983). *Oriomo Papuans: Ecology of Sago-Eaters in Lowland Papua.* Tokyo: University of Tokyo Press.

(1993). Changing food and nutrition of the Gidra in lowland Papua New Guinea. In *Tropical Forests, People and Food: Biocultural Interaction and Application to Development*, ed. C. M. Hladik, O. F. Hladik, H. Pagezy, A. Semple and M. Hadley. Paris: UNESCO, pp. 259–69.

(1994). Subsistence ecology and carrying capacity in two Papua New Guinea populations. *Journal of Biosocial Science* **26**, 395–407.

Ohtsuka, R. and Suzuki, T. (eds.) (1990). *Population Ecology of Human Survival: Bioecological Studies of the Gidra in Papua New Guinea.* Tokyo: University of Tokyo Press.

Pawley, A. K. and Ross, M. (1993). Austronesian historical linguistics and culture history. *Annual Reviews of Anthropology* **22**, 425–59.

Riley, I. D. (1983). Population change and distribution in Papua New Guinea: an epidemiological approach. *Journal of Human Evolution* **12**, 125–32.

Riley, I. D. and Lehmann, D. (1992). The demography of Papua New Guinea: migration, fertility, and mortality patterns. In *Human Biology in Papua New Guinea: The Small Cosmos*, ed. R. D. Attenborough and M. P. Alpers. Oxford: Clarendon Press, pp. 67–92.

Rowland, I. (1992). *Timor: Including Islands of Roti and Ndao.* Oxford: Clio Press.

Serjeantson, S. W., Kirk, R. L. and Booth, P. B. (1983). Linguistic and genetic differentiation in New Guinea. *Journal of Human Evolution* **12**, 77–92.

Sillitoe, P. (1988). *Made in Niugini: Technology in the Highlands of Papua New Guinea.* London: The Trustees of the British Museum.

Simmons, D., Thompson, C. F. and Volklander, D. (2001). Polynesians: prone to obesity and type 2 diabetes mellitus but not hyperinsulinaemia. *Diabetes Medicine* **18**, 193–8.

Spencer, J. E. (1963). The migration of rice from mainland Southeast Asia into Indonesia. In *Plants and the Migration of Pacific Peoples*, ed. J. Barrau. Honolulu: Bishop Museum Press, pp. 83–9.

Spriggs, M. (1997). Recent prehistory (the Holocene). In *The Cambridge History of the Pacific Islanders*, ed. D. Denoon. Cambridge: Cambridge University Press, pp. 52–69.

Stanhope, J. M. (1970). Patterns of fertility and mortality in rural New Guinea. *New Guinea Bulletin* **34**, 24–41.

Suda, K. (1997). Dietary change among the Kubo of Western Province, Papua New Guinea, between 1988 and 1994. *Man and Culture in Oceania* **13**, 83–98.

Swadling, P. (1983). *How Long Have People Been in the Ok Tedi Impact Region?* Papua New Guinea National Museum Record No. 8, Port Moresby.

Tanaka, M., Umezaki, M., Natsuhara, K., *et al.* (2005). No difference in serum leptin concentrations between urban-dwelling Austronesians and Non-Austronesians in Papua New Guinea. *American Journal of Human Biology* **17**, 696–703.

Townsend, P. K. (1971). New Guinea sago gatherers: a study of demography in relation to subsistence. *Ecology of Food and Nutrition* **1**, 19–24.

Ulijaszek, S. J. (2001). Sago, economic change and nutrition in Papua New Guinea. In *New Frontiers of Sago Palm Studies*, ed. K. Kainuma, M. Okazaki, Y. Toyoda and J. E. Cecil. Tokyo: Universal Academy Press, pp. 219–26.

Umezaki, M., Yamauchi, T. and Ohtsuka, R. (1999). Diet among the Huli in Papua New Guinea Highlands when they were influenced by the extended rainy period. *Ecology of Food and Nutrition* **37**, 409–27.

   (2002). Time allocation to subsistence activities among the Huli in rural and urban Papua New Guinea. *Journal of Biosocial Science* **34**, 133–7.

van Baal, J. (1961). Review of peoples of the Tor. *Nieuw Guinea Studiën* **5**, 339–42.

White, J. P. and O'Connell, J. F. (1982). *A Prehistory of Australian, New Guinea and Sahul*. Sydney: Academic Press.

Yamauchi, T., Umezaki, M. and Ohtsuka, R. (2001). Influence of urbanisation on physical activity and dietary changes in Huli-speaking population: a comparative study of village dwellers and migrants in urban settlements. *British Journal of Nutrition* **85**, 65–73.

Yoshida, M., Ohtsuka, R., Nakazawa, M., Juji, T. and Tokunaga, K. (1995). HLA-DRB1 frequencies of non-Austronesian-speaking Gidra in south New Guinea and their genetic affinities with Oceanian populations. *American Journal of Physical Anthropology* **96**, 177–81.

4   *Changing nutritional health in South East Asia*

GEOFFREY C. MARKS

Introduction

Most of the health problems that result from both deficiency and excess of nutrient intake can be found in South East Asia, with those of particular public health significance being protein-energy malnutrition, iodine deficiency disorders, vitamin A deficiency, iron deficiency and the non-communicable diseases (NCDs) for which food intake is an important causal factor. However, the relative importance of each of these has shifted significantly over the last few decades. Social, economic and agricultural developments have brought dramatic achievements in the reduction of poverty and the extent of famine, hunger and starvation in the region, as economies dominated by agricultural production have shifted to those based predominantly on secondary and service industries. These industries have also brought rapid urban growth, dietary and lifestyle changes and demographic ageing of populations, all of which directly or indirectly influence health profiles of nations. But these transitions have not been uniform either across or within countries, and they have features that present significant challenges for governments and agencies in the region.

This chapter focuses on Indonesia, Malaysia, the Philippines, Singapore and Thailand – the five original members of the Association of South East Asian Nations (ASEAN). After describing some of the key features of these countries, I review overall trends in food production, consumption and nutritional health for the region since the 1960s. The situation of Indonesia will be considered in some detail to develop and illustrate some of the key issues facing the region, with a particular focus on the policy and programme challenges facing decision makers.

Social, cultural and economic features of the region

The region is remarkable for its diversity. This exists in terms of history, ethnicity, cultural and religious beliefs, affluence, and political and

Health Change in the Asia-Pacific Region: Biocultural and Epidemiological Approaches, ed. Ryutaro Ohtsuka and Stanley J. Ulijaszek. Published by Cambridge University Press.

economic systems (Table 4.1). Indonesia was colonized by the Dutch from the early seventeenth century, occupied by Japan from 1942 to 1945, and declared its independence in 1945. It is the world's largest archipelagic state and has by far the largest population in the region. It is a multi-ethnic country, the biggest single group being Javanese, and the main religion nationally is Islam. From the 1970s, the Indonesian government invested heavily in agriculture and the social sector, with substantial gains as reflected by the following: (1) rice production increase from 23 million tons in 1977 to 38 million tons by 1984, the country becoming self-sufficient in rice between 1985 and 1990; (2) a doubling of the number of primary schools between 1973 and 1991; and (3) a decline in adult illiteracy among women from 69% to 17% between 1961 and 1990. Since the 1980s, there has been a shift away from employment in agriculture towards employment in manufacturing and labour-intensive industries, although agriculture still employed about 45% of the workforce in the late 1990s (United Nations Development Program 2001). Poverty rates more than halved from those of the mid 1970s but were exacerbated by the 1998 financial crisis that affected most countries in the region. However, poverty remains widespread (Tabor *et al.* 2002).

Malaysia was colonized by the British during the eighteenth and nineteenth centuries, occupied by Japan between 1942 and 1945, and became independent as the Federation of Malaya in 1957. The nation was renamed 'Malaysia' in 1963, when the former British colonies of Singapore and the East Malaysian states of Sabah and Sarawak joined the Federation. Singapore seceded from the Federation in 1965. The country has three major ethnic groups, with Malays comprising just over half of the population, Chinese people about a quarter of it, and Indians of South Asian origin being an important minority at 8% of it. While Malaysia is a Muslim country, with Islam being practised mainly by the Malay population, several other religions are also widely followed. Malaysia is now a middle-income country, transformed – from the 1970s to the 1990s – from a producer of raw materials into an emerging multi-sector economy. This growth was driven almost exclusively by exports, particularly of electronics. Poverty rates are now low, at about 8% of the population below the poverty line (Central Intelligence Agency 2005).

The Philippines became a Spanish colony during the sixteenth century, was ceded to the United States in 1898 following the Spanish–American War, and attained independence in 1946 after Japanese occupation during the war. The population is ethnically Malay, with Roman Catholicism as the dominant religion. While the Philippines has benefited from agricultural and economic development over the past few decades, about 45% of

Table 4.1. *Social, cultural and economic features of South East Asian countries*

	Indonesia	Malaysia	Philippines	Thailand	Singapore
Population[a]	238,452,952	23,522,482	86,241,697	64,865,523	4,353,893
Proportion[a]					
• 0–14 yrs	29.4%	33.3%	35.8%	24.1%	16.5%
• 65 yrs and over	5.1%	4.5%	3.9%	7.3%	7.8%
Infant mortality[a,b]	36.82	18.35	24.24	21.14	2.28
Life expectancy[a]	69.26 yrs	71.95 yrs	69.6 yrs	71.41 yrs	81.53 yrs
Ethnic groups[a]	Javanese 45%	Malay and other	Christian Malay 91.5%	Thai 75%	Chinese 76.7%
	Sundanese 14%	indigenous 58%	Muslim Malay 4%	Chinese 14%	Malay 14%
	Madurese 7.5%	Chinese 24%	Chinese 1.5%	Others 11%	Indian 7.9%
	Coastal Malays	Indian 8%	Others 3%		Others 1.4%
	7.5%	Others 10%			
	Others 26%				
Religions[a]	Muslim 88%	Muslim	Roman Catholic 83%	Buddhism 95%	Buddhist (Chinese)
	Protestant 5%	Buddhist	Protestant 9%	Muslim 3.8%	Muslim (Malays)
	Roman Catholic	Daoist	Muslim 5%	Christianity 0.5%	Christian
	3%	Hindu	Buddhist and others 3%	Hinduism 0.1%	Hindu
	Hindu 2%	Christian		Others 0.6%	Sikh
	Buddhist 1%	Sikh			Toaist
	Others 1%	Shamanism			Confucianist
Literacy[a,c]	87.9%	88.7%	92.6%	92.6%	92.5%
Per capita GDP (USD)[a]	$3,200	$9,000	$4,600	$7,400	$23,700

Population below poverty line[a]	27%	8%	40%	10.4%	Not reported
Labour force by occupation[a]:					
• Agriculture	45%	14.5%	45%	49%	Manufacturing 18%
• Industry	16%	36%	15%	14%	Construction 6%
• Services	39%	49.5%	40%	37%	Transportation and communication 11%
					Financial, business and other services 49%
Human Development Index 1975 – value[d]	0.467	0.6140	0.653	0.613	0.724
Human Development Index 2002 – value (rank)[d]	0.692 (111)	0.793 (59)	0.753 (83)	0.768 (76)	0.902 (25)

[a] Source: Central Intelligence Agency 2005; estimates generally 1999 or later;
[b] Infant mortality – deaths per 1000 live births;
[c] Literacy – age 15 and over who can read and write;
[d] Source: United Nations Development Programme (2005).

the population is still employed in agriculture. Poverty remains a significant national problem with 40% of the population being below the poverty line. The country was less severely affected by the Asian financial crisis of 1998 than its neighbours, as it was aided by remittances from its large number of overseas workers (Central Intelligence Agency 2005).

Singapore was founded as a British trading colony in 1819. It joined the Malaysian Federation in 1963 but separated two years later and became independent. As in Malaysia, the population of Singapore comprises three main ethnic groups, although the majority group is Chinese, comprising about three quarters of the population, with Malays and South Asian Indians being important minority groups (14% and 8% of the population, respectively). The main religions are also similar to that in Malaysia but with different relative importance. Singapore is now one of the world's most prosperous countries, with strong international trading links and per capita GDP equal to that of the leading nations of Western Europe. The service sector is the main employer, although the economy depends heavily on exports, particularly of electronics and manufacturing industries (Central Intelligence Agency 2005).

Thailand was established as a unified kingdom in the mid-fourteenth century. It is the only South East Asian nation never to have been colonized by a European power. A bloodless revolution in 1932 led to a constitutional monarchy. Three quarters of the population are ethnically Thai, with Chinese people forming a significant minority group. Most of the population are Buddhist. Since the 1970s, Thailand has been transforming from a subsistence agrarian society to an industrial one, although about half of the population still remains employed in agriculture (Central Intelligence Agency 2005). Progress was hampered by the Asian economic crisis of 1998, which was triggered by the collapse of the Thai baht in July 1997. However, Thailand recovered much more quickly than either Indonesia or Malaysia (Wolf 2000).

The absolute and relative socioeconomic progress of these five countries in recent decades is reflected by the Human Development Index (HDI), a summary index that combines measures of life expectancy, school enrolment, literacy and income (United Nations Development Program 2005). All countries show significant improvements, but Singapore stands out with by far the highest HDI ranking in 2002. Malaysia and Thailand are middle-ranking nations, with the Philippines and Indonesia being the poorest of these. Socioeconomic progress and diversity of the region are reflected also in food and nutritional changes over the last three decades.

Patterns of food production and consumption between 1960 and the 1990s

National food balance sheet data provide useful information about food availability and population consumption patterns. Table 4.2 summarizes changes in total food availability for each of the countries since the 1960s based on Food and Agriculture Organisation estimates (Khor *et al.* 1990; Nandi 1999). Improved practices for the cultivation of staples, particularly

Table 4.2. *Changes in available calories supply in South East Asian countries, 1961–84 and 1982–92*

A.[a]	Total calories available in 1982–84 per caput per day	Change in total available calories 1961–84 (%)	Animal calories / total calories 1982–84 (%)	Other notable changes
Indonesia	2,433	35.0	2.4	Increase in contribution of cereals.
Malaysia	2,549	8.4	13.9	Drop in contribution of cereals.
Philippines	2,399	28.0	10.1	Drop in contribution of cereals.
Singapore	2,725	12.1	26.2	Drop in contribution of cereals; roots and tubers up by 163%.
Thailand	2,322	9.6	6.3	Drop in contribution of cereals; sugar and honey up by 164%.

B.[b]	Total calories available in 1990–92 per caput per day	Change in total available calories 1961–84 (%)	Animal calories / total calories 1982–84 (%)	Other notable changes
Indonesia	2,696	10.7	3.9	Slight drop in contribution of cereals.
Malaysia	2,817	4.5	16.3	
Philippines	2,292	3.3	11.9	
Singapore				
Thailand	2,374	7.0	10.1	

Sources: [a] Khor *et al.* (1990); [b] Nandi (1999).

rice, resulted in significantly increased production over the period. Increased rice output per hectare accounted for 85% of regional growth in production, with the harvesting of larger areas of land accounting for 15% of the increase. Indonesia and Thailand expanded their areas under cultivation significantly from 1960 to the early 1980s (by 30% and 62% respectively) (Khor *et al.* 1990).

From 1961 to 1984 there was an average increase in the availability of total dietary energy per head of population of 20.9% for the region overall, overcoming much of the famine and seasonal hunger that had previously been common in many areas. Indonesia and Philippines had the largest increases, of 35% and 28% respectively. While the bulk of total calories came from cereals, by 1984, the proportion of dietary energy from cereals was decreasing in all countries except Indonesia. The other major calorie sources were animal products (meat, milk, eggs), roots and tubers (mainly cassava), sugar and honey, and fats and oils. From 1961 to 1984, the proportion of dietary energy from animal products increased by 5% in Malaysia and 56% in Singapore, while it was stable or declined slightly in the other countries. South East Asian countries together were estimated to be importing about a quarter of the world's trade in milk powder in the early 1980s. The proportion of total calories from roots and tubers increased significantly in each of the countries except Indonesia, recording an increase of 163% during the period 1961 to 1984 in Singapore. The proportion of dietary energy from sugar and honey also increased in Malaysia, Philippines and Thailand, was stable in Indonesia, and declined in Singapore. The contribution of fats and oils to total energy intake increased in all countries from 1961 by between 14% and 27% (Khor *et al.* 1990).

Between 1982 and 1992, the contribution of cereals to total calories also declined in Indonesia. In other respects, the trends seen across the region between the 1960s and the 1970s continued, but with lower rates of change. By 1992 national calorie availability for all countries was in a range consistent with a healthy diet, with the Philippines at the lower end. Overall, similar general trends have occurred in all countries of the region since the 1960s – increased total calories, greater consumption of animal food sources and dietary diversification – with the timing and extent of change related to the rate and extent of change in economic development. Singapore now shows some of the dietary changes observed, or at least advocated, in North America and Western Europe – a decline in red meat consumption and increased consumption of vegetables. Results of a national food consumption survey in 1996 showed a ratio of consumption of vegetables and fruits to meats of 1.58, up from 1.45 in 1986 and indicating adoption of more healthy diets (Kiang 1998).

At the household level, the shifts in consumption patterns are more complex. Socioeconomic development has been accompanied by rapid urban growth, with the urban population in the region growing at about twice the rate of the overall population (Nandi, 1999). As a result, an increasing proportion of the population now relies on food purchases rather than on home production to meet their food needs. There is also a trend towards greater consumption of food outside the household. In Thailand, Kosulwat (2002) reports a major shift in expenditure on foods purchased ready-to-eat. Recent National Household Economic Surveys show more than 50% of food expenditure is on ready-to-eat food in greater Bangkok, compared with less than 30% in other regions. But it has also changed in the rural areas, shifting from less than 12.2% of total food expenditure in 1990 to 21.7% of it in 1999. Noor (2002) reports similar changes in Malaysia, although in this case involving significant growth of the fast food industry, including both North American and local companies. Experience elsewhere suggests that this is likely to be accompanied by increased fat, sugar and salt consumption (Popkin 2002).

The evolution of nutritional health

Various kinds of malnutrition have been well recognized as problems of public health significance in most countries of the region since the 1960s. Indeed, addressing malnutrition has been an explicit national development objective in several of the countries of this region, and nutrition programmes have frequently been given high profiles. In spite of this, precise data are not available for most of the nutritional health problems affecting populations, although trends can be inferred from the few national surveys and the numerous regional and community surveys in each country.

Table 4.3 shows representative data available for cross-country comparisons of the extent of underweight in pre-school children. In the 1970s, around half of all Indonesian children below the age of five years were moderately or severely underweight (Soekirman 1997); this declined steadily to 25.8% in 2002. The Philippines had slightly lower rates of underweight in the early 1980s, showed little change between 1982 and 1992, but showed steady improvement since then. Thailand and Malaysia had the lowest rates of underweight, and both showed important reductions during the early 1990s.

The extent of micronutrient malnutrition in the region during the early 1980s is reflected by data compiled by Tontisirin and reported in Khor

Table 4.3. *Proportion of under-fives that are underweight[a] in four South East Asian countries*

	Year	%	Year	%	Year	%	Year	%	Year	%
Indonesia	1987	39.9	1989	37.5	1992	35.6	1998	29.5	2002	25.8
Malaysia					1990	25.0	1995	20.1		
Philippines	1982	33.2	1989–90	33.5	1992	33.4	1998	31.8	2003	27.6
Thailand			1987	25.3	1993	18.6	1995	17.6		

Notes:
[a] Underweight – falls below −2 SD from median weight for age of National Center for Health Statistics/World Health Organization reference population.
Sources: World Health Organization (2003); BPS Statistics (2000, 2003); Food and Nutrition Research Institute, Philippines (2005).

et al. (1990). Selected results are given in Table 4.4, which show that iron deficiency anaemia was widespread in all countries, vitamin A deficiency affected mainly Indonesia and Thailand, and iodine deficiency was localized in endemic areas such as the mountainous interiors of Sabah and Sarawak in East Malaysia, and the rural northern and northwestern regions in Thailand. The public health impact of mild to moderate vitamin A deficiency on child mortality was not recognized at the time, and its occurrence was described mainly in terms of xerophthalmia as a precursor of blindness. As a result, the extent of its occurrence is likely to have been underestimated. In contrast to these results, data from Singapore suggest that overnutrition is more likely to have been the main nutritional health problem. Between 1976 and 1983, a total of 705,511 school children aged 7–12 years were screened for obesity. Based on Body Mass Index (BMI), the prevalence of obesity was reported as 8% in 1983 and was thought to have been increasing, as a reflection of the growing affluence of the country (Khor *et al.* 1990).

Changes in the extent of iron deficiency disorders can be gauged from international reports and some country-level data. In their Second Report on the World Nutrition Situation, the United Nations Administrative Committee on Coordination – Sub-Committee on Nutrition (1992) reported that the prevalence of anaemia among women aged 15 to 49 years in the 1980s was estimated to be approximately 48% in South East Asia, second only to South Asia. The dietary iron supply was assessed as having dropped slightly between 1970 and 1990, especially from vegetable sources since 1980. The prevalence of anaemia among non-pregnant women was estimated to have increased from approximately 40% in 1970–80 to 57% in 1980–90. While it is difficult to judge national trends

Table 4.4. *Prevalence of anaemia, vitamin A deficiency and goitre in four South East Asian countries in the 1980s and 1990s*

A.[a]	Indonesia 1980–81	Malaysia 1979–83	Philippines 1982	Thailand 1985
Iron deficiency anaemia				
Infants	–	34%	50%	–
Preschool children	40%	33%	32% (preschool and school children)	10–40%
School age children	9%	37% (adolescent girls)		30–60%
Adults	–	25% (18–45 yr females)	–	17–37%
Pregnant women	70%	–	49%	30–70%
Vitamin A deficiency				
Clinical assessment	50% with night blindness	–	1.8% with night blindness	–
Preschool children			1.4% with Bitot's spots	
Iodine deficiency				
Goitre	10–90% in endemic areas	1.5% of total population in Sarawak; endemic in Sabah and Northern Peninsula Malaysia	7.6% women of child-bearing age, 7.5% lactating mothers, 3.2% pregnant women in endemic areas	10% school children in North and North East regions

Iron deficiency anaemia:
Malaysia: less than 15% serum transferrin saturation
Thailand: less than 33% haematocrit level

Table 4.4. (*cont.*)

B.[b]	Indonesia	Malaysia	Philippines	Thailand
Iron deficiency anaemia				
Pregnant women	1975 37%	–	1978 53%	1986 36%
	1980 64%		1980 54%	1996–97 22%
	1980 70%		1982 34%	
	1982 68%		1986 48%	
	1986 72%		1993 (national) 44%	
	1991 50%		1998 (national) 51%	
	1995 50%			
Vitamin A deficiency				
Serum retinol	1978, preschool, 67%	–	(all national)	Preschool children
	1991, preschool		Preschool children	(subnational)
	(subnational) 58%		1993 35%	1990 20%
	1995, preschool		1998 38%	
	(national) 50%		Lactating women	
			1993 16%	
			1998 16%	
			Pregnant women	
			1993 16%	
			1998 22%	
Iodine deficiency				
Goitre	School aged children	–	School aged children	School aged children
	(national)		(national)	1989 19%(15 provs)
	1982 37%		1987 boys 0.8%,	1990 17%(15 provs)
	1988 28%		girls 6.4%	1992 12%(39 provs)
	1996 10%		1993 boys 0.6%,	1994 7.9%(57 provs)
			girls 4.8%	1996 4.3%(75 provs)
				1997 3.3%(75 provs)
				1998 2.6%(75 provs)
				1999 2.2%(75 provs)

Sources: [a] Khor *et al.* (1990); [b] Deitchler *et al.* (2004).

from regional surveys, the information presented in Table 4.4B lends support for this trend in Indonesia and the Philippines. In contrast, the iron supplementation programme in Thailand is reported to have been effective in reducing levels of anaemia (Winichagoon *et al.* 2001); information in Table 4.4B supports this view.

One might expect to see reductions in the prevalence of vitamin A deficiency as a result of the largely successful introduction of supplementation programmes with high-dose vitamin A capsules. This is not supported by the data available, although it is difficult to judge, given only two or fewer nationally representative surveys in each country. Universal coverage with vitamin A capsules in Indonesia has been difficult to sustain, high-risk groups being frequently missed over time (Helen Keller International 1998–2004). Similarly, the widespread introduction of iodine fortification of salt in the region is likely to have had an important impact on levels of iodine deficiency. This is shown for Indonesia and Thailand in Table 4.4B; the national prevalence of goitre was already low for the Philippines. The outcome of these types of programmes is often that micronutrient malnutrition is reduced in the general population, but with pockets of hard-to-reach communities remaining, in which significant nutritional deficiency problems persist. This is the case for vitamin A deficiency in Indonesia, iodine deficiency in Thailand (Chaseling *et al.* 2002), and undernutrition more generally among the indigenous population of Malaysia (Cuthberson *et al.* 1999).

Historically, South East Asia is also known for other types of micronutrient malnutrition. Beri-beri, caused by thiamine deficiency, was endemic in many parts of the world in the 1920s, including Malaysia, Indonesia and the Philippines (Hardy 1995). Although thought to be largely a disease of the past by many, recent reports suggest an ongoing vulnerability in particular groups, including a selected group of urban Indonesian elderly people (Juguan *et al.* 1999); patients with malaria or other febrile illness in a provincial Thai hospital (Krishna *et al.* 1999); and a Karen displaced population (McGready *et al.* 2001); and it could possibly be a cause of sudden death among Thai workers in Singapore (Phua *et al.* 1990). Beri-beri has generally occurred in rice-eating populations when the diets were undiversified and based on milled (polished) rice. The report of the United Nations Administrative Committee on Coordination – Sub-Committee on Nutrition (1992) commented on outbreaks of scurvy (vitamin C deficiency), pellagra (niacin deficiency) and beri-beri (thiamine deficiency) in refugee camps, presumably because of their long-term total or near-total dependence on the inadequate rations provided. Indonesia, the Philippines and Thailand all contain numerous camps with refugees or communities that have been internally displaced due to natural disasters or civil conflicts, and that are potentially at risk for these deficiencies.

This mixed progress in addressing malnutrition must now be considered in a context where there is rapid growth in the occurrence of diet-related NCDs. Yusuf *et al.* (2001) have described global trends with decreasing proportions of deaths from infections and concomitant increases in cardio-vascular and other chronic diseases. They attribute the global trends to 'overarching factors' such as urbanization (from 36.6% of the world's population living in urban areas in 1970, to 44.8% in 1994, and projected to increase to 61.1% in 2025) and the patterns of lifestyle change typically associated with urbanization; that is, a marked increase in consumption of energy-rich foods, a decrease in energy expenditure and a loss of tradi-tional support structures. They also point to variations in ethnicity and religion as being factors explaining variability in these trends. All of these factors are relevant to the countries of South East Asia. The extent of change in disease patterns in South East Asia is illustrated in Table 4.5. Overweight and obesity is included as an indicator of the extent of change in type 2 diabetes and cardiovascular disease morbidity, as well as being an important risk factor in its own right. Because of differences in the criteria used for defining overweight and obesity in children, between-country comparisons of prevalence are meaningless, although subgroup compar-isons within countries are valid, as are all comparisons for adults.

Recent surveys in Indonesia show that both underweight and over-weight affect significant proportions of the population, but that there are distinct differences in who is most affected. Soekirman *et al.* (2002) report on a representative survey of school children in West Jakarta and Bogor, both large urban areas. Thinness was slightly more prevalent than over-weight, showing that while overweight and obesity is an emerging issue, undernutrition continues to affect large numbers of children. Importantly, the extent of overweight was much higher and underweight lower in private schools, indicating a gradation by socioeconomic status. This is the inverse of the trends shown in North America and Western Europe, where overweight and obesity are usually more common among people of low socioeconomic status. The authors note that the extent of overweight in the urban samples of West Jakarta and Bogor is greater than that observed in surveys of rural children. Adult patterns of overweight and obesity are shown by the Indonesian Family Life Survey (Strauss *et al.* 2004), which was conducted in 13 provinces and is likely to be representa-tive of national trends. More women than men are either overweight or obese, levels of both increasing with age for both men and women. Overweight hardly exists among young women, while almost a third of

Table 4.5. *The extent of obesity, diabetes mellitus and cardiovascular diseases in South East Asian countries in the late 1990s[a]*

	Indonesia 1998	Malaysia 1996	Philippines 2000–03	Singapore 1998	Thailand 1996–97
Overweight and obesity Children – obesity, variously defined	(8–10 yrs. Jakarta. Bogor) *Boys* Overwt 17.8% Underwt 24% *Girls* Overwt 15.3% Underwt 17.4% Urban more overwt Private schools more overwt	(~7–10 yrs. Kuala Lumpur & rural) Overall 4.1% Urban 7.7% Rural 1.9% Boys 4.8% Girls 2.7%	(Preschool, national) Overwt 1.4% Underwt 27.6% (6–10 yrs, national) Overwt 1.3% Underwt 26.7% (~8–10 yrs; Manila) Overall: Public school 3.3% Private school 12.0%	School children, 12% obese	(Preschool; Saraburi) Urban 22.7% Rural 7.4% (7–9yrs, Khon Kaen; obesity) Overall 10.8% Boys > Girls
Adults	(13 provinces) *Males 15–19 yrs* Overwt 2.0% Underwt 40.8% *Males 40–59 yrs* Overwt 13.9% Underwt 16.3% *Females 15–19 yrs* Overwt 5.1% Underwt 26.1% *Females 40–59 yrs* Overwt 30.3% Underwt 14.3%	(National, > =20 yrs) Overall overwt 26.5% Urban = Rural *Males* Malay 24.3% Chinese 28.7% Indian 27.5% *Females* Malay 33.5% Chinese 22.7% Indian 34.6%	(National) Overall: Overwt 23.9% Underwt 12.3% *Overwt by age* 11–19 yrs 3.5% 20–39 yrs 20.6% 40–59 yrs 30.8% 60 and over 19.1%	(National, 18–69 yrs) Overall obesity 6.0% Not different from 1992 Men 5.3% Women 6.7% Chinese 3.8% Malay 16.2% Indian 12.2%	(National) *Males:* Overwt 13.2% Underwt 32.4% *Females:* Overwt 25.0% Underwt 26.1% Overweight in Bangkok > other regions

Table 4.5. (cont.)

	Indonesia 1998	Malaysia 1996	Philippines 2000–03	Singapore 1998	Thailand 1996–97
Diabetes mellitus	Estimated 1.2–2.3% above 15 yrs; variable across locations – up to 6.1% in Manado	(>=30 yrs) Overall 8.3% diabetic Urban slightly higher Highest amongst Indians	(20–65 yrs, Luzon) Overall 5.1% diabetic Urban = rural Women higher	(18–69 yrs) Overall 9.0% diabetic Not different from 1992 Women slightly greater Chinese 8.0% Malay 11.3% Indian 15.8%	(>=35 yrs) Overall 9.6% diabetic Urban greater
Cardiovascular diseases	1972, 11th leading cause of death; 1986, 3rd; now leading cause	Mortality rate doubled from 1970 to 54.8/100,000 Hospitalization up 77% from 1985 to 493/100,000	–	Regular physical activity significantly increased from 1992, but so did prevalence of high total cholesterol and hypertension	Leading cause of death since 1989

Sources: Indonesia: Soekirman *et al.* (2002); Sutanegara *et al.* (2000); Strauss *et al.* (2004); *Malaysia:* Tee *et al.* (2002); Ismail *et al.* (2002); Zaini (2000); Noor (2002); *Philippines:* Florentino (2002); Florentino *et al.* (2002); *Thailand:* Sakamoto *et al.* (2001); Langendijk *et al.* (2003); Kosulwat (2002); Aekplakorn *et al.* (2003); *Singapore:* Yian (2000).

women aged between 40 and 59 years are overweight; conversely a quarter of young women are underweight while a smaller but still significant number (14%) of the older age group are underweight. Diabetes is an emerging health problem, but there are few population-level data; reported rates range from 1.2% to 6.1% (Sutanegara *et al.* 2000). Cardiovascular diseases have shifted from being the 11th leading cause of death in 1972 to being the leading cause by the late 1990s (Soekirman *et al.* 2002).

The prevalence of overweight and NCDs in Malaysia is greater than for Indonesia, consistent with the more pronounced and long-standing changes in food consumption patterns there. Tee *et al.* (2002) report on a representative survey of school children in Kuala Lumpur and compare it with survey data from rural populations (69 villages and 7 estates across Peninsula Malaysia). The extent of overweight among urban children is about four times that of their rural counterparts; conversely the prevalence of stunting and underweight is about four times higher among rural children. Data from the 1996 National Health and Morbidity Survey (NHMS) show that obesity is well established as a health issue among adults, 20.7% of them being overweight and 5.8% obese, with approximately equal prevalences in urban and rural populations (Ismail *et al.* 2002). Overweight and obesity prevalence is clearly greater among women than men. In women, obesity prevalence is higher in Malays and Indians of South Asian origin, while among men, the Chinese and South Asian Indians have the highest prevalences. The NHMS also assessed diabetes status, showing 8.3% of adults over the age of 30 years to be diabetic, the prevalence increasing with age (Zaini 2000). Increases in the occurrence of cardiovascular diseases are reflected in the mortality rate, which doubled from 1970 to the 1990s, and the hospitalization rate, which increased by 77% between 1985 and the mid 1990s (Noor 2002). Hospitalization for hypertension and diabetes also increased by about 40% between 1985 and 1994. While this is likely to reflect true increases, part of the increase is also likely to reflect patients seeking treatment for these conditions at earlier stages because of increased awareness of these disorders among the general public.

National surveys in the Philippines show that the prevalence of obesity is still relatively low among preschool and primary school children, and that undernutrition remains the more important problem in this age range (Food and Nutrition Research Institute, Philippines 2005). However, a survey of school children in the city of Manila has shown a clear association with socioeconomic status, with 12% of children attending private schools being overweight or obese; this is four times greater than among those attending public schools (Florentino *et al.* 2002). The situation for adults is quite different, with about a quarter of adults being either over-weight or obese, and only 12.3% of them being underweight (Food and Nutrition Research Institute, Philippines 2005). Results of the Second National Diabetes Survey have shown about 5% of adults to be diabetic, an increase of 54% from the prevalence observed in the first survey in 1982 (Baltazar *et al.* 2003). Perhaps surprisingly, the extent of diabetes was found to be similar in both rural and urban areas.

In Thailand, Sakamoto *et al.* (2001) estimated the extent of obesity and overweight among preschool children in a province north of Bangkok to be almost a quarter of that among urban children, and three times that found in rural areas. They also found a strong positive association between socioeconomic status and childhood obesity. An obesity survey of school children in Khon Kaen, a relatively poor area in North East Thailand, found about 11% of them to be obese (using different growth reference values from the Sakamoto study), but importantly showed higher preva-lence in boys than in girls, and found differences between the schools surveyed, with greater prevalence among those with higher socioeco-nomic status (Langendijk *et al.* 2003). Among adults, National Health Examination Surveys (NHES) showed the prevalence of overweight to increase almost twofold between 1991 and 1996 for both males and females. In both years, obesity was much more common among women than men. However, overweight was still less common than underweight. Most cardiovascular disease risk factors also increased significantly between surveys (Kosulwat 2002). The 1996 NHES also showed that 9.6% of adults aged 35 years or more were diabetic – a prevalence similar to that of Malaysia (Aekplakorn *et al.* 2003).

The nutritional health of the population of Singapore is not changing as rapidly as in other South East Asian countries. The overall prevalence of obesity among adults as determined in the 1998 National Health Survey was 6.0%, similar to the prevalence found in the 1992 survey (Yian 2000). Rhen and Lee (2000) report that diabetes also appeared to plateau in the 1990s, increasing from about 2% in 1975 to 4.7% in 1984, 8.6% in 1992 and 9.0% in 1998. While the prevalence of some risk factors for cardiovascular disease continued to rise (National Health Survey; Yian 2000), Singapore had an annual decline in incidence of myocardial infarction of 2.3% from 1991 to 1999, and a drop in case-fatality rate comparable to those observed in many Western countries (Ounpuu and Yusuf 2003). In 1998, the three leading causes of death were cancer, coronary heart disease (CHD) and stroke, accounting for nearly 60% of all deaths in Singapore. As has occured in Malaysia, the disease burden in Singapore varies among ethnic groups, with the Chinese having lower rates of obesity, diabetes and cardiovascular disease than either Malays or South Asian Indians. The ethnic differences in cardiovascular disease morbidity are not explained by differences in traditional risk factors (Ounpuu and Yusuf 2003) or socioeconomic status (Thumboo *et al.* 2003).

Considering the region overall, there are clear differences in the evolution of nutritional issues between countries, but also many similarities. For example, NCDs have become the leading cause of death, while undernutrition remains a significant problem in all countries except Singapore, particularly among children. Furthermore, rates in rural areas are almost as high as in urban areas, while socioeconomic status and ethnicity are generally and independently associated with obesity and NCD risk. The transition has been rapid, although the pace of change is different across countries.

There is ongoing debate in the literature about the potential roles of fetal and childhood malnutrition in programming children for increased risk of NCDs as adults (Reddy 2002). Reports from a longitudinal study in the Philippines show that maternal energy status and diet are predictive of blood pressure in their offspring (Adair *et al.* 2001); mothers with low energy status during pregnancy give birth to male offspring with high cardiovascular disease risk in adolescence, as reflected by lipid profiles (Kuzawa and Adair 2003). These observations suggest that even if malnutrition is eliminated in the near future, there will be at least another generation in most countries of the region with heightened risk for NCDs as a result of malnutrition among children in recent decades.

Indonesia: transitions and challenges

The economic, dietary and nutrition transitions outlined above have been accompanied by other changes in society that present challenges for understanding and addressing nutritional health issues. These are considered here with respect to Indonesia, but are illustrative of the situation in most nations of this region.

The urban growth and shift away from employment in agriculture towards manufacturing and labour-intensive industries in Indonesia since the 1980s have altered the way economic externalities affect food consumption and how governments must now respond to certain nutritional issues. The widespread drought through the mid 1990s was followed by the Asian economic crisis, both events having important impacts on food security. Tabor *et al.* (2002) made the following observations:

> The late 1990s experience also changed the way in which the Indonesian authorities came to understand food security, ... food security was understood as having rather little to do with domestic food production per se and much more to do with household incomes. After the 1996 El Niño, Indonesia's most severe drought in nearly a century, food prices remained relatively stable and there were just scattered signs of hardship, mainly in isolated parts of Eastern Indonesia. Conversely, food riots erupted after the exchange rate devaluation caused rice prices to double in the first half of 1997, pushing the cost of basic foods beyond the reach of many low-income urban families.

Changes have also occurred in the attitudes and expectations of the population. The *Posyandu* is Indonesia's main national community nutrition programme. The major activities of a *Posyandu* include the 'weighing post' (growth monitoring), other primary health services and family planning. These are undertaken by village volunteers, with assistance from health and nutrition professionals from sub-district health centres. Coverage of *Posyandus* grew strongly through the 1980s, so that by 1989 one or more were implemented by almost every village (more than 65,000) in all 27 provinces. In 1990, there were about 250,000 *Posyandus*, there being two to three of them per village (Soekirman *et al.* 1992).

Anecdotal reports suggest that the *Posyandu* began to decline in coverage and attendance from the early 1990s. The 2002 Indonesian Family Life Survey shows changes in its usage by the community and its quality during the late 1990s. While Strauss *et al.* (2004) report a decline in the use of outpatient health care facilities generally between 1997 and 2000, the decline in *Posyandu* use was the most marked (from 52% to 40%). The decline in *Posyandu* use took place in both urban and rural areas, but with a larger proportional decline in rural areas. There was also a decline in provision of *Posyandu* services such as child growth monitoring, which dropped from 50% to 36% between 1997 and 2000, and a decline in availability of KMS (growth monitoring) cards by 24%. Over the same period there was an increase in the utilization of private health facilities by young children, and an improvement in some health indicator outcomes. For example, there was a significant increase in the proportion of children receiving all of their vaccinations (from 35% to 55%).

The 2001 Indonesia Human Development Report (United Nations Development Program 2001) states that 'The decline in usage of the *Posyandu*, is partly because of a reduction in public support for the PKK, the Family Welfare Movement, but also because the *Posyandu* rely on volunteer helpers who as a result of the crisis will have less time to volunteer.' These observations reflect fundamental changes in attitudes and expectations from health services in the population – a shift in consumer preferences for services and fewer village residents being prepared to volunteer – as well as a general decline in service quality. There have been attempts to 'revitalize' the *Posyandu*, but to date these have met with limited success.

Finally, following 30 years of a highly centralized government, Indonesia implemented a policy of decentralization in 2001, devolving most decision-making responsibilities (including for food and nutrition programmes) to the 400 or so districts. This has brought both significant opportunities and challenges. In terms of opportunities, it has allowed districts to move away from the uniform programme models that have typically been used in the past and to introduce nutrition activities that meet expectations of the local population, and directly address the major causes of poor nutrition that are specific to their district. Experience has shown that a cluster of districts with similar levels of malnutrition can have quite different profiles of factors that contribute to it. Consequently, situation analyses need to be done separately for each district. The major challenge is the development of organizational and workforce capacity that is able to undertake such assessments at a district level, to choose cost-effective programme models and to implement them effectively. This will also involve significant institutional changes at provincial and national levels.

This example illustrates how urbanization and social and political change can have important implications for strategies to improve nutritional health. There is broad recognition in Indonesian nutrition and health agencies of these issues, and that new approaches to them are needed. But there are also barriers to overcome, with a need to engage in a different way with district heads and legislators and to develop better linkages with non-government organizations and the private sector.

Conclusion

Much has been written about the epidemiologic and nutrition transitions in South East Asia, as well as elsewhere in the world. Authors tend to highlight those aspects that are most relevant to their audience, generally the field they represent. Those concerned with food production and agriculture focus on food security and food safety, the particular challenges associated with urbanization and on changes associated with patterns of food purchasing. From this perspective, water shortage and environmental pollution are also emerging issues. Those writing for the health sector highlight demographic changes, the shift in disease priorities and the need to reorient health services to address the rapidly growing issue of NCDs. Those concerned with nutritional health are at the intersection of these approaches, needing to understand both of them, as well as other perspectives, to be able to respond effectively to changes in nutritional health.

Significant socioeconomic developments in South East Asia in recent decades have brought changes at many levels, including food consumption and nutritional disease patterns. But there are also changes in community expectations, governance and other factors that impact on how agencies respond to improve population health. The type and extent of change have often been influenced by particular characteristics of population subgroups. In most countries of the region, socioeconomic status and ethnicity appear to have been particularly influential in this. The traditional divide between urban and rural populations is becoming less important as a classifier of distinct types of health problem in the region. As a consequence of these factors, the understanding of nutritional health and the programmes needed to improve it have become more complex in all countries.

Tensions might be expected to grow between the need to address the needs of different groups in the identification of health priorities and allocation of funding by different agencies: undernourished versus overnourished people; children versus adults versus the aged. With improved understanding of the different risk profiles for NCDs in population subgroups will come the need for tailored and targeted interventions. Increasingly, this will also be needed to respond to the re-emergence of some old nutritional issues such as beri-beri and to the likely development of pockets of malnutrition among the hard-to-reach. This segmentation of the nutrition market will present important challenges both in terms of the information needed to understand changing situations and in the development of appropriate policy and programme responses.

References

Adair, L. S., Kuzawa, C. W. and Borja, J. (2001). Maternal energy stores and diet composition during pregnancy program adolescent blood pressure. *Circulation* **104**, 1034–9.

Aekplakorn, W., Stolk, R. P., Neal, B., *et al.* (2003). The prevalence and management of diabetes in Thai adults. *Diabetes Care* **26**, 2758–63.

Baltazar, J. C. C., Ancheta, C. A., Aban, I. B., Fernando, R. E. and Baquilod, M. M. (2003). Prevalence and correlates of diabetes mellitus and impaired glucose tolerance among adults in Luzon, Philippines. *Diabetes Research and Clinical Practice* **64**, 107–15.

BPS Statistics (2000). *End of Decade Statistical Report: Data and Descriptive Analysis.* Jakarta: BPS Statistics Indonesia.

(2003). *Susenas Household Iodised Salt Consumption Survey.* Jakarta: BPS Statistics Indonesia.

Chaseling, M., Frank, V., Rongavilla, E. and Vallejos, H. (2002). An exploratory study of iodine deficiency disorders: a public health problem in Baan Bon Khoa Kang-Rieng, Sri Sawat, Kanchanburi, Thailand. Thesis for Master of Public Health, University of Queensland, Brisbane, Australia.

Central Intelligence Agency (CIA) (2005). *World Fact Book.* http://www.cia.gov/cia/publications/factbook. (sighted January 2005.)

Cuthberson, C., Naemiratch, B. and Thompson, L. (1999). Food sources, dietary intakes, cultural and other factors associated with the prevalence of iodine deficiency and iron deficiency anaemia in Orang Asli women of childbearing age in Sungai Lalang, Chemong village, Hulu Langat, Selangor, Malaysia. Thesis for Master of Community Nutrition, University of Queensland, Brisbane, Australia.

Deitchler, M., Mason, J., Mathys, E., Winichagoon, P. and Tuazon, M. A. (2004). Lessons from successful micronutrient programs Part 1: Program initiation. *Food and Nutrition Bulletin* **25**, 5–29.

Florentino, R. F. (2002). The burden of obesity in Asia: challenges in assessment, prevention and management. *Asia-Pacific Journal of Clinical Nutrition* **11**(Supplement), S676–80.

Florentino, R. F., Villavieja, G. M. and Lana, R. D. (2002). Regional study of nutritional status of urban primary schoolchildren. 1. Manila, Philippines. *Food and Nutrition Bulletin* **23**, 24–30.

Food and Nutrition Research Institute, Philippines (FNRI) (2005). *The 6th National Nutrition Surveys: Initial Results.* http://www.fnri.dost.gov.ph/nns/6thanthrop.pdf (sighted April 2005).

Hardy, A. (1995). Beriberi, Vitamin B1 and World Food Policy, 1925–1970. *Medical History* **39**, 61–77.

Helen Keller International (1998–2004). *Indonesia Crisis Bulletins.* http://www.hkiasiapacific.org/Resources/Downloads/bulletins_indonesia.htm (sighted April 2005).

Ismail, M. N., Chee, S. S., Nawawi, H., *et al.* (2002). Obesity in Malaysia. *Obesity Reviews* **3**, 203–8.

Juguan, J. A., Lukito, W. and Schultink, W. (1999). Thiamine deficiency is prevalent in a selected group of urban Indonesian elderly people. *Journal of Nutrition* **129**, 366–71.

Khor, G. L., Tee, E. S. and Kandiah, M. (1990). Patterns of food production and consumption in the ASEAN region. In *Aspects of Food Production, Consumption and Energy Values*, ed. G. H. Bourne. *World Review of Nutrition and Dietetics* **61**, 1–40.

Kiang, Y. P. (1998). *Statistics Singapore Newsletter: Trends in food consumption.* Statistics Singapore. http://www.singstat.gov.sg/ssn/feat/contents.html (sighted March 2005).

Kosulwat, V. (2002). The nutrition and health transition in Thailand. *Public Health Nutrition* **5**, 183–9.

Krishna, S., Taylor, A. M., Supanaranond, W., *et al.* (1999). Thiamine deficiency and malaria in adults from southeast Asia. *Lancet* **353**, 546–9.

Kuzawa, C. W. and Adair, L. S. (2003). Lipid profiles in adolescent Filipinos: relation to birth weight and maternal energy status during pregnancy. *American Journal of Clinical Nutrition* **77**, 960–6.

Langendijk, G., Wellings, S., van Wyk, M., *et al.* (2003). The prevalence of child-hood obesity in primary school children in urban Khon Kaen, Northeast Thailand. *Asia Pacific Journal of Clinical Nutrition* **12**, 66–72.

McGready, R., Simpson, J. A., Cho, T., *et al.* (2001). Postpartum thiamine deficiency in a Karen displaced population. *American Journal of Clinical Nutrition* **74**, 808–13.

Nandi, B. K. (1999). Nutritional transition. In *Nutritional Security: Asian Perspective Beyond 2000*. Food and Agriculture Organisation. http:www.fao.org//docrep/005/ac622e/ac622e04.htm (sighted March 2005).

Noor, M. I. (2002). The nutrition and health transition in Malaysia. *Public Health Nutrition* **5**, 191–5.

Ounpuu, S. and Yusuf, S. (2003). Singapore and coronary heart disease: a population laboratory to explore ethnic variations in the epidemiologic transition. *European Heart Journal* **24**, 127–9.

Phua, K. H., Koh, K., Tan, T. C., *et al.* (1990). Thiamine deficiency and sudden deaths: lessons from the past. *Lancet* **335**, 1471.

Popkin, B. M. (2002). The dynamics of the dietary transition in the developing world. In *The Nutrition Transition: Diet and Disease in the Developing World*, ed. B. Caballero and B. M. Popkin. Sydney: Academic Press, pp. 111–28.

Reddy, K. S. (2002). Cardiovascular diseases in the developing countries: dimensions, determinants, dynamics and directions for public health action. *Public Health Nutrition* **5**, 231–7.

Rhen, W. and Lee, W. (2000). The changing demography of diabetes mellitus in Singapore. *Research and Clinical Practice* **50**(Supplement 2), S35–9.

Sakamoto, N., Wansorn, S., Tontisirin, K. and Marui, E. (2001). A social epidemiology study of obesity among preschool children in Thailand. *International Journal of Obesity and Related Metabolic Disorders* **25**, 389–94.

Soekirman, Tarwotjo, I., Jus'at, I., Sumodiningrat, G. and Jalal, F. (1992). *Economic Growth, Equity and Nutritional Improvement in Indonesia*. Geneva: United Nations Administrative Committee on Coordination – SubCommittee on Nutrition.

Soekirman (1997). Indonesia continues groundbreaking model for fighting hunger and poverty. *2020 Vision News and Views March 1997*. Washington: International Food Policy Research Institute.

Soekirman, Hardinsyah, Jus'at, I. and Jahari, A. B. (2002). Regional study of nutritional status of urban primary schoolchildren. 2. West Jakarta and Bogor, Indonesia. *Food and Nutrition Bulletin* **23**, 31–40.

Strauss, J., Beegle, K., Dwiyanto, A., *et al.* (2004). *Indonesian Living Standards Three Years after the Crisis: Evidence from the Indonesian Family Life Survey*. Santa Monica, California: Rand Corporation.

Sutanegara, D., Darmono and Budhiarta, A. A. G. (2000). The epidemiology and management of diabetes mellitus in Indonesia. *Diabetes Research and Clinical Practice* **50**(Supplement 2), S9–16.

Tabor, S. R., Dillon, H. S. and Sawit, M. H. (2002). *Child Growth, Food Insecurity and Poverty in Indonesia.* Mimeo. Paper prepared for 'International Expert Meeting on Child Growth', Jakarta. 10–13 November 2002.

Tee, E. S., Khor, S. C., Ooi, H. E., *et al.* (2002). Regional study of nutritional status of urban primary schoolchildren. 3. Kuala Lumpur, Malaysia. *Food and Nutrition Bulletin* **23**, 41–7.

Thumboo, J., Fong, K. Y., Machin, D., *et al.* (2003). Quality of life in a urban Asian population: the impact of ethnicity and socio-economic status. *Social Science and Medicine* **56**, 1761–72.

United Nations Administrative Committee on Coordination – Sub-Committee on Nutrition (1992). *Second Report on the World Nutrition Situation,* Vol. 1: *Global and Regional Results.* Geneva: United Nations Administrative Committee on Coordination – Sub-Committee on Nutrition.

United Nations Development Program (UNDP)/BAPPENAS/BPS (2001). *Indonesian Human Development Report 2001, Towards a New Consensus, Democracy and Human Development in Indonesia.* Jakarta: BPS Statistics Indonesia.

United Nations Development Program (UNDP) (2005). *Human Development Reports Database.* http://hdr.udnp.org/statistics (sighted January 2005).

Yian, T. B. (2000). *Statistics Singapore Newsletter: Highlights of the 1998 National Health Survey.* Statistics Singapore. http://www.singstat.gov.sg/ssn/feat/contents.html (sighted March 2005).

Yusuf, S., Reddy, S., Ounpuu, S. and Anand, S. (2001). Global burden of cardiovascular diseases, Part 1: General considerations, the epidemiologic transition, risk factors, and impact of urbanization. *Circulation* **104**, 2746–53.

Winichagoon, P., Yhoung-aree, J. and Pongchareon, T. (2001). *The Current Situation and Status of Micronutrient Policies and Programs in Thailand.* One of the country case studies presented at a workshop on 'Successful Micronutrient Programs' held at the International Union of Nutritional Sciences, Vienna, August 2001. http://www.inffoundation.org/NatlCountry CaseStudies.html (sighted April 2005).

Wolf, C. (2000). *Asian Economic Trends and Their Security Implications.* Santa Monica, California: Rand Corporation.

World Health Organization (2003). *WHO Global Database on Child Growth and Malnutrition.* Geneva: World Health Organization. http://www.who.int/nutgrowthdb/ (sighted January 2006).

Zaini, A. (2000). Where is Malaysia in the midst of the Asian epidemic of diabetes mellitus? *Diabetes Research and Clinical Practice* **50**(Supplement 2), S23–8.

5 Obesity and nutritional health in Hong Kong Chinese people

GARY T. C. KO

Introduction

Obesity is a major health problem in the Asia-Pacific region and elsewhere because of its increasing prevalence and the morbidity and mortality associated with it (Seidell *et al.* 1996; World Health Organization 1998; Behn and Ur 2006). The prognostic significance of obesity in cardiovascular disease and total mortality is mainly due to its close associations with hypertension, dyslipidaemia, hyperinsulinaemia and glucose intolerance (Sowers 2003; Bray 2004; Behn and Ur 2006). There are also close associations between various anthropometric measures of body fatness, including body mass index (BMI), waist–hip ratio (WHR) and waist circumference (WC), and multiple cardiovascular risk factors. Most of these relationships are independent of age and smoking, and some of the risk factors increase in severity exponentially with increase in fatness and obesity.

Western definitions of obesity, if used to determine levels in Asian populations, give much lower prevalences than among European and North American populations. For example, if a $BMI \geq 30\,kg/m^2$ is used to define obesity, up to 25% of Europeans but only 2% to 5% of Hong Kong Chinese will be classified as obese (Ko *et al.* 1999a; Rossner 2002). However, there is now a wealth of data showing that the prevalences of both type 2 diabetes and hypertension are reaching epidemic proportions among Asians, despite currently low prevalence rates of obesity, as defined as BMI greater than $30\,kg/m^2$, in all nations of Asia. In 1990, the prevalence of diabetes in Hong Kong was reported to be 4.5% using the 1985 World Health Organization (WHO) criteria (Cockram *et al.* 1993). By 1996, this had increased to 9.5% and 9.8% in Hong Kong Chinese men and women respectively (Janus 1997); these prevalences are similar to those of populations in Europe and the United States, which lie between 5% and 15% (Harris *et al.* 1998; International Diabetes Federation 2000).

Health Change in the Asia-Pacific Region: Biocultural and Epidemiological Approaches, ed. Ryutaro Ohtsuka and Stanley J. Ulijaszek. Published by Cambridge University Press. © Cambridge University Press 2007.

Cut-offs for obesity in Hong Kong Chinese people

Obesity is characterized by an excess of body fat (BF), which is defined conventionally as more than 25% of body weight as fat in males and more than 35% of weight as fat in females (World Health Organization 1995; Lohman *et al.* 1997; Deurenberg *et al.* 1998). The BMI corresponding to such BF content is approximately $30 \, kg/m^2$ in young European adults (World Health Organization 1995). There is a significant positive association between BMI, when above $30 \, kg/m^2$, and risk of morbidity and mortality. While the WHO recommends a BMI cut-off for obesity of $30 \, kg/m^2$ (World Health Organization 1995), there is increasing evidence that this cut-off cannot be readily applied to Asians (Ko *et al.* 1999a,b). In a study involving 5,153 Hong Kong Chinese subjects, the optimal BMI to predict BF which is (1) elevated ($\geq 30\%$ of body weight in women; $\geq 20\%$ in men); (2) high ($\geq 35\%$ of body weight in women; $\geq 25\%$ in men); and (3) very high ($\geq 40\%$ of body weight in women, $\geq 30\%$ in men) were 22.5, 24.2 and $26.1 \, kg/m^2$ in women, and 23.1, 23.8 and $25.4 \, kg/m^2$ in men, respectively (Ko *et al.* 2001a), much lower than conventional cut-offs for fatness and obesity in Europeans.

Obesity is a strong predictor for diabetes and hypertension (Chan *et al.* 1996; Ko *et al.* 1997; Sowers 2003). In a study of the likelihood of Hong Kong Chinese adults having type 2 diabetes or hypertension at various cut-off values of BMI, WHR and WC, increases in anthropometric indexes of one unit ($1 \, kg/m^2$ in BMI; 0.01 in WHR; 1 cm in WC) were shown to be associated with a 1.05- to 1.27-fold increased likelihood of having diabetes or hypertension (Ko *et al.* 1999b). If a likelihood ratio (LR) of 2.5 or above is used to identify Chinese subjects at high risk of developing these diseases, the corresponding BMI cut-offs for type 2 diabetes risk are $29.5 \, kg/m^2$ (men) and $25.5 \, kg/m^2$ (women). Corresponding values for hypertension are 26.5 and $25.3 \, kg/m^2$ for males and females respectively. Using a mean BMI of $25 \, kg/m^2$ to predict diabetes or hypertension in Hong Kong Chinese subjects gives an LR of around 2 in men and about 3 in women (Fig. 5.1). Similarly, using the mean levels of WC of European populations (93 cm for men and 82 cm for women) to predict diabetes or hypertension in Hong Kong Chinese gave an LR of between 2.0 and 3.5 in both men and women (Ko *et al.* 1999b).

In another cohort involving 702 Hong Kong Chinese adults recruited from a medical out-patient clinic, there were similar significant trends between obesity indices, the severity of cardiovascular risk factors, and the prevalence of morbidity conditions. Using $19.0–20.9 \, kg/m^2$ and $<70 \, cm$ as reference values for BMI and WC respectively, subjects with $BMI \geq 25.0 \, kg/m^2$

(A) Men

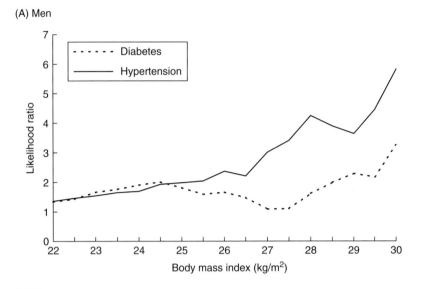

(B) Women

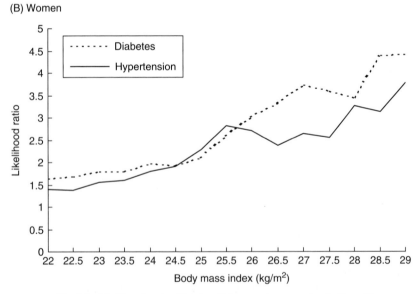

Fig. 5.1. Likelihood ratio of having diabetes or hypertension in Hong Kong Chinese subjects with various BMI cut-off values.

(in both sexes) and/or a WC ≥ 85 cm (males) or ≥ 75 cm (females) had age-adjusted odds ratios of between 3.2 and 4.4 for the occurrence of at least one of the following morbidity conditions: type 2 diabetes, hypertension, dyslipidaemia and albuminuria (Lee *et al.* 2002).

In the year 2000, the World Health Organization Western Pacific Region, the International Association for the Study of Obesity (IASO) and the International Obesity Task Force (IOTF) published a joint statement in which the classification of overweight and obesity for people in the Asia-Pacific region was revised (World Health Organization Western Pacific Region, International Association for the Study of Obesity and International Obesity Task Force 2000). The BMI cut-offs suggested for the classification of overweight and obesity for Asians were 23 and 25 kg/m^2 respectively. Subsequently, the Cooperative Meta-Analysis Group of the Working Group on Obesity in China recommended a further revision to this classification for the identification of overweight and obesity in China, to BMI cut-offs of $\geq$24 and $\geq$28 kg/m^2 respectively (Zhou 2002). Nevertheless, for international comparison, a BMI of $\geq$30 kg/m^2 is still frequently adopted to define obesity even among Asians.

With respect to central obesity, the WHO, IASO and IOTF Asian-Pacific guideline put forward WC $\geq$ 90 cm in men and $\geq$80 cm in women as cut-offs for central obesity in adult Asians. The Hong Kong Society for Endocrinology, Metabolism and Reproduction (HKSEMR) has also issued a statement to health care professionals in Hong Kong to the effect that female patients with type 2 diabetes and high WC ($\geq$90 cm for males and $\geq$80 cm for females) should be considered to be in an unsatisfactory medical state (HKSEMR 2000).

Prevalence of obesity in Hong Kong

In 1990, 28.2% and 28.0% of Hong Kong men and women were either overweight and obese, with BMI $\geq$ 25 kg/m^2 (Ko *et al.* 1999a), while only 2.2% and 4.8% of men and women had BMI $\geq$ 30 kg/m^2, and could thus be considered obese. A subsequent survey involving 17,121 subjects measured in 1996 to 1997 showed 28.9% of them to be overweight and obese (BMI $\geq$ 25 kg/m^2), while 3.8% of them were obese (BMI $\geq$ 30 kg/m^2) (Table 5.1). The age-standardized prevalence rates of overweight and obesity were 30.5% in men and 22.1% in women (Ko *et al.* 2001b). The prevalence of obesity (BMI $\geq$ 30 kg/m^2) is much higher among Euro-American populations, where values vary from 7% in France, 8.9% in Spain, 18% in Finland to 22.4% in the United States (Pietinen *et al.* 1996; Flegal *et al.* 1998; Gutierrez-Fisac *et al.* 1999; Maillard *et al.* 1999; Rossner 2002). There has also been an increase in excess of twofold in the prevalence of childhood overweight and obesity (Leung *et al.* 1998), in line with other Asian countries (Kotani *et al.* 1997; Luo and Hu 2002; Likitmaskul *et al.* 2003).

Table 5.1. *Body mass index (BMI), overweight and obesity among 17,121 Hong Kong Chinese subjects*

Age group (years)	n	BMI (kg/m^2) Mean ± SD	Overweight n (%)	Obese n (%)
(a) Women				
≥20–30	1230	20.2 ± 3.1	76 (6.2)	18 (1.5)
≥30–40	2700	21.8 ± 3.4	411 (15.2)	65 (2.4)
≥40–50	2311	23.1 ± 3.4	584 (25.3)	82 (3.6)
≥50–60	1367	24.1 ± 3.4	508 (37.2)	71 (5.2)
≥60–70	3100	24.3 ± 3.5	1233 (39.8)	184 (5.9)
≥70–80	1468	24.1 ± 3.8	560 (38.2)	95 (6.5)
≥80	164	23.2 ± 3.6	47 (28.7)	3 (1.8)
Total	12340	23.1 ± 3.7	3419 (27.7)	518 (4.2)
(b) Men				
≥20–30	385	22.2 ± 3.6	62 (16.1)	12 (3.1)
≥30–40	1021	23.9 ± 3.5	347 (34.0)	44 (4.3)
≥40–50	949	24.1 ± 3.2	363 (38.3)	33 (3.5)
≥50–60	510	23.7 ± 3.1	174 (34.1)	11 (2.2)
≥60–70	1206	23.6 ± 3.2	390 (32.3)	26 (2.2)
≥70–80	634	23.2 ± 3.0	172 (27.1)	11 (1.7)
≥80	76	21.9 ± 3.2	12 (15.8)	1 (1.3)
Total	4781	23.6 ± 3.3	1520 (31.8)	138 (2.9)

Notes:
Overweight: BMI 25.0–29.9 kg/m^2; Obese: BMI ≥ 30.0 kg/m^2.

Central obesity

There are now many data confirming the importance of central adiposity as a cardiovascular risk factor (Richelsen and Pedersen 1995; Gasteyger and Tremblay 2002; Behn and Ur 2006). This is mainly due to the increased lipolytic activity of visceral adipose tissue, which can lead to increased production of free fatty acids. Free fatty acids can induce insulin resistance by enhancing gluconeogenesis and reducing glucose uptake in muscle, with metabolic and vascular consequences (Björntorp 1991). After adjustment for age, BMI and plasma lipid, WHR has been shown to remain a significant cardiovascular disease risk factor in both European men and women confirmed angiographically to have coronary atherosclerosis (Thompson *et al.* 1991). Other workers have reported similar independent associations between central obesity and blood pressure, and plasma lipid and glucose concentrations in European, Chinese and Japanese subjects (Masuda *et al.* 1993; Lyu *et al.* 1994; Seidell *et al.* 1996).

In the United Kingdom in the 1990s, the mean WHR and WC in men were 0.93 and 93.3 cm, and in women, 0.80 and 82.0 cm, respectively (Lean *et al.* 1995). These values are higher than those for Hong Kong Chinese adults, for whom Ko *et al.* (1999b) reported values for WHR and WC of 0.87 and 90.8 cm for men and 0.80 and 74.9 cm for women, respectively. Using WHRs of ≥0.95 for men and ≥0.80 for women as cut-off values for overweight as suggested by Lean *et al.* (1995), 37.6% of men and 48.2% of women in the UK are considered to have central obesity (Lean *et al.* 1995). This compares with 9.3% and 48.1% of Hong Kong Chinese men and women, using the same cut-offs (Ko *et al.* 1999b). Similarly, using a WC of ≥94 cm for men and ≥80 cm for women as cut-off values, 46.7% of men and 50.9% of women in the UK (Lean *et al.* 1995) are considered to have central obesity, as compared to only 4.5% of men and 25.5% of women in Hong Kong (Ko *et al.* 1999b).

Obesity, morbidity and mortality

Over 50% of all deaths in the United States have been linked to obesity; this accounts for 3% to 7% of total health costs for the nation. In a 12-year prospective study, Seidell *et al.* (1996) followed up 48,287 Dutch subjects and found a threefold increase in coronary heart disease deaths in those with a BMI ≥25 kg/m^2. However, intentional weight loss reduces blood pressure, improves lipid patterns and reduces the risk of diabetes. These relationships are linear and become evident with weight loss of 5% to 10% (Bray 1998). Increasing BMI is strongly associated with increased risk of developing diabetes (Chan *et al.* 1994; Adler 2002). A prospective study of oral glucose tolerance carried out annually among 115 Hong Kong Chinese subjects with normal glucose tolerance compared the relative effects of various BMI groups on progression to diabetes. After a mean follow-up of 2.1 years (range one to eight years), the 50% median survival time for subjects with normal glucose tolerance to progress to diabetes was 5.97 years. This gave a crude rate of progression to diabetes of 8.4% per year (Ko *et al.* 2004). Using BMI cut-offs for overweight and obesity of 23 and 25 kg/m^2 (WHO 2000 revised guideline) or BMI of 24 and 28 kg/m^2 (China 2002 guideline), the progression rates to diabetes in subjects with normal glucose tolerance were significantly different among different BMI groups (Fig. 5.2). Those with BMI ≥ 25 kg/m^2 had a rate of conversion to diabetes five times greater than those with BMI < 23 kg/m^2 (20.4% against 4.1%) and those with BMI ≥ 28 kg/m^2 had a rate ten times higher than those with BMI < 24 kg/m^2 (21.0% against 2.1%). For subjects with BMI ≥ 25 kg/m^2, the crude

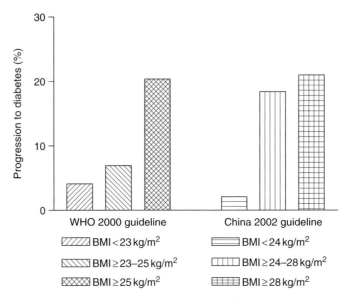

Fig. 5.2. Progression to diabetes in Hong Kong Chinese people with normal glucose tolerance after a mean follow-up of two years.

rate of progression to diabetes was 12.5% per year. The corresponding rate for subjects with BMI ≥ 28 kg/m^2 was 14.6% per year.

Metabolic Syndrome

Reaven (1988) observed a clustering of elevated plasma insulin concentration, obesity, hypertension, dyslipidaemia and hyperglycaemia, and described the condition as Syndrome X or the Metabolic Syndrome. Obesity-inducing insulin resistance is now believed to be the main pathogenic factor underlying the Metabolic Syndrome. Subjects with insulin resistance have impaired insulin-stimulated glucose uptake by peripheral tissues that leads to impaired glucose tolerance and eventually diabetes. Associated hyperinsulinaemia has stimulatory effects on the adrenergic and renin-aldosterone systems and promotes renal tubular sodium and water re-absorption. These factors may contribute to elevation of blood pressure. In normal subjects, insulin inhibits lipolysis and production of very low density lipoprotein cholesterol (VLDL-C) and promotes the catabolism of VLDL-C to high density lipoprotein cholesterol (HDL-C). Hence, insulin resistance is associated with dyslipidaemia, characterized particularly by increased triglyceride and VLDL-C, and reduced HDL-C

(Reaven 1988; Williams 1994). Visceral obesity is an important linking factor in the Metabolic Syndrome (Björntorp 1991). Hormonal abnormalities such as age-related declines in growth hormone and sex steroids, and increased cortisol levels due to psychosocial stress, can all enhance the deposition of visceral fat. These fat depots within the peritoneal cavity are more responsive to catecholamine-stimulated lipolysis than subcutaneous fat, with increased formation of free fatty acids. In the presence of excess free fatty acids, insulin becomes less effective at promoting glucose uptake at various tissue sites, including liver, muscle and adipose tissue, for energy production or storage (Randle *et al.* 1963; Björntorp 1991). Insulin resistance increases by 30% to 40% when body weight increases by 35% to 40% above the ideal. On the other hand, exercise and weight reduction can improve insulin sensitivity. In Hong Kong Chinese people, age and obesity explain most of the variance in the components of the Metabolic Syndrome and abnormal lipid patterns even after correction for insulin resistance (Chan *et al.* 1996).

Nutritional health of Hong Kong Chinese people

The problem of overnutrition is believed to be one of the major factors related to the rising prevalence of obesity and diabetes in Hong Kong (Chan *et al.* 1996). Although obesity occasionally presents as a feature of some endocrine or metabolic disorders, it is usually a consequence of energy imbalance due to excessive energy intake and/or inadequate energy expenditure. In the 1990s, the mean daily energy intake of Hong Kong Chinese people was 2,400 kcal in men and 1,800 kcal in women, similar to those of Australians and higher than those of Singapore Chinese people (Woo *et al.* 1998). Energy intake has increased and the level of physical activity decreased among Hong Kong Chinese people over past decades.

Hong Kong Chinese people have sodium intakes 50% to 80% higher than Australians and 20% to 50% higher than Singapore Chinese people, with only 20% of Hong Kong Chinese subjects having a daily intake of sodium below the desirable maximum level of 2,300 mg/day (Woo *et al.* 1998). High salt intake is strongly associated with the development of hypertension (Intersalt Cooperative Research Group 1988), and increasing intake of sodium with age may also partly contribute to the increasing prevalence of hypertension with age in Hong Kong Chinese people. Most Hong Kong Chinese people have lipid intakes which are optimal for cardiovascular health (Woo *et al.* 1998; Lee *et al.* 2000), although serum lipid profiles are similarly high when compared to many Western populations (Lau *et al.* 1993).

Dietary patterns in Hong Kong are becoming Westernized, especially among children. For Hong Kong Chinese children aged seven years or less, daily energy intakes are 82–98 kcal/kg (boys) and 73–100 kcal/kg (girls), a little below levels of energy intake of Australian and Finnish children; however, protein intakes are higher (Leung *et al.* 2000a). Children in Hong Kong consume less milk fat and more meat, and are less physically active than Australian or Finnish children. About 30% of the total daily energy intake of Hong Kong Chinese children comes from fat. Although this is lower than in many Western populations, it is much more than the traditional Chinese diet (Leung *et al.* 2000b). With emerging problems of overnutrition in Hong Kong children, the future prevalences of obesity, type 2 diabetes and other diseases of urbanism are only likely to increase.

Physical activity in Hong Kong Chinese people

Low physical activity is a strong predictor of weight gain in Europeans as well as Chinese people (Esparza *et al.* 2000; Bell *et al.* 2001). The observation that fewer than one third of Hong Kong Chinese adults have regular exercise (S. Hui *et al.*, 2005) suggests that this may be a major factor contributing to the emergence of overweight and obesity. Woo *et al.* (2002) examined Hong Kong Chinese subjects aged 70 years and over and found that a higher level of physical activity was associated with decreased mortality and hospitalization. A study of young Hong Kong Chinese adults has shown that after a three-month exercise programme, reduction in body weight and BMI was significant in the aerobic group but not the anaerobic group, while BF percentage declined for both groups (Ko 2004). Thus, anaerobic exercise may increase the lean to fat body mass ratio. Plasma glucose was reduced in the aerobic group but not in the anaerobic group, while plasma insulin level remained similar before and after exercise in both groups. This suggests an improved tissue sensitivity to insulin and a steady insulin hormone secretion after aerobic exercise (Rodnick *et al.* 1987).

On the other hand, obesity and physical activity are related not only to medical conditions but mental health as well. In a cross-sectional study involving 876 Hong Kong Chinese subjects, health-related quality of life as assessed by SF-36 questionnaire, was noted to be lower in obese subjects (Ko 2006). In addition, as the level of physical activity decreases, the scores on most of the SF-36 scales also drop. Obese women without physical activity had a lowest score while obese women with physical activity, despite the obesity, had a high score on the SF-36.

Conclusion

Obesity should be considered a chronic disease due to the morbidity and mortality associated with it. With similar BF contents, the BMI cut-off values corresponding to the categories of overweight and obesity are lower in Hong Kong Chinese people than in Europeans. Using the cut-off of $BMI \geq 25 \, kg/m^2$, the age-standardized prevalence of overweight and obesity in Hong Kong adults is 30.5% in men and 22.1% in women, while the prevalence of obesity ($BMI \geq 30 \, kg/m^2$), at between 2% and 5%, is low in comparison with most European nations and the United States and Canada. Overnutrition and inadequate physical activity are the major causes of the increasing prevalence of overweight and obesity in the Hong Kong population.

References

Adler, A. (2002). Obesity and target organ damage: diabetes. *International Journal of Obesity and Related Metabolic Disorders* **26** (Supplement 4), S11–14.

Behn, A. and Ur, E. (2006). The obesity epidemic and its cardiovascular consequences. *Current Opinion in Cardiology* **21**, 353–60.

Bell, A. C., Ge, K. and Popkin, B. M. (2001). Weight gain and its predictors in Chinese adults. *International Journal of Obesity and Related Metabolic Disorders* **25**, 1079–86.

Björntorp, P. (1991). Metabolic implications of body fat distribution. *Diabetes Care* **14**, 1132–43.

Bray, G. A. (1998). Obesity: a time bomb to be defused. *Lancet* **352**, 160–1.

Bray, G. A. (2004). Medical consequences of obesity. *Journal of Clinical Endocrinology and Metabolism* **89**, 2583–9.

Chan, J. M., Stampfer, M. J., Ribb, E. B., Willett, W. C. and Colditz, G. A. (1994). Obesity, fat distribution and weight gain as risk factors for clinical diabetes in man. *Diabetes Care* **17**, 961–9.

Chan, J. C., Cheung, J. C., Lau, E. M., *et al.* (1996). The metabolic syndrome in Hong Kong Chinese – the inter-relationships amongst its components analyzed by structural equation modeling. *Diabetes Care* **19**, 953–9.

Cockram, C. S., Woo, J., Lau, E., *et al.* (1993). The prevalence of diabetes mellitus and impaired glucose tolerance among Hong Kong Chinese adults of working age. *Diabetes Research Clinical Practice* **21**, 67–73.

Deurenberg, P., Yap, M. and van Staveren, W. A. (1998). Body mass index and percent body fat: a meta-analysis among different ethnic groups. *International Journal of Obesity and Related Metabolic Disorders* **22**, 1164–71.

Esparza, J., Fox, C., Harper, I. T., *et al.* (2000). Daily energy expenditure in Mexican and USA Pima Indians: low physical activity as a possible cause of obesity. *International Journal of Obesity and Related Metabolic Disorders* **24**, 55–9.

Flegal, K. M., Carroll, M. D., Kuczmarski, R. J. and Johnson, C. L. (1998). Overweight and obesity in the United States: prevalence and trends, 1960–1994. *International Journal of Obesity and Related Metabolic Disorders* **22**, 39–47.

Gasteyger, C. and Tremblay, A. (2002). Metabolic impact of body fat distribution. *Journal of Endocrinology Investigation* **25**, 876–83.

Gutierrez-Fisac, J. L., Rodriguez Artalejo, F., Gulallar-Castillon, P., Banegas, J. R. and del Rey Calero, J. (1999). Determinants of geographical variations in body mass index and obesity in Spain. *International Journal of Obesity and Related Metabolic Disorders* **23**, 342–7.

Harris, M. I., Flegal, K. M., Cowie, C. C., *et al.* (1998). Prevalence of diabetes, impaired fasting glucose, and impaired glucose tolerance in U.S. adults: the Third National Health and Nutrition Examination Survey, 1988–1994. *Diabetes Care* **21**, 518–24.

Hong Kong Society for Endocrinology, Metabolism and Reproduction, Diabetes Division (HKSEMR) (2000). A statement for health care professionals on type 2 diabetes mellitus in Hong Kong. *Hong Kong Medical Journal* **6**, 105–7.

Hui, S. S. C., Thomas, N. and Tomlinson, B. (2005). Relationship between physical activity, fitness, and CHD risk factors in middle-age Chinese. *Journal of Physical Activity and Health* **3**, 307–23.

International Diabetes Federation (2000). Diabetes Atlas. Brussels: International Diabetes Federation.

Intersalt Cooperative Research Group (1988). Intersalt: an international study of electrolyte excretion and blood pressure. Results for 24 hour urinary sodium and potassium excretion. *British Medical Journal* **297**, 319–28.

Janus, E. D. (1997). *Hong Kong Cardiovascular Risk Factor Prevalence Study 1995–1996*. Hong Kong: Hong Kong University Press.

Ko, G. T. (2004). Short-term effects after a 3-month aerobic or anaerobic exercise program in Hong Kong Chinese. *Diabetes Nutrition Metabolism* **17**, 124–7.

Ko, G. T., Chan, J. C., Woo, J., *et al.* (1997). Simple anthropometric indexes and cardiovascular risk factors in Chinese. *International Journal of Obesity and Related Metabolic Disorders* **21**, 995–1001.

Ko, G. T., Cockram, C. S., Critchley, J. A. and Chan, J. C. (1999a). Obesity – definition, aetiology and complications. *Medical Progress* **26**, 10–14.

Ko, G. T., Chan, J. C., Cockram, C. S. and Woo, J. (1999b). Prediction of hypertension, diabetes, dyslipidaemia or albuminuria using simple anthropometric indexes in Hong Kong Chinese. *International Journal of Obesity and Related Metabolic Disorders* **23**, 1136–42.

Ko, G. T., Tang, J., Chan, J. C., *et al.* (2001a). Lower BMI cut-off value to define obesity in Hong Kong Chinese: an analysis based on body fat assessment by bioelectrical impedance. *British Journal of Nutrition* **85**, 239–42.

Ko, G. T., Wu, M. M., Tang, J., *et al.* (2001b). Body mass index profile in Hong Kong Chinese adults. *Annals of the Academy of Medicine Singapore* **30**, 393–6.

Ko, G. T., Chan, J. C., Chow, C. C., *et al.* (2004). Effects of obesity on the conversion from normal glucose tolerance to diabetes in Hong Kong Chinese – implication of different body mass index cutoffs for obesity. *Obesity Research* **12**, 889–95.

Ko, G. T. (2006). Both obesity and physical activity are associated with a less favourable health-related quality of life in Hong Kong Chinese. *American Journal of Health Promotion* **21**, 49–52.

Kotani, K., Nishida, M., Yamashita, S., *et al.* (1997). Two decades of annual medical examinations in Japanese obese children: do obese children grow into obese adults? *International Journal of Obesity and Related Metabolic Disorders* **21**, 912–21.

Lau, E., Woo, J., Cockram, C. S., *et al.* (1993). Serum lipid profile and its association with some cardiovascular risk factors in an urban Chinese population. *Pathology* **25**, 344–50.

Lean, M. E., Han, T. S. and Morrison, C. E. (1995). Waist circumference as a measure for indicating need for weight management. *British Medical Journal* **311**, 158–61.

Lee, H. Y., Woo, J., Chen, Z. Y., Leung, S. F. and Peng, X. H. (2000). Serum fatty acid, lipid profile and dietary intake of Hong Kong Chinese omnivores and vegetarians. *European Journal of Clinical Nutrition* **54**, 768–73.

Lee, Z. S., Critchley, J. A., Ko, G. T., *et al.* (2002). Obesity and cardiovascular risk factors in Hong Kong Chinese. *Obesity Review* **3**, 173–82.

Leung, S. S., Chan, Y. L., Lam, C. W., *et al.* (1998). Body fatness and serum lipids of 11-year-old Chinese children. *Acta Paediatrica* **87**, 363–7.

Leung, S. S., Chan, S. M., Lui, S., Lee, W. T. and Davies, D. P. (2000a). Growth and nutrition of Hong Kong children aged 0–7 years. *Journal of Paediatric Child Health* **36**, 56–65.

Leung, S. S., Lee, W. T., Lui, S. S., *et al.* (2000b). Fat intake in Hong Kong Chinese children. *American Journal of Clinical Nutrition* **72**(5 Supplement), 1373S–8S.

Likitmaskul, S., Kiattisathavee, P., Chaichanwatanakul, K., *et al.* (2003). Increasing prevalence of type 2 diabetes mellitus in Thai children and adolescents associated with increasing prevalence of obesity. *Journal of Pediatric Endocrinology and Metabolism* **16**, 71–7.

Lohman, T. G., Houtkoper, L. and Going, S. B. (1997). Body fat measurements goes high tech. Not all are created equal. *ACSM's Health & Fitness Journal* **1**, 30–5.

Luo, J. and Hu, F. B. (2002). Time trends of obesity in pre-school children in China from 1989 to 1997. *International Journal of Obesity and Related Metabolic Disorders* **26**, 553–8.

Lyu, L. C., Shieh, M. J., Bailey, S. M., *et al.* (1994). Relationship of body fat distribution with cardiovascular risk factors in healthy Chinese. *Annals of Epidemiology* **4**, 434–44.

Maillard, G., Charles, M. A., Thibult, N., *et al.* (1999). Trends in the prevalence of obesity in the French adult population between 1980 and 1991. *International Journal of Obesity and Related Metabolic Disorders* **23**, 389–94.

Masuda, T., Imai, K. and Komiya, S. (1993). ·Relationship of anthropometric indices of body fat to cardiovascular risk in Japanese women. *Annals of Physiological Anthropology* **12**, 35–144.

Pietinen, P., Vartiainen, E. and Mannisto, S. (1996). Trends in body mass index and obesity among adults in Finland from 1972 to 1992. *International Journal of Obesity and Related Metabolic Disorders* **20**, 114–20.

Randle, P. J., Hales, C. N., Garland, P. B. and Newsholme, E. A. (1963). The glucose fatty-acid cycle: its role in insulin sensitivity and the metabolic disturbances of diabetes mellitus. *Lancet* **2**, 785–9.

Reaven, G. M. (1988). Banting lecture 1988: role of insulin resistance in human disease. *Diabetes* **37**, 1595–607.

Richelsen, B. and Pedersen, S. B. (1995). Associations between different anthropometric measurements of fatness and metabolic risk parameters in non-obese, healthy, middle-aged men. *International Journal of Obesity and Related Metabolic Disorders* **19**, 169–74.

Rodnick, K. J., Haskell, W. L., Swislocki, A. L., Foley, J. E. and Reaven, G. M. (1987) Improved insulin action in muscle, liver and adipose tissue in physically trained human subjects. *American Journal of Physiology* **253**, E489–95.

Rossner, S. (2002). Obesity: the disease of the twenty-first century. *International Journal of Obesity* **26** (Supplement 4), S2–4.

Seidell, J. C., Verschuren, W. M., van Leer, E. M. and Kromhout, D. (1996). Overweight, underweight, and mortality. A prospective study of 48,287 man and women. *Archives of Internal Medicine* **156**, 958–63.

Sowers, J. R. (2003). Obesity as a cardiovascular risk factor. *American Journal of Medicine* **115** (Supplement 8A), 37S–41S.

Thompson, C. J., Ryu, J. E., Craven, T. E., Kahl, F. R. and Crouse, J. R. (1991). Central adipose distribution is related to coronary atherosclerosis. *Arteriosclerosis and Thrombosis* **11**, 327–33.

Williams, B. (1994). Insulin resistance: the shape of things to come. *Lancet* **344**, 521–4.

Woo, J., Leung, S. S., Ho, S. C., Lam, T. H. and Janus, E. D. (1998). Dietary intake and practices in the Hong Kong Chinese population. *Journal of Epidemiology and Community Health* **52**, 631–7.

Woo, J., Ho, S. C. and Yu, A. L. (2002). Lifestyle factors and health outcomes in elderly Hong Kong chinese aged 70 years and over. *Gerontology* **48**, 234–40.

World Health Organization (1995). *Physical Status: The Use and Interpretation of Anthropometry*. Technical Report Series 854. Geneva: World Health Organization.
    (1998). *Obesity: Preventing and Managing the Global Epidemic*. Report on a WHO Consultation on Obesity, Geneva, 3–5 June 1997 WHO/NUT/NCD/98. Geneva: World Health Organization.

World Health Organization Western Pacific Region, International Association for the Study of Obesity and International Obesity Task Force (2000). *The Asia-Pacific Perspective: Redefining Obesity and Its Treatment*. Australia: Health Communications Australia Pty Limited.

Zhou, B. F., for the Cooperative Meta-Analysis Group of the Working Group on Obesity in China (2002). Predictive values of body mass index and waist circumference for risk factors of certain related diseases in Chinese adults – study on optimal cut-off points of body mass index and waist circumference in Chinese adults. *Biomedical and Environmental Science* **15**, 83–96.

6 Modernization, nutritional adaptability and health in Papua New Guinea Highlanders and Solomon Islanders

TARO YAMAUCHI

Introduction

Economic modernization is global in outreach, and the countries in the South Pacific are not exempt. The Melanesian populations of Papua New Guinea (PNG) and the Solomon Islands are experiencing increasing involvement in the cash economy, and associated changes in diet, which now includes more purchased foods. Modernization and urbanization have also led to sedentary lifestyles, due to increased mechanization of work and a shift from subsistence activities (including traditional agriculture, fishing, hunting and gathering) to paid jobs. Dietary changes, such as increases in fat and sugar intakes and decreased fibre intake, together with reduced physical activity, have contributed to the increasing prevalence of obesity and degenerative diseases, such as type 2 diabetes and cardiovascular diseases.

Melanesians are broadly divided into two language groups – Austronesian (AN) and non-Austronesian (NAN). The AN-language speakers within PNG are part of a much larger AN-language family, which stretches from South East Asia to Hawaii and from Micronesia to New Zealand (Foley 1992). By contrast, NAN-language speakers are found predominantly on the mainland of PNG and have a much longer history than AN-language speakers (Foley 1992). It has been argued that the two groups differ genetically, this being reflected in their differing susceptibilities to type 2 diabetes and cardiovascular diseases (being less prevalent in NAN- than in AN-language speakers) (King 1992; Hodge et al. 1996). In this chapter, the influences of modernization on body composition of PNG Highlanders (NAN-language speakers) and Solomon Islanders (AN-language speakers) are compared, and lifestyle changes in terms of daily time allocation, physical activity level

Health Change in the Asia-Pacific Region: Biocultural and Epidemiological Approaches, ed. Ryutaro Ohtsuka and Stanley J. Ulijaszek. Published by Cambridge University Press.

(PAL), diet and nutritional intake of these populations are examined in relation to modernization.

Study populations and participants

PNG Highlands

PNG is located north of the Australian continent and east of Indonesia. The Huli, one of the largest language groups in PNG, inhabit the Tari basin, located about 1,600 m above sea level, between 142°70′ and 143°30′ east longitude, and between 5°70′ and 6°20′ south latitude, in the central part of the Southern Highlands Province. Traditionally, they practised subsistence agriculture that involved the production of sweet potatoes and pig husbandry (Glasse 1968; Wood 1985; Ballard 1995). Details of the villages involved in this study (in the Tari basin) are provided elsewhere (Kuchikura 1999; Umezaki *et al.* 2000; Yamauchi *et al.* 2000). Migration of Highlanders, including the Huli, to the capital, Port Moresby, has increased steadily since the 1970s, with a spurt around the independence of PNG in 1975. Migrants from rural villages to Port Moresby reside in settlements, clustering according to ethnicity and language group. They build houses themselves, using second-hand or makeshift materials. Most migrants are unable to secure formal jobs and depend on income from informal sector jobs, such as collecting empty bottles and selling betel nut, cigarettes, or cooked meat in markets or settlements (Hodge *et al.* 1996; Yamauchi *et al.* 2001a).

Solomon Islands

The Solomon Islands lie east of PNG and stretch for some 900 miles (between 155°30′ and 170°30′ east longitude and between 5°10′ and 12°45′ south latitude) across the South West Pacific. The six biggest islands are Choiseul, New Georgia, Santa Isabel, Guadalcanal, Malaita and Makira, there being in addition approximately 992 smaller islands, atolls and reefs. The Western Province is located in the northwest part of the country and borders PNG and Bougainville Island. One traditional village and three semi-modernized villages formed the study populations in the Solomon Islands. The traditional village was a coastal one, 32 km north of Munda (one of the largest population centres in the Solomon Islands), accessible only by canoe. Villagers engage in subsistence farming and fishing, participation in the cash economy remaining very limited. By contrast, the semi-modernized population of this study was involved in the cash economy to some extent, largely in respect of commercial logging on their lands by foreign companies. Cash income is from the royalties that logging companies pay to the village, wages paid by the logging companies and trade in marine resources, such as shells, fish and sea cucumber. Males living in the semi-modernized village belonged to the same church as the traditional village.

Survey methods

Studies were carried out in rural and urban PNG in 1994 and 1995 respectively, and in the Solomon Islands in 2001. Anthropometric measurements were obtained for more than 90% of village residents during the survey period, and included 213 rural villagers and 140 urban migrants in PNG, and 374 rural and 204 semi-modernized villagers in the Solomon Islands. Daily time allocation, physical activity and food consumption surveys involved smaller samples of 12 to 37 individuals, based on married couples in each of the 8 groups (Table 6.1).

Anthropometry was carried out following a standard protocol (Weiner and Lourie 1981). Each participant was weighed on a portable scale (Tanita model 1597, Japan) to a precision of ±0.1 kg. Stature was measured to the nearest 1 mm using a portable field anthropometer (GPM,

Table 6.1. *Survey participants*

	Males				Females			
	Papua New Guinea		Solomon Islands		Papua New Guinea		Solomon Islands	
	Rural	Urban	Traditional[a]	Semi-modern[b]	Rural	Urban	Traditional[a]	Semi-modern[b]
Anthropometry	86	101	191	103	127	39	183	101
Time allocation and physical activity	15	14	14	30	12	15	14	33
Dietary survey	15	14	14	35	12	15	14	37

Notes:
[a] Traditional village;
[b] Semi-modernized village.

Switzerland). Triceps and subscapular skinfold thicknesses were measured to the nearest 0.2 mm, using Holtain skinfold callipers (Holtain, Briberian, UK). The two-site skinfold equation of Durnin and Womersley (1974) was used in combination with the equation of Siri (1956) to estimate body fat percentage (%fat). Body mass index (BMI) was calculated as body weight, in kilograms, divided by height in metres, squared.

Time allocation survey

For each participant in PNG, minute-by-minute heart rate (HR) was recorded for a single 24-hour period using a portable HR monitor (Polar Vantage XL HR Monitor; Polor Electro, Kempele, Finland). While monitoring HR, the behaviour of each participant was observed across the day. The exact time at which physical activity changed was recorded for each participant. Observed physical activities were placed into 19 different categories (Table 6.2), and a physiology-based classification that considered posture and physical exertion level was used to calculate the energy cost of each activity from the mean HR values (Yamauchi *et al.* 2000). 'Strolling' was defined as slow-paced walking in and around the house or in the marketplace, when the subject alternated between standing and walking a few steps at a time. 'Vehicle-riding' represented riding (mostly sitting) in a bus known locally as public motor vehicle or PMV (the common mode of transportation in Port Moresby). This was observed only in urban subjects. 'Recreation' activity included both less energetic activities, such as playing cards (sitting) and watching a rugby match (sitting or standing), and more energetic activities, such as playing billiards or darts.

By contrast, a household-based survey was performed in the Solomon Islands. All households were visited hourly between 7:00 a.m. and 8:00 p.m., and participants' behaviour observed. If participants were not near their houses, neighbours were asked what they had done during that period, and this report was confirmed when the participant was encountered on a subsequent round. This method has been carried out among Melanesian populations elsewhere (Kuchikura 1999; Umezaki *et al.* 2002), and its reliability confirmed for a Huli population against an individual tracing method (Yamauchi *et al.* 2000). Daytime activities (14 hours/day) were classified into 18 categories (Table 6.2). The activity classification for the Solomon Islanders considered the nature of activities performed, rather than posture and physical exertion level as in the Huli of PNG.

Table 6.2. *Mean time (minutes) spent in activity categories per day for Papua New Guinea Highlanders*

	Male					Female					
	Rural (n = 15)		p^a	Urban (n = 14)		p^a	Rural (n = 12)		Urban (n = 15)		p^a
	Mean	SD		Mean	SD		Mean	SD	Mean	SD	
Sleeping	**536**	**69**		**471**	**94**	*	**523**	**36**	**521**	**48**	NS
Resting	**536**	**124**		**227**	**154**	**	**495**	**66**	**281**	**138**	**
Lying (daytime)	16	24		31	44	NS	0	–	26	40	–
Sitting	456	117		147	121	**	415	65	221	133	**
Standing	64	31		49	31	NS	80	56	34	18	**
Travelling	**141**	**60**		**204**	**99**	*	**132**	**71**	**163**	**81**	NS
Strolling	22	26		86	45	**	15	16	101	38	**
Walking	118	45		52	28	**	116	66	29	26	**
Running	1	1		1	3	NS	1	1	0	1	NS
Vehicle-riding	0	–		65	16		0	–	33	15	
Occupational	**59**	**65**		**272**	**220**	**	**145**	**90**	**179**	**202**	NS
Gardening	43	60		0	–		121	81	0	–	
Gathering	8	19		0	–		8	5	0	–	
Pig husbandry	8	13		0	–		16	18	0	–	
Cash earning	0	–		272	220		0	–	179	201	
Recreation	**31**	**84**		**91**	**92**	NS	**0**	–	**25**	**83**	
Meals	**82**	**22**		**44**	**28**	**	**104**	**35**	**134**	**73**	NS
Food preparation	32	18		8	17	**	48	24	85	64	NS
Eating	50	19		36	16	*	56	22	49	21	NS
Miscellaneous	**55**	**46**		**131**	**70**	**	**41**	**39**	**137**	**83**	**
Total	**1440**			**1440**			**1440**		**1440**		

Notes:
^{a} NS: Not significant;
*$p < 0.05$;
**$p < 0.01$.

Physical activity level and total energy expenditure (TEE)

For the PNG sample, basal metabolic rate (BMR) was estimated individually from body weight and according to sex using predictive equations which had been developed for the Huli population (Yamauchi and Ohtsuka 2000). For the Solomon Islands, BMR was estimated using predictive equations developed for international use, and based on body weight, sex and age groups (Food and Agriculture Organization/World Health Organization/United Nations University (FAO/WHO/UNU) 1985). Total energy expenditure (TEE) was calculated for the standard male and female from the energy costs of categorized activities, and the time spent in them by members of the Solomon Islands group. For the PNG Highlanders, the flex-HR method was used to estimate TEE (Spurr *et al.* 1988; Leonard 2003); additional details of the procedure are published elsewhere (Yamauchi *et al.* 2000). For HR values that were equal to, or below the flex-HR point, the resting metabolic rate (RMR) – determined as the mean energy expenditure (EE) for lying, sitting and standing – was assumed to represent the energy costs of very low physical exertion (Yamauchi and Ohtsuka 2000). The EE during sleep was considered to be equal to BMR (Goldberg *et al.* 1988). Values above the flex-HR point (representing higher physical exertion) were converted into energy costs using a regression line calibrated for each subject. The TEE was then calculated using the following formula:

$$\text{TEE} = \Sigma(\text{sleep EE}) + \Sigma(\text{sedentary EE}) + \Sigma(\text{activeEE})$$

The PAL (= TEE/BMR), an index of overall daily exertion, was also determined for each participant.

For the Solomon Islanders, the factorial method was used to estimate TEE (Ohtsuka *et al.* 2004). Based on the time spent in the 18 activity categories, the daily PALs of the standard adult male and female were estimated. Using the FAO/WHO/UNU (1985) equations, the BMR of the 'standard' traditional male, who weighed on average 64.3 kg, was calculated to be 1,663 kcal/day (using the equation for the 18 to 30 years age group) or 1,625 kcal/day (for the 30 to 60 years age group). The present analysis used the mean of these two values (1,644 kcal/day). The BMRs of the standard semi-modernized male and two female groups were calculated in similar fashion.

Following this, the physical activity ratios (PARs) of the corresponding activities in the FAO/WHO/UNU (1985) report were applied to the observed activities, and the integrated energy indices (IEI) calculated from the empirical data after considering pause time (James and Schofield 1990) (Table 6.3). The equations for the IEI are shown in

Table 6.3. *Observed activities and corresponding activity categories given by FAO/WHO/UNU (1985), with corresponding physical activity ratio (PAR) and integrated energy indices (IEI)*

Activities observed	Corresponding activities in FAO/WHO/UNU report	Male			Female		
		PAR	Activity category	IEI	PAR	Activity category	IEI
Subsistence activities							
Agriculture and planting	[Male] Mean: planting tree /[Female] Mean: five categories of agriculture[a]	4.1	Heavy	3.08	3.7	Moderate	3.20
Fishing	Mean: paddling canoe + fishing from canoe	2.8	Moderate	2.49	2.8[b]	Moderate	2.52
Hunting	Hunting pig	3.3	Moderate	2.86		Not observed	
Gathering	Agriculture and planting	3.3	Moderate	2.86	3.6	Moderate	3.12
Market activities	Mean: sedentary in recreation + strolling	2.4	Light	1.76	2.3	Light	1.84
Livestock	Feeding	3.6	Moderate	3.10	3.6[b]	Moderate	3.12
Domestic activities							
Food preparation	Cooking	1.8	Light	1.61	1.8	Light	1.71
Domestic work	Mean: sedentary in recreation + strolling	2.4	Light	1.76	2.3	Light	1.84
Childcare	Childcare	2.2[b]	Light	1.71	2.2	Light	1.81
Meals	Sitting quietly	1.2	Light	1.46	1.2	Light	1.56
House-building	Mean: seven categories of house-building	3.8	Moderate	3.24	3.8[b]	Moderate	3.27
Collecting firewoods	Mean: cutting trees + walking at nomal pace	4.0	Heavy	3.02	4.0[b]	Heavy	3.07
Making tools	Mean: carving plates, combs, etc.	2.1	Light	1.68	2.1[b]	Light	1.79
Resting and recreation	Sedentary in recreation	2.2	Light	1.71	2.1	Light	1.79
Church activities	Mean: sedentary in recreation + sitting quietly + strolling	2.0	Light	1.66	2.0	Light	1.74
Meeting	Mean: sitting at desk + strolling	1.9	Light	1.63	1.9[b]	Light	1.74
Visit town or other villages	Mean: sedentary in recreation + strolling	2.4	Light	1.76	2.3	Light	1.84
Miscellaneous	Mean: sitting quietly + standing quietly + strolling	1.7	Light	1.58	1.7	Light	1.69
Sleeping (8 h)	Sleeping	1.0	Light	1.00	1.0	Light	1.00
Resting (2 h)	Mean: sitting quietly + standing quietly	1.3	Light	1.30	1.3	Light	1.30

Notes:

[a] Clearing ground + digging ground + digging holes for planting + weeding + harvesting root crops;

[b] Applying male/female values to female/male.

Table 6.4. *Equations for estimating the integrated energy indices (IEI), according to James and Schofield (1990)*

	Males	Females
Light activities		
(PAR: 1.0–2.5)	IEI = 75% of 1.54 + 25% of PAR	IEI = 75% of 1.68 + 25% of PAR
Moderate activities		
(PAR: 1.0–2.5)	IEI = 25% of 1.54 + 75% of PAR	IEI = 25% of 1.68 + 75% of PAR
Heavy activities		
(PAR: 4.0 +)	IEI = 40% of 1.54 + 60% of PAR	IEI = 40% of 1.68 + 60% of PAR

Table 6.4. Interviews showed that the hours between 21:00 and 7:00 were generally devoted to sleep and rest. Across this period, eight of the ten hours involved sleep and two hours involved resting. The PAR values for sleeping and resting were used as the IEI. The PAL was determined as ΣIEI × time (minutes) for the 20 categorized activities (the 18 observed activities plus sleeping and resting) divided by 1440 (the number of minutes in a day). The TEE was then calculated by adding up BMR and PAL values, and multiplying by BMR, thus obtaining a value for the 24-hour period.

Dietary survey

Dietary consumption surveys were conducted using two different methods, one for each country. In PNG, all foods consumed throughout the day of HR monitoring were weighed directly before cooking, for each subject, using a portable beam scale. In addition, in the morning, participants were asked about the types and amounts of foods consumed the previous night. Sweet potatoes and green leafy plants, which were most frequently eaten in the villages, were sampled in the Huli villages and their food compositions analysed in the laboratory in Japan (Umezaki *et al.* 1999). For other foodstuffs, the energy, protein and fat contents were estimated using the food composition tables of Hongo and Ohtsuka (1993), and other available data for PNG and surrounding areas (Yamauchi *et al.* 2001a).

By contrast, a semi-quantitative method (one-hour recall) was used to assess the dietary intake of the Solomon Islands group. Over seven consecutive days, we visited all the households hourly between 7:00 a.m. and 8:00 p.m., and checked all food items and the amount of food consumed during the one-hour period between visits. Basic information on food and meals (food type, individual portions, recipes and waste of raw food) was also obtained on other occasions. These data were useful for the estimation of the amount of each food item consumed by each subject. Energy, protein and fat contents of foods were estimated using food composition tables for Pacific Island countries (Dignan *et al.* 1994).

Results and Discussion

Anthropometry

Table 6.5 shows the anthropometric characteristics for the PNG and Solomon Islands groups. There were significant differences in body weight between the rural and urban PNG groups for each sex, urban subjects being heavier than their rural counterparts. The difference was particularly marked in PNG females (11.3 kg). By contrast, there were no significant differences among the Solomon Islanders of either sex. Solomon Islanders of both sexes were taller than PNG Highlanders. Significant differences between rural and urban groups were found in the PNG sample but not between the traditional and the semi-modernized groups in the Solomon Islands. The finding in PNG that urban subjects who had been born and raised until puberty or adolescence in their home villages had significantly larger bodies, especially height, suggests that well-nourished individuals were more likely to have migrated to Port Moresby, the capital city, than undernourished ones. Urban subjects in PNG had higher BMIs than their rural counterparts, the difference being particularly marked for females. The mean BMI of urban PNG females was above 25.0, categorizing them as 'overweight' on average, according to World Health Organization (2000) criteria. Conversely, no significant urban–rural difference was found for the Solomon Islander groups of either sex. Urban subjects in PNG had significantly greater %BF than their rural counterparts. As with the differences in body weight and BMI, the difference was particularly notable in females. There were no such differences between traditional and semi-modernized groups in the Solomon Islands.

Body composition

To further understand the differences in BMI between the two PNG subgroups and those of the Solomon Islands, BMI was divided into two

Table 6.5. *Body composition of rural versus urban subjects in Papua New Guinea, and traditional versus semi-modernized subjects in the Solomon Islands*

	Males										Females										
	Papua New Guinea						Solomon Islands						Papua New Guinea						Solomon Islands					
	Rural $n=86$		Urban $n=101$			Traditional $n=191$		Semi-modernized $n=103$			Rural $n=127$		Urban $n=39$			Traditional $n=183$		Semi-modernized $n=101$		
	Mean	SD	Mean	SD	p	Mean	SD	Mean	SD	p	Mean	SD	Mean	SD	p	Mean	SD	Mean	SD	p
Body weight (kg)	59.8	9.5	65.5	7.2	<0.01	64.3	7.6	63.0	9.6	NS	50.8	7.1	62.1	11.3	<0.01	59.4	8.6	58.1	13.3	NS
Height (cm)	157.5	6.7	161.8	5.3	<0.01	163.7	5.4	163.7	6.4	NS	149.5	5.4	151.5	5.3	<0.05	153.6	5.4	152.8	5.8	NS
BMI (kg/m^2)[a]	24.0	2.8	25.0	2.4	<0.01	24.0	2.3	23.5	3.3	NS	22.7	2.5	27.0	4.5	<0.01	25.1	3.1	24.8	5.2	NS
Triceps skinfold (mm)	6.1	1.5	6.4	1.5	NS	8.2	3.2	8.7	4.7	NS	9.0	3.4	13.0	6.4	<0.01	20.6	7.0	20	9.2	NS
Subscaplar skinfold (mm)	12.6	4.5	14.7	4.8	<0.01	12.5	4.9	14.1	8.1	<0.05	16.6	6.4	25.1	9.4	<0.01	29.6	8.8	27.5	12.6	NS
Body fat (%)[b]	15.9	3.9	17.9	3.9	<0.01	17.3	4.8	17.9	6.9	NS	25.1	5.3	31.1	6.3	<0.01	35.9	5.2	34.2	7.7	<0.05

Notes:
[a] Body mass index;
[b] Estimated from the sum of two skinfold measurements (Durnin and Womersley 1974; Siri 1956).

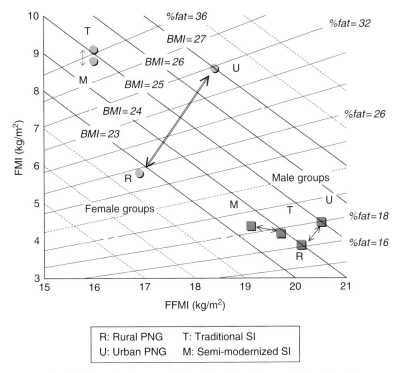

Fig. 6.1. Mean fat-free mass index (FFMI) and fat mass index (FMI) in the Papua New Guinea and Solomon Islands groups, by sex.

indices, the fat-free mass index (FFMI) and fat mass index (FMI), and then plotted in a scatter diagram (Fig. 6.1) (Van Itallie *et al.* 1990; Hattori *et al.* 1997). Body weight was divided into fat mass (FM) and fat-free mass (FFM) using %BF estimated from skin-fold measurements at two sites, with FMI and FFMI being calculated as follows:

$$FMI = FM/height^2 \quad (kg/m^2)$$

$$FFMI = FFM/height^2 \quad (kg/m^2)$$

The data for the four female groups (urban and rural in PNG; traditional and semi-modernized in the Solomon Islands) are located in the top left of Fig. 6.1, while the data for the four male groups are in the bottom right. This analysis examines the effect of modernization and urbanization on body composition by looking at changes in the FM and FFM of rural people when they move from village to city; it also examines how, in the PNG samples, body composition changes when a traditional subsistence people begin to adapt to an urban environment. In addition, differences in body composition between traditional and semi-modernized villagers in the Solomon Islands are described.

Urban PNG females had FFMI and FMI which were 1.5 and 2.8 kg/m^2 greater than their rural PNG counterparts. Both FMI and FFMI were 0.6 kg/m^2 greater in urban PNG males, as compared to their rural counterparts. The difference was statistically significant for FMI but not for FFMI. No such trend was observed in the Solomon Islands groups.

Time allocation

Table 6.6 shows the daily activity patterns of rural and urban subjects in PNG, and traditional and semi-modernized subjects in the Solomon Islands. Urban subjects in PNG spent less time per day resting (309 and 214 minutes for males and females, respectively ($p < 0.01$)), but more time travelling (63 minutes more for males ($p < 0.05$); 31 minutes more for females (not statistically significant)) and in occupational activities (213 minutes more for males ($p < 0.01$); 34 minutes more for females (not statistically significant)) than did their rural counterparts. Urban participants spent considerably less time sitting than their rural counterparts. Of the categories of travel activity used in this study, time spent strolling was significantly greater in the urban group, but time spent walking was much lower than in the rural group (118 minutes against 52 minutes in males ($p < 0.01$), and 116 minutes against 29 minutes in females ($p < 0.01$)). Time spent in food preparation and eating was lower in urban men, but greater among urban women. Rural Huli males generally cook for themselves in their men's house, while in urban settlements, people live together as families, and women do most of the cooking.

Differences in daily time allocation were not prominent between traditional and semi-modernized female Solomon Islanders, but were pronounced between traditional and semi-modernized males, especially with respect to subsistence activities. Traditional village men spent two hours longer per day in subsistence activities than do their semi-modernized counterparts, including more than two hours per day spent in community work, organized by the Christian Fellowship Church (Yamauchi 2002). Replanting was the major activity undertaken during community work during the study period (January–March 2001). Males living in the

Table 6.6. *Mean time (minutes) spent in activity categories per day for Solomon Islanders*

	Males		Females	
	Traditional $n = 14$	Semi-modernized $n = 30$	Traditional $n = 14$	Semi-modernized $n = 33$
Subsistence activities[a]	368	239	190	206
Agriculture and planting	238	145	165	169
Fishing	110	35	21	14
Hunting	9	4	0	0
Gathering	0	2	1	5
Market activities	5	51	1	17
Livestock	6	3	2	1
Domestic activities				
Food preparation	13	19	76	89
Domestic work	7	22	62	102
Childcare	2	16	31	51
Meals	28	36	24	38
House-building	15	36	0	8
Collecting firewoods	1	7	1	1
Making tools	13	17	12	18
Resting and recreation	353	280	393	200
Church activities	24	48	36	63
Meeting	0	23	0	6
Visit town or other villages	7	58	0	31
Miscellaneous	9	38	15	27
Total	840	840	840	840

Note:
[a] Sum of time spent in six subsistence activities.

semi-modernized village spent 225 minutes per day in agriculture and planting (Yamauchi 2002), which was comparable to that of traditional village males (238 minutes per day).

Time spent at work by the different groups is shown in Fig. 6.2. For the PNG groups, working time included time spent (1) gardening and gathering; (2) in pig husbandry and earning cash (Table 6.2); (3) travelling to and from gardens or workplaces; and (4) in other activities in the work situation, including resting. For the Solomon Islanders, the working time was the sum of the time spent in the six subsistence activities shown in

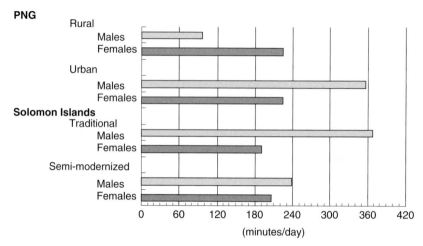

Fig. 6.2. Time spent in subsistence and occupational activities by Papua New Guinea Highlanders and Solomon Islanders.

Table 6.6. Working time was similar across all the female groups (range: 190 to 224 minutes per day). While men in urban PNG worked about four times longer than their rural counterparts, those in the traditional villages in the Solomon Islands had work times which were about 50% longer than their semi-modernized counterparts.

Only in rural PNG did females work longer per day than males (Fig. 6.2). A number of studies have reported gender inequalities in the division of labour in PNG Highland societies, with females working longer than males (Waddell 1972; Strathern 1982; Allen *et al.* 1995; Yamauchi *et al.* 2000, 2001b; Sillitoe 2002). Table 6.7 shows gender differences in the time spent in agricultural work (subsistence gardening and cash cropping) for ten PNG Highland societies, including the Huli (Kuchikura 1999; Yamauchi *et al.* 2001b). The figures show that women generally work one to two hours longer than men, the gender difference among Huli being within this range (100 minutes per day).

In both traditional and semi-modernized villages in the Solomon Islands, males spent longer in subsistence activities, including agriculture and planting, fishing, hunting, gathering, market activities and livestock activities, than did females (Table 6.6). In the traditional village, men spent three hours longer in subsistence activities than did women. By contrast, females spent more time in domestic activities such as preparing food, cleaning and childcare. When the time spent in these three domestic activities was combined with the time spent in subsistence activities, the

Table 6.7. *Time spent by men and women in subsistence gardening and cash cropping for populations in the Papua New Guinea Highlands*

Population	Agriculture-related working (minutes/day)			Author	Published year
	Male	Female	Difference[a]		
Saine[b]	156	201	45	Norgan *et al.*	1974
Duna	164	233	69	Modjeska	1982
Bomagai-Angoiang Maring	141	211	70	Clarke	1971
Sinasina[b]	101	186	85	Hide	1981
Wola Mendi	134	221	87	Sillitoe	1983
Raiapu Enga[b]	146	234	88	Waddell	1972
Tairora[b]	150	239	89	Grossman	1984
Yamiyufa[b]	102	192	90	Sexton	1986
Huli	88	188	100	Yamauchi *et al.*	2001b
Karinje Enga	8	132	124	Wohlt	1978

Notes:
[a] Male–female;
[b] Sum of the time spent on subsistence and cash cropping.
Source: Kuchikura (1999).

work time of women was about two and a half hours longer per day than that of men in semi-modernized villages, whereas in the traditional village, work time of women was about half an hour shorter than that of men.

Physical activity

Table 6.8 summarizes the mean body weight, BMR, TEE, PAL, and interpretation of PAL as light, moderate, or heavy activity from the FAO/WHO/UNU (1985) criteria. In the PNG samples, urban females, but not males, were significantly heavier than their rural counterparts ($p < 0.05$); however, there was no significant difference in BMR between urban and rural groups. Similarly, there were no significant differences in TEE and PAL between rural and urban males; significant differences, however, were found in these indices in females (TEE: $p < 0.05$, PAL: $p < 0.01$).

The PAL values of the PNG and Solomon Islands groups cannot be compared with each other due to the different methodologies used;

Table 6.8. *Body weight, basal metabolic rate, total daily energy expenditure, physical activity level and interpretation of activity levels*[a]

	Papua New Guinea										Solomon Islands[b]			
	Males					Females					Males		Females	
	Rural n=15		Urban n=14			Rural n=12		Urban n=15			Traditional n=14	Semi-modernized n=30	Traditional n=14	Semi-modernized n=33
	Mean	SD	Mean	SD	p	Mean	SD	Mean	SD	p	Mean	Mean	Mean	Mean
Weight (kg)	63.6	7.3	70.0	9.5	NS	53.3	7.6	59.7	6.4	<0.05	64.3	63.0	59.4	58.1
BMR (MJ/day)	7.13	0.82	7.13	0.97	NS	5.82	0.83	5.71	0.61	NS	6.89	6.81	5.69	5.62
TEE (MJ/day)	13.13	2.12	12.20	2.01	NS	11.04	2.60	9.31	1.27	<0.05	12.04	11.16	9.41	9.36
PAL	1.84	0.22	1.71	0.21	NS	1.88	0.26	1.63	0.19	<0.01	1.75	1.64	1.65	1.66
Level[c]	M–H		M			H		M			M	L–M	M	M

Notes:

[a] BMR, predicted basal metabolic rate; TEE, total energy expenditure; PAL, physical activity level (TEE/BMR);

[b] No statistical test was performed;

[c] Level of work classified by PAL as light (L); moderate (M); and heavy (H) (FAO/WHO/UNU 1985).

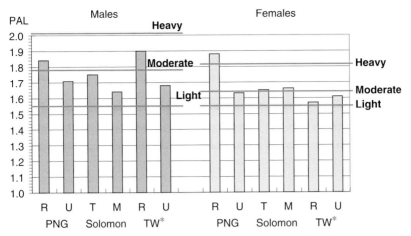

Fig. 6.3. Comparison of physical activity levels in Papua New Guinea, the
Solomon Islands and Third World (TW) populations.
Note: * Weighted average of the study populations (Ferro-Luzzi and Martino
1996; Yamauchi *et al.* 2001a).

however, within-sample comparisons are valid. According to the FAO/
WHO/UNU (1985) work level criteria, PAL of the PNG sample was heavy
for rural females, moderate-to-heavy in rural males and moderate in urban
males and females. Unsurprisingly, urban subjects were less active than
their rural counterparts. Conversely, no significant difference was found in
the PAL values of traditional and semi-modernized villagers in the
Solomon Islands, although semi-modernized village men tended to be
slightly less active than their traditional counterparts. The PAL values
were moderate in all Solomon Islander groups apart from males in the
semi-modernized village, where it was light to moderate.

Figure 6.3 shows the average PAL values for all eight groups, together
with averages for rural and urban Third World (TW) populations, as
compiled by Yamauchi *et al.* (2001b), based on data from Ferro-Luzzi
and Martino (1996). Samples from both males and females in the urban
and modernized populations in both PNG and the Solomon Islands had
similar values to the average of a number of urban populations in the
developing world. While the PALs of rural and traditional males in PNG
and the Solomon Islands respectively were below the Third World average
for rural populations, the PAL of rural Huli females (1.88) was very high
relative to both traditional Solomon Islanders and the rural Third World
more generally. Rural Huli females had PALs similar to those of Indian
farmers, rural Gambians and Upper Volta farmers in the dry season
(Ferro-Luzzi and Martino 1996).

Modelling activity profiles: how a desirable level of physical activity can be achieved

Several studies have found evidence of precursors for chronic degenerative disease in urban PNG (World Health Organization 1985; Dowse *et al.* 1994; Inaoka *et al.* 1996; Natsuhara *et al.* 2000). These precursors are likely to be associated with reduced levels of physical activity (Yamauchi *et al.* 2001a,b), as well as obesity. In terms of public health, it is worth examining what constitutes a desirable level of physical activity, and how it may be achieved by modifying lifestyles.

There are two possible ways to determine a desirable level of physical activity: one is based on epidemiological analyses of contemporary populations, and the other involves an evolutionary perspective, based on estimating the physical activity of Palaeolithic stone-age people. With respect to the former, analyses of over 40 national physical activity studies worldwide show a significant inverse relationship between the average BMI of adult men and PAL (Ferro-Luzzi and Martino 1996); groups with PAL $>= 1.8$ are much less likely to be overweight (BMI ≥ 25). The relationship for women, although not statistically significant, is similar. Based on this analysis, the WHO (2000) has suggested that people should remain physically active throughout life, and sustain a PAL ≥ 1.75 to avoid excessive weight gain. The average PAL of stone-age humans has been estimated to have been approximately 1.8 (range: 1.6–2.0) (Leonard and Robertson 1992; Cordain *et al.* 1998). Therefore, it is reasonable to consider a PAL of 1.75–1.80 as a desirable level of daily physical activity. These values correspond to a moderate level of physical activity in males and moderate-to-heavy physical activity in females, according to the WHO criteria (FAO/WHO/UNU 1985).

Figure 6.4 shows the average activity patterns of rural and urban Huli participants. While the rural Huli participants achieved PAL values above the 1.75 to 1.8 range, the urban groups did not. Urban men could increase their PAL from 1.71 to 1.80 if they stopped using motorized transport (65 minutes per day, Table 6.2) and increased their walking time to the level of their rural counterparts (118 minutes per day). Women could increase their PAL from 1.63 to 1.76 by increasing their walking time to the level of their rural counterparts (an additional 87 minutes per day). However, it is insufficient for women simply to avoid riding buses (33 minutes per day), as a further 54 minutes more of moderate physical activity would be needed per day to achieve a level that would minimize the risk of becoming overweight.

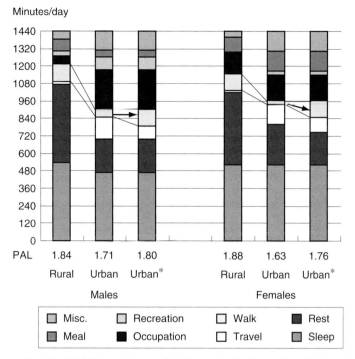

Fig. 6.4. Modelling activity profiles for rural and urban subjects in Papua New Guinea: increasing the walking time of the urban group to the level of the rural group in order to achieve an overall mean physical activity level of between 1.75 and 1.8.
Note: Urban*: hypothesized cases, in which urban subjects spend same time walking as their rural counterparts.

Dietary intake

Daily dietary intakes of all groups are shown in Figure 6.5. Papua New Guinea Highlanders of both sexes had higher energy intakes than did Solomon Islanders. When total energy intake (TEI) was compared with TEE, the energy balance (TEI – TEE) was within ±1.0 MJ per day in three of the four groups; the exception was among urban females in PNG, whose TEE exceeded TEI by 1.12 MJ per day. Thus in three of the four groups, the difference between measured energy intake and expenditure was within an acceptable range (Ulijaszek 1995). Among the Solomon Islands groups, large energy imbalances exceeding 1.0 MJ per day were found in both traditional (−1.84 MJ per day) and semi-modernized (−1.96 MJ per day) village males, indicating that the TEI of Solomon Islander males is likely to have been underestimated. In contrast, TEI and TEE were

(A)

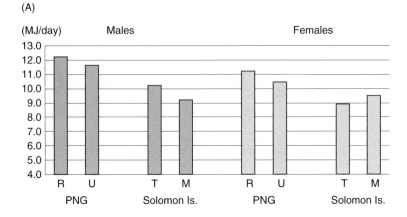

(B)

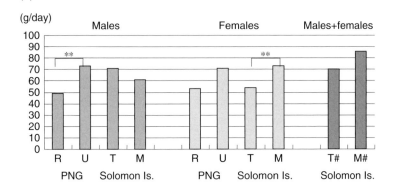

(C)

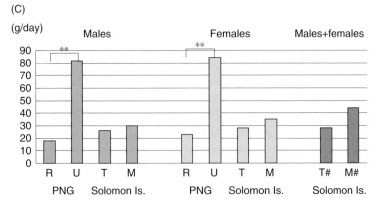

Fig. 6.5. Dietary intakes of Papua New Guinea Highlanders and Solomon Islanders. (A) Daily energy intake of the PNG and Solomon Islands groups, including results from previous studies. (B) Daily protein intake of the PNG and Solomon Islands groups, including results from previous studies. (C) Daily fat intake of the PNG and Solomon Islands groups, including results from previous studies. *Notes:* R: rural PNG; U: urban PNG; T: traditional Solomon Islands; M: semi-modernized Solomon Islands; T#: traditional Solomon Islands (the same village as this study, 1985; Eason *et al.* 1987); M#: semi-modernized Solomon Islands (not the same villages as this study, 1985; Eason *et al.* 1987). *$p < 0.05$; **$p < 0.01$.

Table 6.9. *Contribution of purchased food[a] to daily nutrient intake, expressed as percentages*

	Papua New Guinea		Solomon Islands	
	Rural $n = 27$	Urban $n = 29$	Trad $n = 28$	Semi-mod $n = 72$
Energy intake	21.1	approaching 100	5.0	42.5
Protein intake	24.8	approaching 100	3.5	26.5
Fat intake	61.4	approaching 100	1.5	32.0

Note:
[a] Purchased food: bread, rice, flour, tinned fish, meat, etc.

reasonably balanced in both traditional (-0.51 MJ per day) and semi-modernized (0.14 MJ per day) village Solomon Islander females.

Figure 6.5A shows that there were no regional differences in TEI for PNG males and females or for Solomon Islander males and females. However, the dietary patterns of the subgroups differed markedly in PNG and the Solomon Islands. Table 6.9 shows the proportional contribution of purchased foods (including rice, bread, biscuits, tinned fish, tinned meat and lamb chops) to total daily energy, protein and fat intakes. The dependence on purchased food was lowest among Solomon Islanders in the traditional village, followed by rural villagers in PNG. Solomon Islanders in semi-modernized villages had higher dependence on bought food, while urban dwellers in PNG were totally dependent on it.

The daily protein and fat intakes for each group are shown in Figs. 6.5B and 6.5C, respectively, together with results of previous studies in the Solomon Islands (Eason *et al.* 1987). Urban PNG males consumed significantly more protein than their rural counterparts (Fig. 6.5B), as did semi-modernized Solomon Islander females in comparison with their traditional counterparts. With the exception of Solomon Islander males, urban PNG and semi-modernized Solomon Islander groups had greater protein intakes than their rural and traditional counterparts, respectively. Both traditional and semi-modernized Solomon Islands groups had slightly lower daily protein intakes than was observed there previously (Eason *et al.* 1987). Nevertheless, the average amount of protein consumed per day in all eight groups exceeded the WHO safe level (FAO/WHO/UNU 1985).

Significant differences were found in daily fat intake between rural and urban PNG subjects of both sexes, whereas fat intakes were similar for both the traditional and semi-modernized Solomon Islander groups (Fig. 6.5C). Fat intakes were much higher in both male and female urban

PNG groups than in any of the other groups. Urban Huli fat intakes were comparable to those of urban Wanigela people in Port Moresby, whose fat intakes in the 1990s were 71 g/day for men and 65 g/day for women, predominantly of saturated fat (Hodge *et al.* 1996). Since more than 60% of the fat consumed by the Huli migrants in Port Moresby came from sheep meat, butter and chicken (Yamauchi *et al.* 2001a), their fat intake is also likely to be predominantly of a saturated nature.

Changing patterns of diet and nutrient intakes

Intakes of dietary energy, protein and fat have been stable for at least 18 years in the Solomon Islands. During this period (1985 to 2003), great changes in terms of modernization and urbanization have taken place. Despite this, villagers in the Western Province of the Solomon Islands have maintained traditional subsistence lifestyles under the guidance of an indigenous Christian sect that began there in 1960. Although the nutrient intakes of the traditional and semi-modernized village Solomon Islander groups of this study were similar, dietary patterns, including contributions of purchased food, differed markedly between them (Table 6.9). By contrast, the effects of urbanization on dietary patterns and nutrient intake were evident in both rural and urban PNG groups. Protein and fat intakes increased greatly in urban subjects, due to increased reliance on purchased foods.

Although the relative contribution of purchased foods to nutrient intake was lower than among rural villagers in PNG (Table 6.9) and the daily nutrient intake may have been underestimated, traditional villagers in the Solomon Islands consumed more protein and fat than their rural counterparts in PNG (Figs. 6.5B, 6.5C). The most marked difference between the subsistence diet of Solomon Islanders and PNG Highlanders is that the former eat fish and coconut, while the latter do not. Of the traditional local foods in the Solomon Islands, fish seemed to be the most important source of protein and fat. Indeed, in the traditional village, fish and other seafood formed more than 50% of the protein intake in males and 40% in females, as well as 45 and 25% of the fat intake, respectively.

Summary

Urban PNG Highlanders were significantly heavier than their rural counterparts, probably because of their greater intakes of protein and fat from market foods and considerably lower PAL, especially in women. There was great gender inequality in the division of labour in rural but not in urban Huli, the differences in physical activity patterns between the rural and urban groups being larger for females than males.

Like other Pacific countries, the incidence of chronic degenerative diseases in PNG has increased recently with modernization and urbanization (WHO 1985; Inaoka *et al.* 1996). For example, Dowse *et al.* (1994) found a high prevalence of non-insulin-dependent diabetes mellitus (NIDDM) among Wanigela dwellers in a Port Moresby settlement. Furthermore, the hospital admission rate for NIDDM has increased rapidly over the past 20 years, while the proportion of patients with cardiovascular diseases admitted to Port Moresby General Hospital has increased steadily between 1960 and 1997 (Yamauchi *et al.* 2001b). Urban migrant Huli-speaking people have a higher risk of cardiovascular disease than their rural counterparts (Natsuhara *et al.* 2000).

In contrast, body size of the semi-modernized Solomon Islanders was similar to that of the traditional population. In the traditional village, the subjects ate mostly subsistence foods, such as root crops, fish and coconut, while the populations of the semi-modernized villages depended on subsistence food, with the addition of purchased food, such as rice, tinned fish and flour. The local diet in the Solomon Islands, based on fish, root crops and coconut, is nutritionally good, and it is perhaps unsurprising that the nutritional status of the rural villagers was similar to that of the more modernized villagers. Unlike the groups seen in PNG, no clear difference was found in PAL between traditional and semi-modernized villagers of either sex; however, traditional village men had a higher PAL than did their semi-modernized counterparts. This may reflect the fact that the former spent more than 2 hours longer in subsistence activities per day than did the latter.

In addition, it is speculated here that the semi-modernized Solomon Islander population of this study are still in a state of transition between subsistence and cash economies. Females living in the most modernized of the three villages had a relatively high mean BMI of $28.5 \, \text{kg/m}^2$, which classifies the majority of the population as being overweight by WHO (2000) criteria. Furthermore, 13% of males and 16% of females in these communities were categorized as hypertensive. These percentages are higher than those of traditional villagers (4% of males and 9% of females respectively). These data indicate that the full influences of modernization are not yet fully manifest, but that they constitute a continuing and future health risk.

References

Allen, B. J., Bourke, R. M. and Hide, R. L. (1995). The sustainability of Papua New Guinea agricultural systems: the conceptual background. *Global Environmental Change* **5**, 297–312.

Ballard, C. (1995). The death of a great land: ritual, history and subsistence revolution in the Southern Highlands of Papua New Guinea. Ph.D. thesis, Australian National University.

Clarke, W. C. (1971). *Place and People: An Ecology of a New Guinean Community.* Berkeley: University of California Press.

Cordain, L., Gotshall, R. W. and Eaton, S. B. (1998). Physical activity, energy expenditure and fitness: an evolutionary perspective. *International Journal of Sports Medicine* **19**, 328–35.

Dignan, C. A., Burlingame, B. A., Arthur, J. M., *et al.* (1994). *The Pacific Islands Food Composition Tables.* Palmerston North: South Pacific Commission.

Dowse, G. K., Spark, R. A., Mavo, B., *et al.* (1994). Extraordinary prevalence of non-insulin-dependent diabetes mellitus and bimodal plasma glucose distribution in the Wanigela people of Papua New Guinea. *Medical Journal of Australia* **160**, 767–74.

Durnin, J. V. G. A. and Womersley, J. (1974). Body fat assessed from total body density and its estimation from skinfold thickness: measurements on 481 men and women aged from 16 to 72 years. *British Journal of Nutrition* **32**, 77–97.

Eason, R. J., Pada, J., Wallace, R., Henry, A. and Thornton, R. (1987). Changing patterns of hypertension, diabetes, obesity and diet among Melanesians and Micronesians in the Solomon Islands. *Medical Journal of Australia* **146**, 465–73.

FAO/WHO/UNU (1985). *Energy and Protein Requirements.* WHO Technical Report Series 724, Geneva: World Health Organization.

Ferro-Luzzi, A. and Martino, L. (1996). Obesity and physical activity. In *The Origins and Consequences of Obesity*, ed. D. J. Chadwick. Ciba Foundation Symposium 201. Chichester: Wiley, pp. 207–27.

Foley, W. A. (1992). Language and identity in Papua New Guinea. In *Human Biology in Papua New Guinea: The Small Cosmos*, ed. R. D. Attenborough and M. P. Alpers. Oxford: Clarendon Press, pp. 136–49.

Glasse, R. M. (1968). *Huli of Papua: A Cognatic Descent System.* Paris: Mouton.

Goldberg, G. R., Prentice, A. M., Davies, H. L. and Murgatroyd, P. R. (1988). Overnight and basal metabolic rates in men and women. *European Journal of Clinical Nutrition* **42**, 137–44.

Grossman, L. S. (1984). *Peasant, Subsistence Ecology, and Development in the Highlands of Papua New Guinea.* Princeton: Princeton University Press.

Hattori, K., Tatsumi, N. and Tanaka, S. (1997). Assessment of body composition by using a new chart method. *American Journal of Human Biology* **9**, 573–8.

Hide, R. (1981). Aspects of pig production and use in colonial Sinasina, Papua New Guinea. Unpublished Ph.D. dissertation, Department of Anthropology, Columbia University.

Hodge, A. M., Montgomery, J., Dowse, G. K., *et al.* (1996). Diet in an urban Papua New Guinea population with high levels of cardiovascular risk factors. *Ecology of Food and Nutrition* **35**, 311–24.

Hongo, T. and Ohtsuka, R. (1993). Nutrient compositions of Papua New Guinea foods. *Man and Culture in Oceania* **9**, 103–25.

Inaoka, T., Ohtsuka, R., Kawabe, T., Hongo, T. and Suzuki, T. (1996). Emergence of degenerative diseases in Papua New Guinea islanders. *Environmental Sciences* **4**, S79–S93.

James, W. P. T. and Schofield, E. C. (1990). *Human Energy Requirements: A Manual for Planners and Nutritionists.* Oxford: Oxford University Press.

King, H. (1992). The epidemiology of diabetes mellitus in Papua New Guinea and the Pacific: adverse consequences of natural selection in the face of sociocultural change. In *Human Biology in Papua New Guinea: The Small Cosmos*, ed. R. D. Attenborough and M. P. Alpers. Oxford: Clarendon Press, pp. 363–72.

Kuchikura, Y. (1999). The cost of diet in a Huli community of Papua New Guinea: a linear programming analysis of subsistence and cash-earning strategies. *Man and Culture in Oceania* **15**, 65–90.

Leonard, W. R. (2003). Measuring human energy expenditure: what have we learned from the flex-heart rate method? *American Journal of Human Biology* **15**, 479–89.

Leonard, W. R. and Robertson, M. L. (1992). Nutritional requirements and human evolution: a bioenergetics model. *American Journal of Human Biology* **4**, 179–95.

Modjeska, N. (1982). Production and inequality: perspectives from central New Guinea. In *Inequality in New Guinea Highlands Societies*, ed. A. Strathern. Cambridge University Press, pp. 50–108.

Natsuhara, K., Inaoka, T., Umezaki, M., *et al.* (2000). Cardiovascular risk factors of migrants in Port Moresby from the highlands and island villages, Papua New Guinea. *American Journal of Human Biology* **12**, 655–64.

Norgan, N. G., Ferro-Luzzi, A. and Durnin, J. V. G. A. (1974). The energy and nutrient intake and the energy expenditure of 204 New Guinea adults. *Philosophical Transactions of the Royal Society of London B* **268**, 309–48.

Ohtsuka, R., Sudo, N., Sekiyama, M., *et al.* (2004). Gender difference in daily time and space use among Bangladeshi villagers under arsenic hazard: application of the compact spot-check method. *Journal of Biosocial Science* **36**, 317–32.

Sexton, L. (1986). *Mothers of Money, Daughters of Coffee: The Wok Meri Movement.* Ann Arbor: UMI Research Press.

Sillitoe, P. (1983). *Roots of the Earth: Crops in the Highlands of Papua New Guinea.* Kensington: New South Wales University Press.

 (2002). After the 'affluent society': cost of living in the Papua New Guinea highlands according to time and energy expenditure-income. *Journal of Biosocial Science* **34**, 433–61.

Siri, W. E. (1956). The gross composition of the body. *Advances in Biological and Medical Physics* **4**, 239–80.

Spurr, G. B., Prentice, A. M., Murgatroyd, P. R., *et al.* (1988). Energy expenditure from minute-by-minute heart-rate recording: comparison with indirect calorimetry. *American Journal of Clinical Nutrition* **48**, 552–9.

Strathern, A. (1982). *Inequality in Highlands Societies.* Cambridge: Cambridge University Press.

Ulijaszek, S. J. (1995). *Human Energetics in Biological Anthropology*. Cambridge: Cambridge University Press.

Umezaki, M., Yamauchi, T. and Ohtsuka, R. (1999). Diet among the Huli in Papua New Guinea Highlands when they were influenced by the extended rainy period. *Ecology of Food and Nutrition* **37**, 409–27.

(2002). Time allocation to subsistence activities among the Huli in rural and urban Papua New Guinea. *Journal of Biosocial Science* **34**, 133–7.

Umezaki, M., Kuchikura, Y., Yamauchi, T. and Ohtsuka, R. (2000). Impact of population pressure on food production: an analysis of land use change and subsistence pattern in the Tari basin in Papua New Guinea Highlands. *Human Ecology* **28**, 359–81.

Van Itallie, T. B., Yang, M., Heymsfield, S. B., Funk, R. C. and Boileau, R. A. (1990). Height-normalized indices of the body's fat-free mass and fat mass: potentially useful indicators of nutritional status. *American Journal of Clinical Nutrition* **52**, 953–9.

Waddell, E. (1972). *The Mound Builders: Agricultural Practices, Environment, and Society in the Central Highlands of New Guinea*. Seattle: Washington University Press.

Weiner, J. S. and Lourie, J. A. (1981). *Practical Human Biology*. London: Academic Press.

Wohlt, P. B. (1978). Ecology, agriculture and social organization: the dynamics of group composition in the Highlands of Papua New Guinea. Unpublished Ph.D. dissertation, Department of Anthropology, University of Minnesota.

World Health Organization (1985). *Diabetes Mellitus: Report of a WHO Study Group*. Technical Report Series 727. Geneva: World Health Organization.

(2000). *Obesity: Preventing and Managing the Global Epidemic*. WHO Technical Report Series 894, Geneva: World Health Organization.

Wood, A. W. (1985). *The Stability and Permanence of Huli Agriculture*. Occasional paper no. 5. new series. Port Moresby: Department of Geography, University of Papua New Guinea.

Yamauchi, T. (2002). Development and basic human needs in the local villages of Western Province, the Solomon Islands: impact of commercial logging on nutrition and health. *Environment, Development and Culture in Asia-Pacific Societies* **4**, 10–26 (in Japanese).

Yamauchi, T. and Ohtsuka, R. (2000). Basal metabolic rate and energy costs at rest and during exercise in rural- and urban dwelling Papua New Guinea Highlanders. *European Journal of Clinical Nutrition* **54**, 494–9.

Yamauchi, T., Umezaki, M. and Ohtsuka, R. (2000). Energy expenditure, physical exertion and time allocation among Huli-speaking people in the Papua New Guinea Highlands. *Annals of Human Biology* **27**, 571–85.

(2001a). Influence of urbanisation on physical activity and dietary changes in Huli-speaking population: a comparative study of village dwellers and migrants in urban settlements. *British Journal of Nutrition* **85**, 65–73.

(2001b). Physical activity and subsistence pattern of the Huli, a Papua New Guinea highland population. *American Journal of Physical Anthropology* **114**, 258–68.

7 Tongan obesity: causes and consequences

TSUKASA INAOKA, YASUHIRO MATSUMURA
AND KAZUHIRO SUDA

Introduction

The World Health Organization (WHO 2000) warns that obesity has become a global endemic since it has increased over the past several decades in both developed and developing countries. One of the regions showing the highest prevalences is Polynesia, where much obesity-related nutritional and health research has been carried out (Inaoka 2002; Inaoka *et al.* 2003). The frequency of obesity, as judged by body mass index (BMI) (body weight (kg)/body height (m)^2), is higher in Polynesians than in other Pacific Islanders; in urban rather than rural dwellers, and in females rather than males (Hodge *et al.* 1995a,b). In addition, Polynesian migrants in the United States (Hawaii and California, in particular), New Zealand and Australia are characterized by high obesity rates, causing serious health and social problems in those countries (Kumanyika 1993; McAnulty and Scragg 1996; Simmons 1996). Prevalence of obesity among Polynesians has been increasing (Ulijaszek 2000, 2001), in association with elevated incidence of degenerative diseases such as non-insulin-dependent diabetes mellitus (NIDDM) (Hodge *et al.* 1995a, 1996; Bell *et al.* 2001; Colagiuri *et al.* 2002), although their mortalities have been considerably masked due to the availability of reasonably advanced health care systems in such countries. As elsewhere in the world, obesity among Polynesians is due to combined effects of genetics and environment (Fig. 7.1). Since Neel (1962, 1982) suggested the 'thrifty genotype' hypothesis, many obesity-related genes have been reported (Joffe and Zimmet 1998; Chukwuma and Tuomilehto 1998; de Silva *et al.* 1999; Kagawa *et al.* 2002). With respect to environment, it has been suggested that changes in dietary habits and daily activity patterns that have come with lifestyle modernization play fundamental roles in the causation of obesity (Koike *et al.* 1984; McAnulty and Scragg 1996; Bell *et al.* 2001; Lako, 2001; Ball *et al.* 2002; Cameron *et al.* 2003). Furthermore, psychosocial factors, including perceptions favouring

Health Change in the Asia-Pacific Region: Biocultural and Epidemiological Approaches,
ed. Ryutaro Ohtsuka and Stanley J. Ulijaszek. Published by Cambridge University Press.
© Cambridge University Press 2007.

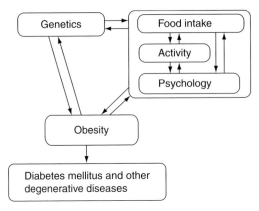

Fig. 7.1. Major factors causing obesity, type 2 diabetes and other degenerative diseases.

large body physique, may have played some part in the emergence of obesity (Brewis *et al.* 1998; Craig *et al.* 1999; Metcalf *et al.* 2000).

There have been debates concerning appropriate indicators for obesity as a risk factor for cardiovascular diseases; some indicators other than BMI, such as waist–hip ratio, having been considered (Hodge and Zimmet 1994; Dalton *et al.* 2003). However, body composition of Polynesians markedly differs from that of Europeans; at any given BMI, the former are more muscular and carry a lower proportion of body fat than the latter (Craig *et al.* 2001); this adds complexity to the assessment of obesity and its functional implications among Pacific Islanders.

In this chapter, Tongan obesity is examined from several viewpoints. First, the time when the general Tongan population became obese is considered in relation to changes from traditional to Westernized life-styles. Secondly, reasons why dietary and other lifestyle changes may have led to obesity are explored. Thirdly, associations between obesity and the prevalence of health disorders, NIDDM in particular, are discussed.

Subjects and methods

Polynesians are genetically very homogeneous, the nation-states to which they belong differing to a lesser degree than nations in the South Pacific region more generally in extents of modernization. Current nation-based health status also markedly differs in the Pacific area. For instance, WHO

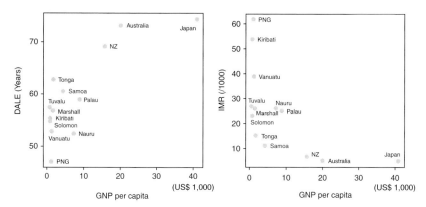

Fig. 7.2. The relationships of DALE (disability-adjusted life expectancy: left) and IMR (infant mortality rate: right) with gross national product (GNP) per capita in selected Pacific countries in 1996. (Source: World Health Organization 2006.)

statistics reveal that disability-adjusted life expectancy (DALE) dramatically increases and infant mortality rate (IMR) exponentially decreases as gross national product per capita, as a development index, increases (Fig. 7.2). Tonga and Samoa are located in median positions in these relationships.

The Kingdom of Tonga is located approximately 650 km southeast of Fiji and approximately 1,850 km northeast of New Zealand (Fig. 7.3). Archaeologically, the Lapita people, or the ancestors of Polynesians, reached the Tongan islands about 3,500 years ago. The first contact of Tongan people with Europeans came with Captain Cook's visit in 1773. He nicknamed Tonga the 'friendly islands', although the people engaged in perpetual civil war at that time. The Tongan population is estimated to have been around 20,000 to 30,000 strong when Captain Cook visited, with fluctuations due to warfare and occasional epidemics.

When did Tongans become obese?

The pathway to obesity must include an historical approach, made possible with various types of information, including historical pictures and photographs of the people (During 1990), and records of trade of modern foods (Campbell 1992; Inoue 2001), although no data of their body physique are available. Pictures of Tongan people painted in the time around Captain Cook's visit suggest negligible European influence, their body physique being well muscled but not obese.

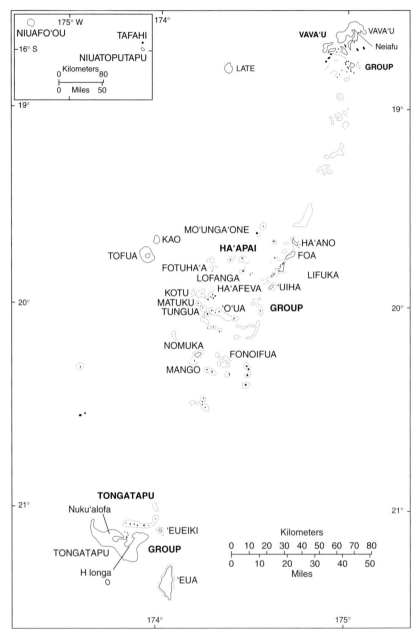

Fig. 7.3. The Kingdom of Tonga.

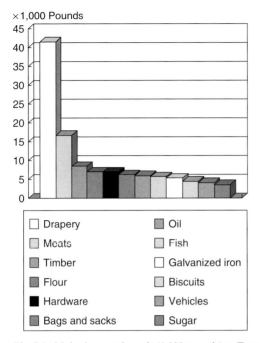

Fig. 7.4. Major imported goods (1,000 pounds) to Tonga in 1909.

The Tongan Civil War ended in 1845, when George Tupou I was crowned. Many Christian missionaries (mostly Catholic ones) came to Tonga in the 1850s, after which time small-scale cash-based trades started to displace barter-based trade. However, cash income seldom increased, because high-return cash crops such as coffee and cotton had not been introduced. Rather, coconuts, which drew a low commodity price, were grown in plantations and sold. By the late 1860s, copra trade, managed by a German trader, flourished, although the average income of many Tongans remained very low.

An important event triggering change in economic life was incorporation of Tonga into a British protectorate in the year 1900, a political situation which continued until the nation's complete independence in 1970. During this period, the Tongan economy developed and its population grew tremendously; moreover, education and health systems were established. In 1909, the most important imported goods by quantity were drapery, followed by meat, timber and flour (Fig. 7.4). Food accounted for about 40% of all imports. According to Campbell (1992), this pattern continued until 1945, suggesting that Westernized diet was not popular among the majority of Tongans before the Second World War.

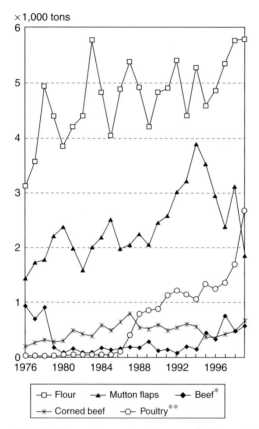

Fig. 7.5. Changes in quantity of imports (1,000 tons) of five selected foodstuffs between 1976 and 1999.

A photograph taken in 1940 shows that among a group of people dancing in a public place, there were but a few of them that were seemingly overweight (Campbell 1992). Many pictures drawn in this period confirm this observation. This suggests that changes in body composition among the general Tongan population began at this time, or after the Second World War, in association with lifestyle modernization. Since 1965, when the present king Tupou IV was crowned, education and health service systems underwent considerable improvement.

Tonga is now an independent nation with 171 islands spreading over approximately 360,000 km^2. It is divided into four island groups, Tongatapu, Ha'apai, Vava'u and Niuas, from south to north. Fewer than 40 islands are inhabited, and of 100,000 Tongan inhabitants, about 70,000 live on Tongatapu island, where the capital city, Nuku'alofa, is located. The Ha'apai island group, about 200 km north of Tongatapu, is more rural than the Va'vau island group, which is located 150 km further north but which has been considerably modernized to provide facilities for the many foreigners that visit its islands to take advantage of one of the best shores for yachting in the world. Niuas, the northernmost island group, is extremely isolated, connected with Tongatapu by weekly airline flights. Apart from city dwellers in Nuku'alofa, all Tongans subsist on horticulture of varieties of bananas and tubers, supplemented by fishing, which is conducted occasionally along the coast. Small-scale cash-earning activities such as kava powder production and tapa cloth making are conducted in many villages.

Figure 7.5 shows changing patterns of flour and meat imports from 1970 (the nation's independence) to 1998 (Inoue 2001). Many items, especially chicken meat, continuously increased in import; however, mutton flap imports sharply decreased after 1994, when the government decided to control the supply of this meat. Consumption of all kinds of meat per person per day is estimated to have been about 120 g in 1990 and 150 g in 1998, while consumption of flour remained constant throughout the period, at about 160 g per person per day. In summary, major dietary changes took place in Tonga in the middle of the twentieth century, when many people migrated to the city areas during the Second World War. Dietary change accelerated since the nation's independence in 1970.

Regional differences in Tongan obesity

The target groups of our survey, which was carried out intermittently between 1999 and 2003, consisted of dwellers in four villages of Ha'ano Island in the northern part of the Ha'apai group, those in the city centre of Nuku'alofa and those in Kolovai village in the western part of the main island of Tongatapu. For the purposes of the present analysis, these are called the 'island', 'urban' and 'sub-urban' populations, respectively. The surveys were carried out in collaboration with staff at the Tongan Ministry of Health, in addition to volunteer participants. Subjects were recruited using the local radio network with the assistance of local leaders. A total of 160 adult subjects, aged 15 years and older, were recruited for the island population, while 155 and 210 adults were recruited for the urban and sub-urban populations respectively. This analysis mainly considers the island and urban groups. Mean ages for males and females were 43.5 (SD: 17.5) and 41.5 (SD: 17.4) years for the former group, and 47.1 (SD: 14.8) and 42.5 (SD: 14.0) years for the latter.

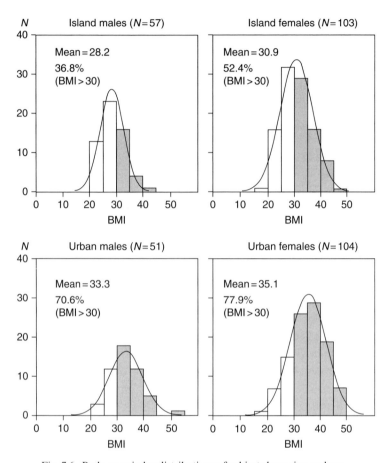

Fig. 7.6. Body mass index distributions of subjects by region and sex.

In the health survey, measurements included urinalysis, systolic and diastolic blood pressures, anthropometry with standard methods, and measurement of blood sugar level using an ordinal kit for a 10 ml spot sample. The serum samples were kept in a freezer to be transferred to Japan for further analysis. In addition, information about each participant's dietary habits and daily activities was obtained by the health assistant by questionnaire.

Figure 7.6 shows the distribution of BMI for males and females of the island and urban groups, respectively. The mean BMI was 28.2 and 30.9 for the island males and females, and 33.3 and 35.1 for the urban males and females, respectively. Following the WHO cut-offs for BMI (undernourished: <18.5; normal: 18.5 to 24.9; overweight: 25 to 29.9; obese: ≥30), the proportion of obese subjects was 36.8% and 52.4% for the island males

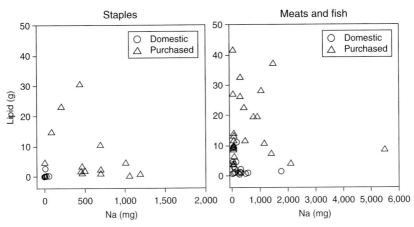

Fig. 7.7. Amounts of fat and sodium contained in staple foods (left) and meats and fish (right) per 100 g edible portion.

and females, and 70.6% and 72.9% for the urban males and females, respectively. The difference in BMI between regions was highly significant in both sexes, largely because of modernized lifestyle in the urban group.

The regional difference in Tongan obesity is likely to have arisen from differences in the timing and nature of modernization. Changes in daily activities, food intakes and psychosocial conditions are the major environmental factors (Fig. 7.1). Traditional Tongan foods are banana, tubers such as taro and yam, and fish cooked with coconut milk, seasonally supplemented by breadfruit. Cassava and sweet potato were added as staples perhaps after European contact. Across the past 50 years or so, these foods have been gradually displaced by varieties of Western foods. Of these, the most commonly available, even in village canteens, have been bread, rice, wheat flour, biscuits, corned beef and tin fish; in addition, frozen chicken, mutton flaps (until 1994) and beef have been sold in stores where electricity has been available.

Fat and sodium contents of commonly consumed staple foods, fish and meats, determined from food composition tables for the South Pacific (Dignan et al. 1994) and Papua New Guinean peoples (Hongo et al. 1989), are shown in Fig. 7.7. Both nutrients are present in much higher quantities in the imported foods than in the domestic foods; the only exception is coconut. It is thus reasonable to assume that the transition from domestic to imported foods caused increased intakes of these nutrients. The increase in fat intake would have raised Tongan daily energy intake because energy content per unit of weight is more than twice as great in fat than that in either protein or carbohydrate.

Similar nutritional changes have been reported for various communities elsewhere in the South Pacific. In Papua New Guinea (Inaoka and Ohtsuka 1995) and Fiji (Lako 2001; Lako and Nguyen 2001), for instance, nutritional status (as determined by increased BMI from previously low levels) initially improved with modernization. However, the continuation of new dietary habits gradually led to excessive energy intakes and increased prevalence of obesity, and of health problems related to the latter.

Concerning psychosocial factors that might affect Tongan obesity, traditional ways of thinking are important. Tongans have considered fatness and obesity in a positive light, especially in respect of adult males. There is no Tongan word which refers to obesity; if a Tongan is asked to give the word for this physical state, the usual answer is 'Sino Lahi', which means a 'big man'. Moreover, Tongans believe that females should not work hard, because their husbands have responsibilities for the household economy. There is a Tongan saying that their way of life is 'Kai Lahe, Inu Lahe, Mohe Lahe', which means, 'Eat a lot, drink a lot, and a sleep a lot'. As pointed out by various researchers (Brewis *et al.* 1998; Craig *et al.* 1999; Metcalf *et al.* 2000), the Tongan perception of appropriate body size and obesity differs from that of Western people, even though new opinions favouring slimness above obesity have been gradually prevailing among young females, particularly in urban areas.

Associations between obesity and health disorders

A summary of the prevalence of diabetes mellitus (DM), as judged from blood spot sugar level (beyond 125 mg/l) after overnight fasting among subjects, is given by sex and region in Fig. 7.8. Most DM among Tongans is non-insulin dependent, or type 2. The prevalences of DM in males and females were 7% and 18% respectively in the island group, and 24% and 20% respectively in the urban group. Crosschecking impaired glucose tolerance (IGT) with the oral glucose tolerance test (OGTT) for the sub-urban sample in 2002 suggests that the IGT prevalence rates shown in Fig. 7.8 might underestimate true values by up to 10%. Colagiuri *et al.* (2002), investigating other Tongan groups, reported that DM prevalence was 7.5% in 1973, increasing to 15.1% around the year 2000. This is consistent with our data in both island and urban areas, particularly in respect of females. These results and those of many other investigations on

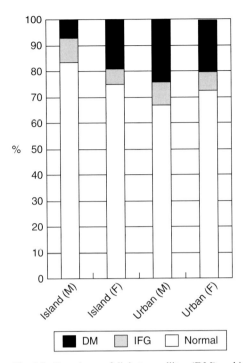

Fig. 7.8. Prevalence of diabetes mellitus (DM) and impaired fasting glucose (IFG) in Tonga.

DM prevalence (including OGTT) in South Pacific countries (Dowse 1996; Dowse *et al.* 1996; Papoz *et al.* 1996) show that Polynesia is a DM-endemic area where the incidence of this disorder continues to increase.

Table 7.1 shows the proportion of subjects according to the three categories of BMI (normal: <25.0; overweight: 25.0 to 29.9; obese: ≥30.0) and three categories of fasting blood glucose level (normal: <110 mg/dl, impaired fasting glucose (IFG): 110 to 125 mg/dl; DM: >125 mg/dl)). There was no statistically significant association between BMI and blood glucose level for any region/sex group, implying the possibility of mixture of two types of people, one with high BMI but undeveloped DM status and another with low BMI but developed DM status. Despite the common view that Polynesians are prone to obesity and NIDDM (Simmons *et al.* 2001) and the knowledge that obesity is an important risk factor for NIDDM (Hodge *et al.* 1996; Bell *et al.* 2001), there are many more factors which modify the relationships between obesity and DM.

Table 7.1. *Island and urban subjects (number) by diabetes and obesity categories*

Island group	Males ($p = 0.39$)[a]				Females ($p = 0.07$)[a]			
	Normal*	IFG**	DM***	Total	Normal*	IFG**	DM***	Total
Obesity								
Normal (BMI < 25.0)	11	2	0	13	13	0	4	17
Overweight (BMI 25.0–29.9)	18	3	2	23	29	2	1	32
Obese (BMI ≥ 30.0)	17	0	2	19	33	4	13	50
Total	46	5	4	55	75	6	18	99

Urban group	Males ($p = 0.85$)[a]				Females ($p = 0.17$)[a]			
	Normal*	IFG**	DM***	Total	Normal*	IFG**	DM***	Total
Obesity								
Normal (BMI < 25.0)	2	0	1	3	2	0	1	3
Overweight (BMI 25.0–29.9)	6	1	5	12	6	1	5	12
Obese (BMI ≥ 30.0)	19	6	11	36	19	6	11	36
Total	27	7	17	51	27	7	17	51

Notes:
[a] *p*: Significance level for chi-square test;
* Normal (<110 mg/dl in fasting blood glucose);
** Impaired fasting blood glucose (110–125 mg/dl in fasting blood glucose);
*** Diabetes mellitus (>125 mg/dl in fasting blood glucose).

Table 7.2 shows the percent distribution of subjects according to three categories (low, normal and high) of BMI, waist–hip ratio, blood pressure and various biochemical markers of lipid metabolism and liver function. The low and high categories indicate levels 20% lower or higher than the normal level, respectively, implying abnormality for either. All groups generally had high values for the zinc turbidity test, γ-glutamyl trans-peptidase (GTP), creatine phosphokinase, total-cholesterol, triglyceride and low density lipoprotein. Low values were found for all groups only for A/G ratio (albumin/globulin ratio) and for highdensity lipoprotein cholesterol among urban males. The proportions of subjects with high values were much greater in the urban group than in the island group for both sexes, indicating that urban inhabitants are at higher risk of health

Table 7.2. *Distribution of subjects (%) by three categories of health indicators*

	Island males (N = 55)			Island females (N = 104)			Urban males (N = 55)			Urban females (N = 100)		
	Low	Normal	High	Low	Normal	High	Low	Normal	High	Low	Normal	High
Waist–hip ratio[a]	0.0	87.7	12.3	0.0	57.3	42.7	0.0	68.6	32.4	0.0	56.7	43.3
Blood pressure	0.0	68.4	31.6	0.0	77.9	22.1	0.0	64.7	35.3	0.0	77.9	22.1
Fasting blood sugar[b]	0.0	60.0	40.0	0.0	64.0	36.0	0.0	40.0	60.0	0.0	47.1	52.9
Total protein	0.0	94.5	5.5	1.0	88.0	11.0	0.0	90.9	9.1	1.0	92.3	6.7
Albumin	1.8	98.2	0.0	8.0	95.0	0.0	0.0	100	0.0	6.7	92.3	0.0
A/G ratio	12.7	87.3	0.0	34.0	66.0	0.0	23.6	76.4	0.0	49.0	51.0	0.0
ZTT	0.0	18.2	81.8	0.0	9.0	91.0	0.0	25.5	74.5	0.0	7.7	92.3
Total bilirubin	0.0	98.2	1.8	0.0	99.0	1.0	0.0	96.4	3.6	0.0	98.1	1.9
ALP	0.0	83.6	16.4	0.0	87.0	13.0	0.0	87.3	12.7	0.0	90.4	9.6
GOT	0.0	90.0	9.1	0.0	84.0	16.0	1.8	74.6	23.6	0.0	79.8	20.2
GPT	0.0	87.3	12.7	0.0	85.0	15.0	0.0	70.9	29.1	0.0	81.7	18.3
LDH	0.0	96.4	3.6	0.0	85.0	15.0	0.0	92.7	7.3	0.0	94.2	5.8
γGTP	0.0	74.5	25.5	0.0	96.0	4.0	0.0	89.1	10.9	0.0	94.2	5.8
Cholinesterase	1.8	90.9	7.3	0.0	96.0	4.0	0.0	89.1	10.9	2.9	91.3	5.8
LAP	0.0	85.5	14.5	1.0	92.0	7.0	0.0	85.5	14.5	1.0	79.8	19.2
CPK	0.0	21.8	78.2	0.0	62.0	38.0	0.0	47.3	52.7	2.9	92.3	4.8
Total cholesterol	3.6	60.0	36.4	1.0	67.0	32.0	3.6	67.3	29.1	2.9	92.3	4.8
HDL cholesterol	3.7	94.4	1.8	2.0	88.8	9.2	20.0	78.2	1.8	0.0	80.8	19.2
Triglyceride	0.0	75.9	24.1	0.0	77.7	22.3	0.0	52.7	47.3	2.9	62.5	34.6
LDL cholesterol	3.6	60.0	36.4	1.0	67.0	32.0	3.6	67.3	29.1	0.0	97.1	2.9
BUN	0.0	85.5	14.5	2.0	91.0	7.0	0.0	89.1	10.9	1.9	97.1	1.0
Creatinine	0.0	90.9	9.1	0.0	97.0	3.0	0.0	94.5	5.5	0.0	97.1	1.0
Uric acid	0.0	83.6	16.4	1.0	94.0	5.0	0.0	50.9	49.1	0.0	87.5	12.5
B2 microglobulin	0.0	85.5	14.5	0.0	86.9	13.1	0.0	85.5	14.5	0.0	89.4	10.6

Notes:

[a] Those exceeding 1.0 (males) or 0.9 (females) were put in the 'High' group;

[b] Including IFG in DM status.

ZTT: zinc turbidity test; ALP: alkali phosphatase; GOT: glutamic oxaloacetic transaminase; GPT: glutamic pyruvic transaminase; LDH: lactic dehydrogenase; γGTP: γ-glutamyl transpeptidase; LAP: leucine aminopeptidase; CPK: creatine phosphokinase; BUN: blood urea nitrogen.

Table 7.3. *Partial correlation coefficients between body mass index and 26 health indicators after controlling for age, sex and region*

	Island		Urban	
	Males	Females	Males	Females
Waist–hip ratio	0.71***	0.33**	0.37**	0.32**
Systolic blood pressure	0.38**	0.53***		
Diastolic blood pressure	0.36**	0.42***		0.27**
Total protein			0.39**	
A/G ratio		#0.23*		#0.21*
ZTT			0.37**	
ALP	#0.32*			
GOT		0.28**	0.42**	
GPT		0.43**	0.46***	
LDH			0.42**	
γGTP		0.32**		
Cholinesterase		0.39***		0.32**
Total cholesterol		0.22*		
HDL cholesterol			#0.41**	#0.23**
Tryglyceride	0.35**	0.40***		
LDL cholesterol		0.21*		
Uric acid		0.45***	0.60***	0.47***
B2 microgloblin	#0.31*			

Notes:
The items which had significant correlations are selected.
*$p < 0.05$; **$p < 0.01$; ***$p < 0.001$.
Inverse relationship.

damage than their rural counterparts, as has been previously reported (Koike *et al.* 1984; Sawata *et al.* 1988).

Table 7.3 shows partial correlation analysis, after controlling for age, of relationships between BMI and biochemical and physiological health parameters. BMI was significantly related to waist–hip ratio in all regions and both sexes. The BMI of island females was related to blood pressure and to many enzymes associated with lipid metabolism and liver function. In the urban group, BMI was related positively with HDL cholesterol and negatively with uric acid for both sexes, and positively with many liver enzymes for males. In no region or either sex was BMI related to blood glucose level. These results are largely consistent with those reported from other populations in the South Pacific (McAnulty and Scragg 1996; Bell *et al.* 2001). The effect of DM status (normal, IFG and DM), as a confounding factor in the relationship between BMI and blood parameters,

Table 7.4. *Proportion of subjects positive for hepatitis-B antigen and antibody respectively, by region and sex*

	Percent positive			
	Island		Urban	
	Males ($N = 55$)	Females ($N = 104$)	Males ($N = 55$)	Females ($N = 100$)
Antigen	5.5	14.4	21.8	14.0
Antibody	72.7	62.5	69.1	63.0

was examined by partial correlation analysis, after controlling for both age and fasting glucose level. The results did not differ from those shown in Table 7.3.

The Pacific region is known as an endemic area for hepatitis-B (Maddrey 2000). Many subjects of our study were positive to hepatitis-B antigen (HbsAg) and hepatitis-B antibody (HbsAb), the proportions ranging from 5.5% to 21.0% for the former, and from 62.5% to 72.7% for the latter among the two regions and both sexes (Table 7.4). Partial correlations between BMI and blood parameters were recalculated after controlling for age, fasting blood glucose level (the first confounding factor) and hepatitis-B status (the second confounding factor). The variables significantly related to BMI remained the same as those shown in Table 7.3. Regarding the observation that many females had high liver activities of the enzymes glutamine oxaloacetic transaminase, glutamic puruvic transaminase and γGTP, further investigations are needed of the habitual consumption of kava and herbal medicine instead of Western alcohol by Tongans in relation to non-alcoholic steatohepatitis, which is very prevalent in the Asia-Pacific region (Farrell 2003; Poordad *et al.* 2003; Chitturi *et al.* 2004).

Concluding remarks

The modernization of Tongan lifestyle, which is tightly associated with the very high prevalence of obesity in them, began several decades ago. Although some specific genetic factors may be particularly important in the development of obesity among Tongans (Duarte *et al.* 2003a,b), two aspects of lifestyle may be just as important. First, the majority of Tongans live in urban areas and even those living in rural areas are not so isolated when compared with many other South Pacific countries such as PNG, the Solomons Islands and Vanuatu. The other is that about 100,000 Tongan citizens are out-migrants to various developed countries, including the United States, Australia and New Zealand. Many Tongans move between their own country and that of overseas residence, and flows of information between Tonga and the larger developed nations to which they migrate are great, as are remittances from out-migrants to Tongan relatives. These factors make it difficult for Tongans to slow down their rate of lifestyle modernization.

Efforts for reducing the risks of developing and for controlling Tongan obesity may be divided into two categories: those that are either governmental or institutionalized; and those that are based upon the efforts of individuals. For the former, a legal attempt to prohibit the importation of lamb flaps, which was seen as a cause of obesity, has been in effect since the mid 1990s (Inoue 2001); however, other regulations for imports of foods have not been put in place. The government also began a campaign to fight against obesity, in which a weight-loss competition with awards of prizes was organized (Englberger 1999; Inoue 2001). This attempt saw effective results in the initial stages, but participants gradually reduced in number. Currently, the elderly still remember their traditional lifestyles before the influences of modernization took hold; their narratives might be helpful in reducing obesity by encouraging aspects of traditional lifestyles that promote good health.

In reducing the risks of Tongan obesity, several strategies are needed, most of which should be undertaken by the government with different levels of cooperation with the people. The first involves regulatory strategies such as legal or penal treatments against import of foreign fat-rich foods. The second involves information-based strategies, focusing on scientific and practical knowledge about healthy dietary habits and other lifestyles. Tongans need to understand that obesity is the most critical factor in causing degenerative diseases if they are to help themselves reduce the prevalence rates of obesity. The third strategy involves incentive-based strategies, related to individual judgement rather than governmental judgement for healthy life.

Acknowledgements

We express our gratitude to Dr K. Takasaka, Dr Y. Kuchikura, Dr R. Ohtsuka, Dr A. Ioue, Dr J. Ohashi, Dr K. Natsuhara, Dr T. Yamauchi, Dr S. Fukuyama and Dr R. Kimura, who participated in these investigations and performed data collection and analysis. We are deeply thankful to Ms Naoko M. Afeaki, who served as coordinator of our research in relation to the needs of the Tongan government, and to Dr T. Palu, Dr S. Foliaki and Dr S. Kupu, Ministry of Health of Tonga, who gave us cooperative advice to help us accomplish our studies. In addition, we are very grateful to residents of Nuku'alofa city and Ha'ano Island, who understood our research purpose and participated in the examinations. This research was financially supported in part by Grant-in-Aid for Scientific Research from the Japanese Ministry of Education, Science and Culture (#13375004; project representative: Y. Kuchikura, and #14255008; project representative: R. Ohtsuka).

References

Ball, K., Mishra, G. and Crawford, D. (2002). Which aspects of socioeconomic status are related to obesity among men and women? *International Journal of Obesity and Related Metabolic Disorders* **26**, 559–65.

Bell, A. C., Swinburn, B. A., Simmons, D., *et al.* (2001). Heart disease and diabetes risk factors in Pacific Islands communities and associations with measures of body fat. *New Zealand Medical Journal* **114**, 364–5.

Brewis, A. A., McGarvey, S. T., Jones, J. and Swinburn, B. A. (1998). Perception of body size in Pacific Islanders. *International Journal of Obesity and Related Metabolic Disorders* **22**, 185–9.

Cameron, A. J., Welborn, T. A., Zimmet, P. Z., *et al.* (2003). Overweight and obesity in Australia: the 1999–2000 Australian Diabetes, Obesity and Lifestyle Study (AusDiab). *Medical Journal of Australia* **178**, 427–32.

Campbell, I. C. (1992). *Island Kingdom: Tongan Ancient and Modern.* Christchurch, New Zealand: Canterbury University Press.

Chitturi, S., Farrell, G. C. and George, J. (2004). Non-alcoholic steatohepatitis in the Asia-Pacific region: future shock? *Journal of Gastroenterology and Hepatology* **19**, 368–74.

Chukwuma, C. Sr and Tuomilehto, J. (1998). The 'thrifty' hypotheses: clinical and epidemiological significance for non-insulin-dependent diabetes mellitus and cardiovascular disease risk factors. *Journal of Cardiovascular Risk* **5**, 11–23.

Colagiuri, S., Colagiuri, R., Na'ati, S., *et al.* (2002). The prevalence of diabetes in the kingdom of Tonga. *Diabetes Care* **25**, 1378–83.

Craig, P., Halavatau, V., Comino, E. and Caterson, I. (1999). Perception of body size in the Tongan community: differences from and similarities to an Australian sample. *International Journal of Obesity and Related Metabolic Disorders* **23**, 1288–94.

(2001). Differences in body composition between Tongans and Australians: time to rethink the healthy weight ranges? *International Journal of Obesity and Related Metabolic Disorders* **25**, 1806–14.

Dalton, M., Cameron, A. J., Zimmet, P. Z., *et al.* (2003). Waist circumference, waist-hip ratio and body mass index and their correlation with cardiovascular disease risk factors in Australian adults. *Journal of Internal Medicine* **254**, 555–63.

de Silva, A. M., Walder, K. R., Aitman, T. J., *et al.* (1999). Combination of polymorphisms in OB-R and the OB gene associated with insulin resistance in Nauruan males. *International Journal of Obesity and Related Metabolic Disorders* **23**, 816–22.

Dignan, C. A., Burlingame, B. A., Arthur, J. M., Quigley, R. J. and Milligan, G. C. (1994). *The Pacific Islands Food Composition Tables.* Auckland, New Zealand: New Zealand Institute for Crop and Food Research Ltd.

Dowse, G. K. (1996). Incidence of NIDDM and the natural history of IGT in Pacific and Indian Ocean populations. *Diabetes Research and Clinical Practice Supplement* **34**, S45–50.

Dowse, G. K., Zimmet, P. Z. and Collins, V. R. (1996). Insulin levels and the natural history of glucose intolerance in Nauruans. *Diabetes* **45**, 1367–72.

Duarte, N. L., Colagiuri, S., Palu, T., Wang, X. L. and Wilcken, D. E. (2003a). A 45-bp insertion/depletion polymorphism of uncoupling protein 2 in relation to obesity in Tongans. *Obesity Research* **11**, 512–17.

  (2003b). Obesity, Type II diabetes and the Ala54Thr polymorphism of fatty acid binding protein 2 in the Tongan population. *Molecular Genetics Metabolism* **79**, 183–8.

During, K. (1990). *Pathways to the Tongan Present.* Nuku'alofa, Tonga: Government Printing Department.

Englberger, L. (1999). Prizes for weight loss. *Bulletin of the World Health Organization* **77**, 50–3.

Farrell, G. C. (2003). Non-alcoholic steatohepatitis: what is it, and why is it important in the Asia-Pacific region? *Journal of Gastroenterology and Hepatology* **18**, 124–38.

Hodge, A. M. and Zimmet, P. Z. (1994). The epidemiology of obesity. *Baillieres Clinical Endocrinology and Metabolism* **8**, 577–99.

Hodge, A. M., Dowse, G. K., Zimmet, P. Z. and Collins, V. R. (1995a). Prevalence and secular trends in obesity in Pacific and Indian Ocean island populations. *Obesity Research* **2**(Supplement), 77S–87S.

Hodge, A. M., Dowse, G. K., Toelupe, P., *et al.* (1995b). Dramatic increase in the prevalence of obesity in western Samoa over the 13 year period 1978–1991. *International Journal of Obesity and Related Metabolic Disorders* **18**, 419–28.

Hodge, A. M., Dowse, G. K., Collins, V. R. and Zimmet, P. Z. (1996). Mortality in Micronesian Nauruans and Melanesian and Indian Fijians is not associated with obesity. *American Journal of Epidemiology* **143**, 442–5.

Hongo, T., Suzuki, T., Ohtsuka, R., *et al.* (1989). Compositional character of Papuan foods. *Ecology of Food and Nutrition* **23**, 39–56.

Inaoka, T. (2002).Tongan obesity: history, genetics and environment. *Abstracts of Inter-Congress of the International Union of Anthropological and Ethnographic Science 2002*, Tokyo, p. 48.

Inaoka, T. and Ohtsuka, R. (1995). Nutritionally disadvantageous effects of small-scale marketing in a lowland Papua New Guinea community. *Man and Culture in Oceania* **11**, 81–93.

Inaoka, T., Matsumura, Y., Ohashi, J., *et al.* (2003). Nutritional and health problems in the South Pacific with special reference to Polynesia: Tongan obesity and diabetes (In Japanese). *Abstract of combined annual meetings of 44th Japanese Tropical Medicine and 18th International Health and Medicine*, Kokura, p. 118.

Inoue, A. (2001). Modernization of food intake and changes of Tongan traditional body perception and health: case study of weight loss competition in Tonga (written in Japanese). *Hokudai Bungakukenkyuka Kiyou* (Bulletin of Faculty of Literature, Hokkaido University) **105**, 1–49.

Joffe, B. and Zimmet, P. Z. (1998). The thrifty genotype in type 2 diabetes: an unfinished symphony moving to its finale? *Endocrine* **9**, 139–41.

Kagawa, Y., Yanagisawa, Y., Hasegawa, K., *et al.* (2002). Single nucleotide polymorphisms of thrifty genes for energy metabolism: evolutionary origins and prospects for intervention to prevent obesity-related diseases. *Biochemical and Biophysical Research Communications* **295**, 207–22.

Koike, G., Yokono, O., Iino, S., *et al.* (1984) Medical and nutritional survey in the Kingdom of Tonga; comparison of physiological and nutritional status of adult Tongans in urbanized (Kolofoou) and rural (Uiha) areas. *Journal of Nutritional Science and Vitaminology* **30**, 341–56.

Kumanyika, S. K. (1993). Special issues regarding obesity in minority populations. *Annals of Internal Medicine* **119**, 650–4.

Lako, J. V. (2001). Dietary trend and diabetes: its association among indigenous Fijians 1952–1994. *Asia Pacific Journal of Clinical Nutrition* **10**, 183–7.

Lako, J. V. and Nguyen, V. C. (2001). Dietary patterns and risk factors of diabetes mellitus among urban indigenous women in Fiji. *Asia Pacific Journal of Clinical Nutrition* **10**, 188–93.

Maddrey, W. C. (2000). Hepatitis B: an important public health issue. *Journal of Medical Virology* **61**, 362–6.

McAnulty, J. and Scragg, R. (1996). Body mass index and cardiovascular risk factors in Pacific Island Polynesians and Europeans in New Zealand. *Ethnicity and Health* **1**, 187–95.

Metcalf, P. A., Scragg, R. K., Willoughby, P., Finau, S. and Tipene-Leach, D. (2000). Ethnic differences in perceptions of body size in middle aged European, Maori and Pacific people living in New Zealand. *International Journal of Obesity and Related Metabolic Disorders* **24**, 593–9.

Neel, J. V. (1962). Diabetes mellitus: a 'thrifty' genotype rendered detrimental by 'progress'? *American Journal of Human Genetics* **14**, 353–62.

(1982). The thrifty genotype revisited. In *The Genetics of Diabetes Mellitus*, ed. J. Kobberling and R. Tattersall. Amsterdam: Academic Press, pp. 137–47.

Papoz, L., Barny, S. and Simon, D. (1996). Prevalence of diabetes mellitus in New Caledonia: ethnic and urban-rural differences. CALDIA Study Group. CALedonia DIAbetes Mellitus Study. *American Journal of Epidemiology* **143**, 1018–24.

Poordad, F., Gish, R., Wakil, A., *et al.* (2003). De novo non-alcoholic fatty liver disease following orthotopic liver transplantation. *American Journal of Transplantation* **3**, 1413–17.

Sawata, S. Hidaka, H., Yasuda, H., *et al.* (1988). Prevalence of cardiovascular diseases in the Kingdom of Tonga. *Japanese Heart Journal* **29**, 11–18.

Simmons, D. (1996). The epidemiology of diabetes and its complications in New Zealand. *Diabetic Medicine* **13**, 371–5.

Simmons, D., Thompson, C. F. and Volklander, D. (2001). Polynesians: prone to obesity and type 2 diabetes mellitus but not hyperinsulinaemia. *Diabetic Medicine* **18**, 193–8.

Ulijaszek, S. J. (2000). Age difference in physique of adult males aged 30–86 years in Rarotonga, the Cook Islands. *International Journal of Food Science and Nutrition* **51**, 229–34.

  (2001). Increasing body size among adult cook islanders between 1966 and 1996. *Annals of Human Biology* **28**, 363–73.

World Health Organization (2000). *Obesity: Preventing and Managing the Global Endemic.* Technical report series No. 894. Geneva: World Health Organization.

  (2006). *Statistical Information System: Statistics by Country or Region.* http://www3.who.int/whosis/menu.cfm?path = whosis,inds&language = english).

8 Nutrition and health in modernizing Samoans: temporal trends and adaptive perspectives

EMBER D. KEIGHLEY, STEPHEN T. MCGARVEY,
CHRISTINE QUESTED, CHARLES MCCUDDIN,
SATUPAITEA VIALI AND UTO'OFILI A. MAGA

Introduction

In 1988, David Tua, a Samoan raised on the island of Upolu, stormed the world's heavyweight boxing circles. He is one of the many Samoans who have become known internationally for their athleticism, whether in the boxing ring, on the rugby field, in American football or in paddling traditional outrigger canoes. While such athletes train and remain fit, Samoan communities in the archipelago and abroad are experiencing the negative health effects of the nutrition transition. Shifts in diet and physical activities are leading to chronic positive energy balance and high prevalence rates of obesity and related non-communicable diseases (NCD). In 2003 in the nation of Samoa, about one-third of men and over half of women aged 25 to 74 years were obese by Polynesian standards for body mass index (BMI). In 2002, 100 km across the ocean in American Samoa, about 60% of men and almost three-quarters of women were obese by the same standard. Along with these marked levels of adulthood overweight and obesity, Samoan people suffer from high levels of cardiovascular disease, type 2 diabetes, and other associated metabolic disorders (McGarvey 2001; McGarvey *et al.* 2005). Recently there has been a distinct rise in childhood and adolescent obesity in Samoa (Roberts *et al.* 2004), which threatens to exacerbate NCDs at earlier ages in the future.

This chapter describes the putative causes and health consequences of the changes in nutritional status among Samoans in the early twenty-first century. The physiology and health of Samoans are intertwined with their evolutionary and demographic history, cultural values and contemporary social roles, and alteration of the traditional subsistence ways of life and patterns of energy balance. We explore these factors and how they may have influenced the emergence of extremely high prevalence rates

Health Change in the Asia-Pacific Region: Biocultural and Epidemiological Approaches,
ed. Ryutaro Ohtsuka and Stanley J. Ulijaszek. Published by Cambridge University Press.
© Cambridge University Press 2007.

of obesity and their associated health outcomes. We employ mainly temporal comparisons to analyse trends in BMI, the proportions of people who are overweight or obese, and dietary patterns. In addition, we examine recent data on physical activities for the years 2002 and 2003. The chapter deals mostly with findings from communities in the Samoan islands, although some data on migrant communities living in Hawaii in the 1970s are used to illustrate changing health patterns with varying degrees of modernization.

The Samoas

The Samoan islands are located between 13° and 15° south latitudes and between 169° and 173° west longitudes in the South Pacific, about halfway between Hawaii and New Zealand. Archeological evidence indicates that Polynesian culture flourished in both Samoa and Tonga about 5,000 years ago (Green *et al.* 1969; Kirch 2000). Demographic, archaeological and genetic evidences suggest that Samoa was originally settled by a small number of founders who reached the islands after long open-ocean voyages against prevailing trade winds (Kirch 2000; Tsai *et al.* 2004).

These voyages were likely to have pushed their participants to the limits of endurance (McGarvey *et al.* 1989; McGarvey 1994); survivorship would have favoured boatmen and women with large body fat (BF) stores and efficient energy metabolism. Excess BF would have protected against both starvation and hypothermia during cold nights on the open water. Once these explorers reached new islands, it may have taken considerable time to establish agricultural crops, extending the duration of possible food shortage. Even after crops had been established, island life was threatened by tropical storms and volcanic activity, which would have created occasional and relatively acute shortages of food. Ancient Polynesians with efficient metabolic mechanisms may have been better able to survive and reproduce under these conditions. Since the end of the Second World War, the Samoan islands have rapidly changed and modernized. Today there is an abundance of food, and cases of starvation are rare to non-existent on the islands.

Although the putative adaptation of Samoan people towards more efficient metabolism may be contributing to the accelerating rates of obesity prevalence, is important not to reify the concept of the thrifty genotype as if it were based on only one or just a few gene systems. The search for a few influential genes causing obesity has been elusive (Damcott *et al.* 2003; Speakman 2004). Any putative thrifty genotype is most likely to be a complex genetic mechanism distributed across several loci which regulate energy intake, extraction, expenditure and storage (Snyder *et al.* 2004). While such a mechanism may increase susceptibility to obesity and its related diseases, it is not deterministic. Obesity is a result of a complex interaction of many lifestyle factors as well as variation in biological susceptibility. Furthermore, there may be negative social implications of overly simplistic thrifty genotype ideas for ethnic groups and families experiencing high levels of obesity and related diseases, leading to fatalistic attitudes about weight control and disease risks.

In 1878, the United States annexed the islands of Tutuila, Aunu'u, Ofu, Olesega and Ta'u, which became the US territory of American Samoa (Meleisea 1987). Germany gained control of Upolu, Savai'i and the small surrounding islands in 1900. From the time when the Samoan islands were divided between these two industrial nations, their history begins to diverge. Like most island nations throughout the Pacific, the colonial history of the islands has had a profound effect on the health of the people.

During the period of US colonial expansion in the late nineteenth century into the Pacific, the presence of the deepwater harbour of Pago Pago was seen as a valuable asset (Philbrick 2003). A US naval coaling station was established in the harbour after an agreement in 1872, and the US naval base was maintained there until 1951. During the Second World War, many US and allied troops were billeted on Tutuila, the main island of American Samoa. In the early 1960s, the US federal government invested substantially in roads, water and sewer services, education and medical care in American Samoa. At the same time, labour costs favourable to industry and other financial advantages led major US companies to build two tuna canneries in Pago Pago harbour. Over the past 50 years, the ways of life and subsistence patterns of many Samoans have changed drastically. Many residents of American Samoa, especially those living on Tutuila, underwent a rapid shift from traditional subsistence lifestyles to wage/salary jobs. For instance, on Tutuila, in 1960, the majority of adult males were engaged in agriculture; by 1974, only 8% remained as full-time farmers (Greksa and Baker 1982).

By contrast, Germany developed Samoa predominately for agriculture. It was made a protectorate of New Zealand after the First World War and gained its independence in 1962. In 1997, the independent nation changed its name from Western Samoa to Samoa. Today, the economy of Samoa is still predominately based on agriculture, with tourism becoming increasingly important (Government of Samoa 2001). In contrast to their American Samoan counterparts, in the 2001 census for Samoa 44% of all men reported agriculture as their primary occupation, with higher proportions in the rural areas (Government of Samoa 2001). The economy in Samoa is also highly dependent upon migrant remittances to their family members. Throughout the Pacific island nations, remittances may contribute up to half of all cash income (Ahlburg 1995). Reliance upon this cash flow from relatives abroad leaves Samoa vulnerable to fluctuations in the world market.

Throughout the nineteenth century, there were documented epidemics of infectious disease throughout the Samoan islands (McArthur 1956; Gilson 1970). These diseases are believed to have been inadvertently brought by Europeans and went on to kill 70% to 90% of the population of the islands (McArthur 1967; Kirch 2000). The Samoas have largely undergone health transitions since the end of the Second World War. Infectious disease has rapidly declined and chronic disease has steadily increased. In 1950, in American Samoa, the death rate from major infectious diseases was 440 per 100,000 (Baker and Crews 1986). By 1960, the death rate from infectious disease had dropped to 128 per 100,000, and there was a decrease in the proportion of deaths due to infectious disease from 42% to 23%. By the 1970s, tuberculosis disappeared as a cause of mortality in American Samoa. In Samoa, the incidence of pneumonia and influenza rapidly dropped from 546 per 10,000 in 1986 to 60.7 per 10,000 in 1994 (Annual Report of the Department of Health 1986 and 1993–94).

Early European visitors to Polynesia noted the health and well-being of the people, and mentioned signs of overweight among high status individuals but not in the general population (Turner 1861, 1884; Beaglehole 1961). Following the Second World War, obesity and associated metabolic disorders became major determinates of morbidity and mortality. As infectious disease decreased in importance, the incidence of hypertensive diseases increased from 10.9 per 10,000 in 1986 to 14.6 per 10,000 in 1994. Type 2 diabetes, which was not among the ten leading causes of mortality in 1986, became the eighth most common cause of death in 1994.

Over the past 100 years many Samoans in both polities have moved away from traditional lifestyles. Today, there are fewer Samoans farming and fishing, and more Samoans participating in sedentary wage jobs. In addition, there has been an alteration in diet towards increased consumption of imported foods, and many Samoans have acquired material possessions which alter lifestyle, such as cars and televisions. With such change has come a shift from physically demanding lifestyles towards more sedentary ones, with increased access to energy-dense, processed foods.

The current obesity epidemic among Samoans in their native islands and abroad may also be partially attributed to cultural values and practices pertaining to social roles, physical activity patterns, dietary intake and body size. Larger body size is often considered attractive, especially among older Samoans, and considered a symbol of wealth, health and prestige in Samoan society more generally (Brewis *et al.* 1998; Brewis and McGarvey 2000; WHO 2002). At *to'onai* (Sunday feasts), *fiafia* (village celebrations) and *fa'alavelave* (larger celebrations for weddings or funerals), food is provided according to each villager's social status. *Matai* (chiefs) and respected elders are fed first and given the largest portions. Once Samoan women have married and raised children, they have completed a large part of their social responsibility, and it becomes the job of their children and younger members of the village to care for them (O'Meara 1990; Holmes 1992). Traditionally, subsistence labour was carried out by young men, and as men aged they took on a more managerial role in *'aiga* (extended family) farming and fishing activities. Thus with increased age, there is often a consequential shift to a more sedentary lifestyle.

Patterns of physical activity and cultural practices have often been tied to changes in body fatness among Samoans. In a study of the diet and physical activity patterns of rural and urban residents in Samoa in 1982, Pelletier (1987) found that higher levels of body fatness among residents of the capital, Apia, could be attributed to adherence to Samoan cultural feasting practices while living a more sedentary lifestyle. Although urban residents consumed fewer calories throughout the week than their rural counterparts, they continued to partake in Sunday *to'onai* and ate an average of 5,930 calories compared to the 5,940 calories consumed by rural villagers. While rural villagers engaged in high levels of physical activity throughout the week, offsetting the positive energy balance from the weekly *to'onai*, urban residents did not.

Nutritional change

Social and economic factors play an integral role in the production of risk and disease. We suggest that the broad nutritional changes that Samoans have experienced are a result of complex interactions among (1) global and regional politico-economic factors such as occupation-related physical activity, agricultural policies and food imports; and (2) local social and cultural factors including nutrition and health education, retention of culturally valued dietary patterns, attitudes about body image and physical activity levels (McGarvey *et al.* 1989; McGarvey 1994; Zimmet 2000; Horgen and Brownell 2002; Popkin 2004; Popkin and Gordon-Larsen 2004).

In what follows, we present findings from research among Samoans living in Hawaii, American Samoa and Samoa between 1975 and 2003.

The emphasis is on adults, although findings on BMI and overweight in children are also presented. Research methods used in surveys in Hawaii in 1975 and 1977, in American Samoa in 1976 and 1978, and in Samoa in 1979 and 1982 are described fully in Baker *et al.* (1986). In all studies, the emphasis was on obtaining nationally representative samples as far as it was possible to do so. In American Samoa in 1976, only adults were recruited, children being recruited in smaller studies there in 1978.

Across the period 1990 to 1994, a representative sample of adults aged 25 to 54 years was recruited from 20 villages and 10 workplaces in American Samoa (McGarvey *et al.* 1993). Between 1991 and 1995, adults aged 25 to 54 years were randomly chosen from nine villages in Samoa, including four from rural Savai'i, three from rural 'Upolu and two from the urban Apia area (Chin-Hong and McGarvey 1996). Children were also studied in Samoa between 1991 and 1995, primarily from the rural 'Upolu and Apia villages (Roberts *et al.* 2004). In the period 1990 to 1995, studies of factors underlying blood pressure variation and change over time were carried out in Samoa, and those with medical diagnoses were excluded from survey. Full descriptions of the sampling and recruitment in the studies carried out in 1990 to 1995 are presented elsewhere (McGarvey *et al.* 1993; Chin-Hong and McGarvey 1996; Galanis *et al.* 1999). Adults were included in the study samples only if they had not received a diagnosis of hypertension or type 2 diabetes, and self-reported all four grandparents as being Samoan. The samples obtained in both American Samoa and Samoa are broadly representative of the adult populations of those polities because of the very low numbers excluded, and the low levels of medically diagnosed NCDs in both locations at the time.

Genetic epidemiology studies were conducted in American Samoa in 2002 and in Samoa in 2003 to estimate genetic and environmental influences on adiposity and glucose metabolic traits, and to identify chromosomal regions with susceptibility genes associated with obesity and type 2 diabetes. Recruitment and sampling was based purely on availability of large families, and probands were not selected specifically for obesity or type 2 diabetes. In American Samoa, probands were randomly selected from prior studies and then supplemented by following extended families to recruit as many individuals as possible. In Samoa, large families were identified in randomly selected villages, adults and children in the same families being recruited. Thus, data collected in 2002 and 2003 are not from pre-selected random surveys, but may be considered representative of the two polities. In addition, many of the individuals are siblings and cousins and no adjustment for familial clustering was performed (although such adjustments change the standard errors of means, not the means or proportions).

The same techniques were used to measure weight, stature and blood pressure in all studies and at all times. In the 2002 and 2003 surveys, body composition was assessed by bioelectrical impedance in all participants. The proportion of body weight as fat was calculated from impedance measures using published equations for Polynesian adults and youths (Swinburn *et al.* 1999; Rush *et al.* 2003). Recent studies have found differences in body composition in Polynesians which affect the validity of the usual World Health Organization (WHO) BMI criteria for classification of overweight and obesity (Swinburn *et al.* 1999). Therefore, we have used standards put forth by Swinburn *et al.* (1999) to define overweight and obesity among Samoans; these define overweight as a BMI of 26–32 kg/m^2 and obesity as a BMI $>32 kg/m^2$.

Adiposity and overweight among Samoan adults

BMI levels and prevalence rates of overweight and obesity in adults were already remarkably high in the 1970s when systematic study of Samoans began (Tables 8.1, 8.2 and 8.3) (McGarvey and Baker 1979; Bindon and Baker 1985; McGarvey 1991). In American Samoa in the 1970s, 69.0% of men and 83.7% of women aged 18 to 74 years were either overweight or obese (as defined by BMI $\geq 26 kg/m^2$) (Table 8.3). In many developing nations, the body composition of women is affected by changes in lifestyle years before that of men (McGarvey *et al.* 1989). This may be explained by several factors. While men in modernizing societies are more likely to continue some of their traditional, physically demanding activities (at least initially), women often experience a more rapid shift towards sedentary life due to reduction in subsistence activities as cash enters communities and households. Women tend to participate in more consistent but fairly low-level physical activities, generally domestic tasks.

Overweight in Samoan women may be influenced by other factors such as body image and weight gain with repeated pregnancies. It is often considered attractive for Samoan women to be larger, and some of the elevated obesity among women may be desired by them (Brewis *et al.* 1998; Brewis and McGarvey 2000). Furthermore, Samoan women are often encouraged to have large families (Brewis *et al.* 1998). In 1974, the average total fertility rate in American Samoa was 6.63 (Filiga and Levin 1988). This rate decreased slightly by 1980 with the introduction of birth control, but remained high, at 5.98. In Samoa, mean completed fertility decreased from approximately 7.0 in the late 1960s and early 1970s, to 6.4 in the mid 1970s, 6.7 in the early 1980s and about 4.8 in 1993 (Brewis *et al.* 1998).

Table 8.1. *Adult body mass index (BMI; kg/m²) in American Samoa by sex and age, 1976–2002*

Age group	1976–78 N	1976–78 Mean (SD)	1990 N	1990 Mean (SD)	2002 N	2002 Mean (SD)
Males						
18–24	138	26.1 (3.8)	–	–	52	30.5 (6.6)
25–34	157	30.3 (4.9)	100	33.7 (5.8)	73	33.6 (8.7)
35–44	133	30.2 (5.4)	126	34.2 (6.7)	79	34.9 (6.3)
45–54	175	31.0 (6.4)	92	34.9 (6.7)	52	35.7 (8.3)
55–64	147	28.7 (5.6)	–	–	46	33.0 (6.2)
65–74	62	28.6 (5.9)	–	–	21	34.1 (5.3)
Females						
18–24	199	27.9 (5.2)	–	–	66	33.1 (8.8)
25–34	227	32.4 (6.5)	151	34.0 (6.3)	93	37.6 (8.6)
35–44	254	34.6 (6.4)	157	36.8 (7.5)	114	38.0 (7.1)
45–54	249	34.9 (6.9)	93	36.3 (5.9)	67	38.7 (8.3)
55–64	153	32.6 (7.2)	–	–	43	35.6 (7.1)
65–74	64	31.8 (5.9)	–	–	27	35.4 (7.1)

The sharp gender difference in overweight and obesity prevalence rates seen in the 1970s in American Samoa appears to diminish in the 1990 and 2002 data, but this may be an artefact of the classifications used. The highest BMI group includes everyone with BMI $>32 kg/m^2$. When the WHO classifications were used, with a slight change (making Class I obesity $32–34.9 kg/m^2$; Class II obesity $35.0–39.9 kg/m^2$; and Class III obesity $\geq 40 kg/m^2$), there remained a large gender disparity in the degree of adiposity among American Samoans in 2002. Inspection of age-specific mean BMI over time of American Samoan men and women (Table 8.1) shows large BMI differences, especially in young adults. For example, in 2002, mean BMI in men aged 25 to 34 years was 33.6 kg/m², while for women it was 37.6 kg/m², a 4 kg/m² difference. Women also had BMIs

Table 8.2. *Adult body mass index (BMI; kg/m^2) in Samoa by sex and age, 1979–2003*

Age group	1979–82		1991		2003	
	N	Mean (SD)	N	Mean (SD)	N	Mean (SD)
Males						
18–24	80	24.0	–	–	69	25.1
		(2.5)		–		(3.5)
25–34	46	25.7	141	27.7	81	27.7
		(3.7)		(3.7)		(4.7)
35–44	36	26.5	108	30.4	105	29.9
		(3.9)		(5.8)		(5.2)
45–54	44	27.4	84	29.0	83	32.9
		(4.2)		(4.8)		(5.7)
55–64	28	28.0	–	–	48	31.9
		(5.7)		–		(5.3)
65–74	13	26.9	–	–	38	29.7
		(2.9)		–		(5.5)
Females						
18–24	79	26.1	–	–	81	28.1
		(3.8)		–		(5.2)
25–34	67	28.0	152	29.2	111	32.2
		(4.5)		(4.5)		(6.2)
35–44	45	29.7	118	31.7	105	34.8
		(4.9)		(5.0)		(6.9)
45–54	48	30.8	105	33.0	91	36.5
		(5.6)		(6.6)		(8.5)
55–64	32	31.1	–	–	60	33.4
		(7.2)		–		(6.1)
65–74	21	28.3	–	–	54	32.2
		(6.1)		–		(7.2)

greater by 3.1 kg/m^2 in the 35 to 44 years age group, and by 3.0 kg/m^2 in the 45 to 54 years age group.

Between 1976 and 2002, BMI levels and prevalence rates of obesity increased greatly among American Samoan men and women. By 1990, the proportion of obese women increased to 70% from 51% in 1976–78, rising only a little thereafter, to 71% in 2002 (Table 8.3). In 2002, less than 10% of women in American Samoa were within the normal range of BMI. In American Samoan men, there was a dramatic increase in overweight and obesity combined, from the 1970s to 1990. The prevalence of obesity rose from 28% to 61%, while there was a smaller decline in overweight, from 41% to 33%. Prevalences of both overweight and obesity declined slightly by the year 2002.

Table 8.3. *Proportion of normal, overweight, and obese*[a] adults, aged 18–74 years, by sex

(A) American Samoa 1976–2002

Age group	1976–78 Overweight % n/N	Obese % n/N	1990[b] Overweight % n/N	Obese % n/N	2002 Overweight % n/N	Obese % n/N
Males						
18–24	35.5 49/138	8.7 12/138	–	–	40.4 21/52	36.5 19/52
25–34	51.6 81/157	29.3 46/157	39.0 39/100	55.0 55/100	34.2 16/79	52.1 16/79
35–44	44.4 59/133	31.6 42/133	32.5 41/126	61.9 78/126	20.3 25/73	70.9 38/73
45–54	34.3 60/175	42.9 75/175	28.3 26/92	67.4 62/92	26.9 14/52	69.2 36/52
55–64	41.5 61/147	25.2 37/147	–	–	30.4 14/46	58.7 27/46
65–74	33.9 21/62	27.4 17/62	–	–	23.8 5/21	66.7 14/21
All ages	40.8 331/812	28.2 229/812	33.3 106/318	61.3 195/318	29.4 95/323	58.8 190/323
Females						
18–24	47.2 94/199	16.1 32/199	–	–	24.2 16/66	50.0 33/66
25–34	37.0 84/227	47.6 108/227	25.8 39/151	62.3 94/151	16.1 15/93	76.3 71/93
35–44	30.3 77/254	63.0 160/254	15.9 25/157	76.4 120/157	18.4 21/114	78.1 89/114
45–54	23.7 59/249	66.7 166/249	23.7 22/93	69.9 65/93	20.9 14/67	77.6 52/67
55–64	28.8 44/153	53.6 82/153	–	–	30.2 13/43	65.1 28/43
65–74	31.3 20/64	51.6 33/64	–	–	25.9 7/27	66.7 18/27
All ages	33.0 378/1146	50.7 581/1146	21.4 86/401	69.6 279/401	21.0 86/410	71.0 291/410

Notes:

[a] Based on Polynesian standards: normal: $BMI < 26 \, kg/m^2$; overweight: $BMI \; 26\text{–}32 \, kg/m^2$; obese: $BMI > 32 \, kg/m^2$;

[b] Includes only adults aged 25–54 years.

Table 8.3. (*cont.*)
(B) Samoa 1979–2003

	1979–82		1991[b]		2003	
	Overweight % n/N	Obese % n/N	Overweight % n/N	Obese % n/N	Overweight % n/N	Obese % n/N
Age group						
Males						
18–24	20.0 16/80	0.0 0/80	–	–	24.6 17/69	4.3 3/69
25–34	41.3 19/46	4.3 2/46	47.5 67/141	13.5 19/141	43.2 35/81	16.0 13/81
35–44	41.7 15/36	11.1 4/36	45.4 49/108	32.4 35/108	44.8 47/105	28.6 30/105
45–54	43.2 19/44	13.6 6/44	46.4 39/84	25.0 21/84	33.7 28/83	53.0 44/83
55–64	39.3 11/28	21.4 6/28	–	–	47.9 23/48	43.8 21/48
65–74	76.9 10/13	0 0/13	–	–	39.5 15/38	34.2 13/38
All ages	36.4 90/247	7.3 18/247	46.5 155/333	22.5 75/333	38.9 165/424	29.2 124/424
Females						
18–24	40.5 32/79	8.9 7/79	–	–	38.3 31/81	22.2 18/81
25–34	43.3 29/67	20.9 14/67	51.3 78/152	23.0 35/152	32.4 36/111	49.5 55/111
35–44	46.7 21/45	28.9 13/45	36.4 43/118	50.0 59/118	23.8 25/105	67.6 71/105
45–54	43.8 21/48	41.7 20/48	32.4 34/105	56.2 59/105	24.2 22/91	67.0 61/91
55–64	25.0 8/32	46.9 15/32	–	–	28.3 17/60	61.7 37/60
65–74	38.1 8/21	23.8 5/21	–	–	42.6 23/54	42.6 23/54
All ages	40.8 119/292	25.3 74/292	41.3 155/375	40.8 153/375	30.7 154/502	52.8 265/502

Notes:
[a] Based on Polynesian standards: normal: BMI $< 26 \,\mathrm{kg/m^2}$; overweight: BMI $26–32 \,\mathrm{kg/m^2}$; obese: BMI $> 32 \,\mathrm{kg/m^2}$;
[b] Includes adults 25–54 years.

Five-year longitudinal studies conducted in American Samoa between 1976 and 1981, and 14-year longitudinal changes observed between 1976 and 1990, show that American Samoans who were already overweight, especially young adults, continued to gain weight across their lives (McGarvey 1991; Gershater and McGarvey 1995).

While levels of overweight and obesity in Samoa were not as severe as those in American Samoa, they too have continued to increase since the 1970s (Table 8.2). In Samoa, BMI has increased most sharply in adults above the age of 35 years, but especially among those aged 45 to 54 years. Between 1979 and 2003, the mean BMI of this age group changed from 27.4 to 32.9 in men and from 30.8 to 36.5 in women. In Samoa, a much larger proportion of women than men are classified as being either overweight or obese at all years of survey. Mean BMIs of women in Samoa in 2003 were lower than those of women in American Samoa in 2002, but are regardless indicative of high levels of body mass and adiposity.

The proportion of adults in Samoa of normal BMI has reduced by almost half between the 1970s and 2003 (Table 8.3). While 56% of men in the 1970s were of normal BMI, less than one-third of them were similarly classified in 1991 and in 2003. The proportion of women of normal BMI fell from 34% in the 1970s to 18% in 1991 and to 17% in 2003. Almost 30% of men and more than 50% of women in Samoa in 2003 were classified as obese (BMI $>32\,\mathrm{kg/m^2}$).

Across both Samoas, both sexes and all years of survey, the greatest temporal increase in average BMI was among young American Samoan women aged 25 to 34 years, despite pre-existing high rates of obesity in the 1970s. While overall obesity rates continue to increase, they are rising most sharply in young adults. Increased prevalence of obesity in younger adults unveils a grim outlook for their health in the future, and suggests not only an increase in, but also an earlier onset of, chronic disease. It also suggests an increase in the incidence of obesity among children. Biologically, high maternal weight both pre-pregnancy and during pregnancy is associated with higher birth weight, and elevated metabolic profiles in young children (Vohr *et al.* 1995, 1999; Vohr and McGarvey 1997). A rise in obesity among women of childbearing age also leads to increased risk for gestational diabetes in women and later risk of type 2 diabetes (Coustan *et al.* 1993). Increasing obesity among young Samoan women may also have social consequences for their children, inasmuch as parents and grandparents are usually models for nutritional behaviour for children within a household (Neumark-Sztainer 1999). Ideas of ideal body image are also transmitted culturally by way of family and age group peers (McKinley 1999; Jones 2004). Samoan children, in both Samoa and especially American Samoa, are likely to be at higher risk of obesity because they are surrounded by adults with high levels of obesity.

There has been a rapid decline over time in older adult Samoans with BMI within the normal range (Tables 8.1 and 8.2). This is due to both the strong cohort effects present in modernizing societies and the differential mortality in later adult life of those with high BMI. Older Samoans studied in the 1970s were born in the early twentieth century, when agricultural and fishing subsistence patterns of diet and physical activity were near universal, long before the cash economy exerted its influence on most Samoans. In American Samoa in the 1970s, men aged 55 to 74 years had lower average BMI than both counterparts of their age in 2002 and younger adults studied at the same time. This cross-sectional age difference largely disappeared by 2002, and any decrease in BMI in the oldest age groups here may be attributed to selective mortality among the obese and sick as they age. There are fewer cross-sectional age differences in mean BMI in men from Samoa in the 1979 to 1982 sample, with a slight increase with increasing age. However, over time there has been an increase in mean BMI in older men and women from Samoa. In both nations, as these older, and putatively healthier, cohorts age and die, their influence as role models of nutritional behaviours and healthy body type may diminish.

The proportions of men and women in the different BMI groups in 2003 in Samoa are remarkably similar to their same-sex counterparts in 1976 in American Samoa (Fig. 8.1). In 1976, 4.3% of American Samoan men were graded as being Class III obese compared to 5.9% of men in Samoa in 2003. Likewise in 1976, 13.2% of American Samoan women were Class III obese compared to 13.7% of women in Samoa in 2003. The strong similarity of the sex-specific distributions of BMI in 2003 in Samoa with those 27 years earlier in American Samoa could be purely coincidental, but could reflect a general trend in modernization and the nutritional transition among Samoans, and perhaps Pacific island populations more generally. Despite a different pace, timing and set of political and economic influences on development in the two Samoan polities, adult Samoans experiencing economic modernization over two to three decades can lead to very high prevalences of overweight and obesity. If the current patterns of increasing adiposity continue in Samoa, as they did in American Samoa over the last 25–30 years, the prevalences of obesity in American Samoan adults in 2002 may foreshadow future adiposity levels among those living in Samoa. However, despite the very high levels of fatness and obesity achieved by contemporary Samoan populations, not all individuals within these populations become obese. In both polities, a proportion of adults retain BMIs within the normal range.

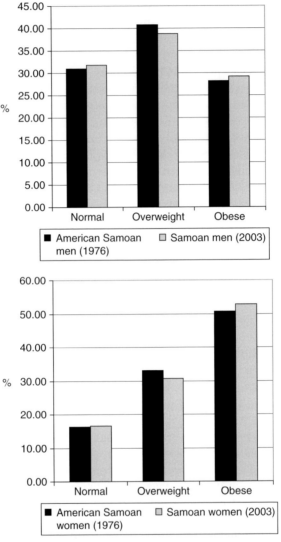

Fig. 8.1. Distribution of normal, overweight and obese adults by sex in American Samoa in 1976 and Samoa in 2003 (normal: BMI $< 26\,\text{kg/m}^2$; overweight: BMI $= 26$–$32\,\text{kg/m}^2$; obese: BMI $> 32\,\text{kg/m}^2$).

Adiposity among Samoan children

Overweight in children is associated with increased levels of childhood cardiovascular disease risk factors as well as an elevated risk of obesity in adulthood (Guo *et al.* 1994; Freedman *et al.* 1999). In this section, we describe adiposity in samples of Samoan children aged 6 to 17 years, from studies conducted in parallel with those carried out among adults and described in the previous section. The data thus collected are compared with those from studies of Samoan children residing in Hawaii during 1975 to 1977. The Hawaii studies are described in detail elsewhere (Bindon and Zansky 1986), the children therein being in families studied in several Samoan residential enclaves on the island of Oahu in 1975 and 1977. The families were chosen by our Samoan contacts in those towns and villages on Oahu without regard for health conditions, social status or recent migration.

All children and adolescents were categorized as normal, overweight or obese using the standards developed by Cole *et al.* (2000). While these standards provide perhaps the best cut-offs currently available for children internationally, if body composition evidence from Samoan adults also applies to children and adolescents, then Samoan children may have lower percent BF at each BMI cut-off. There has been a temporal trend towards higher BMIs among Samoan children, in association with ecological measures of modernization (Table 8.4) (Bindon and Zansky 1986; McGarvey and Schendel 1986). The studies of Samoan children living in Hawaii carried out in the mid 1970s showed them to have high age-specific BMI relative to children in American Samoa and Samoa, also measured around that time. However, by 2002, childhood age-specific BMI levels in American Samoa exceeded those of Hawaii in the years 1975 to 1977 in all but one age and sex group (Table 8.4). Within American Samoa, age- and sex-specific childhood BMI increased between 1978 and 2002 by between 1.9 and 3.8 kg/m^2 in boys and girls aged 6 to 11 years, respectively. In adolescents aged 12 to 17 years, the increases in BMI across the same period ranged from 3.6 to 5.8 kg/m^2 in boys and 4.9 to 5.8 kg/m^2 in girls.

The 1990s data from Samoa were obtained from surveys carried out in villages on 'Upolu island and incorporate few children from the more rural island of Savai'i. General childhood BMI levels for Samoa as a whole may thus be overestimated. The data collected on Samoa in 2003, however, were collected in villages throughout the nation, and may be more representative of the national population. To preserve the national representativeness of the comparison, analysis of temporal trends in childhood BMI in Samoa is limited to the 1979 to 1982 period and data for the year 2003 only. The descriptive data from the 1990s are included in the tables for the sake of completeness.

Between 1979 to 1982 and the year 2003, the average BMI of children living in Samoa increased in every age group. The increases were not as

Table 8.4. *Body mass index (BMI; kg/m²) of Samoan children aged 6–17.9 years, residing in Hawaii, American Samoa and Samoa by sex and age, 1975–2003*

Age (years)	Hawaii 1975–77 N	Hawaii 1975–77 Mean (SD)	American Samoa 1976–78 N	American Samoa 1976–78 Mean (SD)	American Samoa 2002 N	American Samoa 2002 Mean (SD)	Samoa 1979–82 N	Samoa 1979–82 Mean (SD)	Samoa 1991–95 N	Samoa 1991–95 Mean (SD)	Samoa 2003 N	Samoa 2003 Mean (SD)
Males												
6–8	58	18.3	38	17.2	73	19.1	69	16.4	51	16.4	61	17.3
		2.7		2.4		4.5		1.2		1.4		3.6
9–11	51	20.7	46	17.7	86	21.5	58	17.6	44	19.2	77	18.4
		4.3		1.9		5.2		2.7		4.1		2.7
12–14	53	22.0	40	20.3	71	23.9	45	19.2	21	20.7	52	20.1
		3.6		2.6		6.8		2.1		4.1		3.5
15–17	38	23.6	38	22.7	53	28.5	35	20.9	18	23.3	43	22.3
		2.7		2.2		6.1		2.0		5.1		2.6
Females												
6–8	65	18.2	31	16.5	64	18.4	49	16.5	54	16.8	63	17.2
		2.5		1.8		3.1		2.0		2.1		2.8
9–11	50	21.4	41	18.8	77	21.2	55	17.2	34	19.4	74	18.9
		4.3		2.4		5.3		2.1		3.5		3.2
12–14	52	23.8	31	21.8	68	26.7	49	19.1	31	22.4	62	22.0
		4.1		4.5		6.5		2.5		4.7		3.1
15–17	43	26.5	27	24.9	53	30.7	38	22.3	21	26.1	51	24.4
		3.3		3.8		7.7		3.1		4.8		4.2

great as those in American Samoan children, average BMI having increased by between 0.8 and 0.9 kg/m² for boys and 0.7 to 1.7 kg/m² for girls aged 6 to 11 years. Levels of BMI rose more rapidly in adolescents aged 12 to 17 years, with increases of 0.9 to 1.4 kg/m² in boys and 2.1 to 2.9 kg/m² in girls. Using the Cole *et al.* (2000) criteria, large proportions of both American Samoan children and adolescents in 2002 were overweight or obese, there having been a dramatic increase since the mid 1970s (Tables 8.5 and 8.6). Approximately 70% of boys and over 80% of girls aged 15 to 17 years in American Samoa in 2002 were overweight or obese. Of girls aged 12 to 14 years, over 65% were overweight or obese. Even among American Samoan boys and girls aged 6 to 11 years, only 49% to 56% of them had BMIs within the normal range (Table 8.5). In Samoa, the proportions of children who were overweight or obese in the late 1970s and early 1980s were very small (Tables 8.6 and 8.7). Between 1979 and 2003,

Table 8.5. *Prevalence of overweight and obesity among children aged 6–17.9 years in American Samoa and Hawaii, 1975–2002*

Age (years)	Hawaii 1975–77				American Samoa 1976–78				American Samoa 2002			
	N	Normal %N	Over weight %n	Obese %n	N	Normal %n	Over weight %n	Obese %n	N	Normal %n	Over weight %n	Obese %n
Males												
6–8	58	53.4 / 31	39.7 / 23	6.9 / 4	38	73.7 / 28	23.7 / 9	2.6 / 1	73	56.2 / 41	19.2 / 14	24.7 / 18
9–11	51	58.8 / 30	23.5 / 12	17.6 / 9	46	89.1 / 41	10.9 / 5	0 / 0	86	48.8 / 42	27.9 / 24	23.3 / 20
12–14	53	67.9 / 36	20.5 / 11	11.3 / 6	40	82.5 / 33	15.0 / 6	2.5 / 1	71	59.2 / 42	14.1 / 10	26.8 / 19
15–17	38	60.5 / 23	34.2 / 13	5.3 / 2	38	78.9 / 30	21.1 / 8	0 / 0	53	30.2 / 16	20.8 / 11	49.1 / 26
Females												
6–8	65	49.2 / 32	40.0 / 26	10.8 / 7	31	77.4 / 24	22.6 / 7	0 / 0	64	50.0 / 32	34.4 / 22	15.6 / 10
9–11	50	46.0 / 23	38.0 / 19	16.0 / 8	41	75.6 / 31	22.0 / 9	2.4 / 1	77	54.5 / 42	22.1 / 17	23.4 / 18
12–14	52	40.4 / 21	46.2 / 24	13.5 / 7	31	74.2 / 23	16.1 / 5	9.7 / 3	68	32.4 / 22	36.8 / 25	30.9 / 21
15–17	43	30.2 / 13	51.2 / 22	18.6 / 8	27	44.4 / 12	44.4 / 12	11.1 / 3	53	18.9 / 10	37.7 / 20	43.4 / 23

Source: Overweight and obesity classifications based on Cole *et al.* (2000).

Table 8.6. *Percentage of normal, overweight and obese Samoan children aged 6–17.9 years, Hawaii, American Samoa and Samoa, 1975–2003*

	Hawaii 1975–77	American Samoa 1976–78	American Samoa 2002	Samoa 1979–82	Samoa 1991–95	Samoa 2003
Males						
Normal	60.0	81.5	49.8	92.8	84.3	81.1
n/N	120/200	132/162	141/283	192/207	113/134	189/233
Overweight	29.5	17.3	20.8	5.8	11.2	15.0
n/N	59/200	28/162	59/283	12/207	15/134	35/233
Obese	10.5	1.2	29.3	1.4	4.5	3.9
n/N	21/200	2/162	83/283	3/207	6/134	9/233
Females						
Normal	42.4	69.2	40.5	90.1	68.6	70.0
n/N	89/210	90/130	106/262	172/191	96/140	175/250
Overweight	43.3	25.4	32.1	8.9	22.9	21.2
n/N	91/210	33/130	84/262	17/191	32/140	53/250
Obese	14.3	5.4	27.5	1.0	8.6	8.8
n/N	30/210	7/130	72/262	2/191	12/140	22/250

Source: Overweight and obesity classifications based on Cole *et al.* (2000).

however, there was a fourfold increase in the prevalence of obesity among boys and an almost eightfold increase in the prevalence of obesity among girls. In 2003, between 77% and 86% of boys of all ages were of normal BMI. In 2003, the proportions of Samoan girls who were overweight or obese ranged from 21% to 26% in young ages, rising to between 37% and 39% in 12 to 17-year-olds (Tables 8.6 and 8.7).

Childhood obesity in all Samoans has increased with increasing modernization (Table 8.6). In the 1970s, children most exposed to modernization, those living in Hawaii, had the highest BMIs, followed by American Samoans, and finally Samoans living in Samoa. The BMI pattern in children living in Samoa in 1979 is the closest approximation we have of neo-traditional nutritional ecology and body size, although age-specific childhood BMI is likely to have been higher than that of one- or two-hundred years previously. However, they are perhaps the most reasonable targets for ensuring healthy body mass and adiposity in Samoan children in the present day. If present trends continue, it is easy to project that obesity-related NCDs will become important paediatric health problems in Samoans, as well as worrisome antecedents to the health of future Samoan adults.

Table 8.7. *Prevalence of overweight and obesity among children aged 6–17.9 years in Samoa, 1979–2003*

Age (years)	Samoa 1979–82				Samoa 1991–95				Samoa 2003			
	N	Normal %N / N	Overweight %n / n	Obese %n / n	N	Normal %n / n	Overweight %n / n	Obese %n / n	N	Normal %n / n	Overweight %n / n	Obese %n / n
Males												
6–8	69	91.3 / 63	7.2 / 5	1.4 / 1	51	92.2 / 47	7.8 / 4	0 / 0	61	82.0 / 50	14.8 / 9	3.3 / 2
9–11	58	96.6 / 56	0 / 0	3.4 / 2	44	81.8 / 36	11.4 / 5	6.8 / 3	77	85.7 / 66	7.8 / 6	6.5 / 5
12–14	45	91.1 / 41	8.9 / 4	0 / 0	21	76.2 / 16	19.0 / 4	4.8 / 1	52	76.9 / 40	19.2 / 10	3.8 / 2
15–17	35	91.4 / 32	8.6 / 3	0 / 0	18	77.8 / 14	11.1 / 2	11.1 / 2	43	76.7 / 33	23.3 / 10	0 / 0
Females												
6–8	49	91.8 / 45	6.1 / 3	2.0 / 1	54	83.3 / 45	14.8 / 8	1.9 / 1	63	79.4 / 50	6.3 / 4	14.3 / 9
9–11	55	94.5 / 52	3.6 / 2	1.8 / 1	34	64.7 / 22	23.5 / 8	11.8 / 4	74	74.3 / 55	18.9 / 14	6.8 / 5
12–14	49	93.9 / 46	6.1 / 3	0 / 0	31	67.7 / 21	22.6 / 7	9.7 / 3	62	61.3 / 38	37.1 / 23	1.6 / 1
15–17	38	76.3 / 29	23.7 / 9	0 / 0	21	38.1 / 8	42.9 / 9	19.0 / 4	51	62.7 / 32	23.5 / 12	13.7 / 7

Source: Overweight and obesity classifications based on Cole *et al.* (2000).

Table 8.8. *Mean blood pressure[a] (mmHg) of adults in American Samoa by sex and age, 1976–2002*

Age group	N	1976–78 SBP (SD)	DBP (SD)	N	1990 SBP (SD)	DBP (SD)	N	2002 SBP (SD)	DBP (SD)
Males									
18–24	139	122.1	78.5	–	–	–	52	123.4	80.3
		11.4	8.9	–	–	–		10.5	10.5
25–34	158	127.5	84.1	101	124.8	83.3	69	126.6	85.3
		13.9	10.6		12.2	10.0		9.8	9.6
35–44	135	129.1	85.9	127	125.5	84.9	75	130.4	89.8
		17.1	11.0		16.7	13.4		14.7	11.7
45–54	176	132.9	87.0	92	131.4	87.1	43	130.9	88.0
		20.0	12.9		16.5	10.7		15.9	12.1
55–64	147	135.6	85.6	–	–	–	30	126.7	83.6
		23.9	14.9		–	–		16.7	12.1
65–74	64	137.7	84.0	–	–	–	9	136.8	86.3
		23.1	14.6		–	–		18.9	11.7
Females									
18–24	209	115.2	74.2	–	–	–	65	113.4	74.2
		11.5	9.4		–	–		9.8	9.1
25–34	237	119.5	77.8	148	115.6	75.4	90	118.5	80.1
		14.1	10.8		13.1	9.9		12.8	10.0
35–44	261	128.8	84.0	156	121.4	80.8	108	120.2	79.5
		18.6	12.1		14.0	9.9		12.1	10.1
45–54	249	139.2	88.1	92	133.1	86.8	55	129.9	84.3
		25.1	15.6		19.1	11.6		15.6	10.3
55–64	152	142.5	87.5	–	–	–	27	134.0	84.7
		26.0	14.9		–	–		22.7	12.5
65–74	64	148.9	88.6	–	–	–	18	137.9	85.6
		27.1	15.6		–	–		22.8	10.6

Notes:
SBP: systolic blood pressure; DBP: diastolic blood pressure.
[a] Excluding people on blood pressure medication.

Blood pressure and cardiovascular disease

Obesity is the most important risk factor for several NCDs among Samoans, including hypertension and type 2 diabetes (McGarvey and Schendel 1986; McGarvey 1992; Roberts *et al.* 2004). Thus, it is not surprising that age-specific mean blood pressure levels and hypertension (blood pressure $\geq 140/90$ mmHg) prevalence among adults are high (Tables 8.8 to 8.11). Across all age groups, American Samoan men had

Table 8.9. *Mean blood pressure^a (mmHg) of adults in Samoa by sex and age, 1979–2003*

Age group		1979–82			1991			2003	
	N	SBP (SD)	DBP (SD)	N	SBP (SD)	DBP (SD)	N	SBP (SD)	DBP (SD)
Males									
18–24	66	124.5	70.5	–	–	–	70	121.2	75.1
		12.8	9.2		–	–		12.2	8.5
25–34	37	122.4	77.2	142	125.9	78.9	80	122.4	79.5
		11.2	9.4		11.4	9.7		11.4	10.7
35–44	30	120.4	79.0	107	124.1	80.5	105	124.8	84.0
		11.9	9.0		11.1	11.1		13.7	10.6
45–54	34	124.1	79.2	84	125.3	82.2	81	129.5	88.1
		14.2	10.4		14.6	11.4		17.0	12.8
55–64	19	127.6	81.4	–	–	–	44	137.1	89.2
		22.2	11.9		–	–		26.6	13.1
65–74	10	133.6	80.0	–	–	–	32	128.2	81.2
		24.7	14.7		–	–		16.7	8.1
Females									
18–24	56	116.9	73.6	–	–	–	81	114.8	75.4
		10.5	7.5		–	–		10.7	8.0
25–34	43	111.9	72.8	152	116.1	73.2	110	116.2	78.2
		10.9	8.9		11.0	9.1		11.2	10.2
35–44	28	117.9	76.0	120	119.0	76.0	99	117.8	80.9
		12.6	9.6		12.1	9.7		12.7	9.3
45–54	36	124.9	79.7	105	126.9	80.1	78	124.4	82.8
		14.8	9.7		15.7	10.2		16.3	10.1
55–64	18	137.4	83.2	–	–	–	49	138.3	86.1
		32.3	18.1		–	–		23.1	13.7
65–74	10	130.3	80.0	–	–	–	42	133.4	81.6
		12.1	11.4		–	–		19.5	11.2

Notes:
SBP: systolic blood pressure; DBP: diastolic blood pressure.
^{a} Excluding people on blood pressure medication.

prevalence rates of hypertension of almost 40% in the period 1976 to 1978 and 46% in 2002; participants who reported medication taken for high blood pressure were also assumed to have hypertension. In 2002, more than half of all American Samoan men aged 35 years or more were hypertensive (Table 8.10). American Samoan women (all ages combined) had hypertension prevalence rates of about 35% in the 1970s and 31% in the year 2002. However, hypertension increased from 57% in the 1970s to 67% in 2002 in American Samoan women aged 55 to 74 years. In Samoa,

Table 8.10. *Prevalence of hypertension[a] in American Samoa, 1976–2002*

	1976–78		1990		2002	
Age group	%	n/N	%	n/N	%	n/N
Males						
18–24	20.1	28/139	–	–	17.3	9/52
25–34	34.8	55/158	25.7	26/101	38.4	28/73
35–44	42.2	57/135	30.7	39/127	53.2	42/79
45–54	47.2	83/176	35.9	33/92	50.0	26/52
55–64	47.6	70/147	–	–	61.7	29/47
65–74	43.8	28/64	–	–	72.7	16/22
All ages	39.2	321/819	30.6[b]	98/320	46.2	150/325
Females						
18–24	7.2	15/209	–	–	6.1	4/62
25–34	17.3	41/237	8.8	13/148	20.4	19/93
35–44	38.7	101/261	18.6	29/156	23.7	27/114
45–54	51.4	128/249	41.3	38/92	45.6	31/68
55–64	57.2	87/152	–	–	67.4	29/43
65–74	67.2	43/64	–	–	66.7	18/27
All ages	35.4	415/1172	20.2[b]	80/396	31.1	128/411

Notes:
[a] Hypertension: either systolic BP $\geq 140\,mmHg$, diastolic BP $\geq 90\,mmHg$, or current use of anti-hypertensive medication;
[b] Ages 25–54 years only.

hypertension prevalence was high but not as high as in American Samoa (Table 8.11). In Samoa in 1979 to 1982, 17% of men and 12% of women aged 18 to 74 years were hypertensive. By 2003, hypertension prevalence rates had increased to 30% in men and 29% in women, with the highest proportions being among those aged 55 to 74 years.

Blood pressure levels and hypertension in modernizing populations are not related to overweight and obesity only (McGarvey and Schendel 1986). Dietary intake (especially of sodium), psychosocial stress, and physical inactivity are also related to blood pressure levels and risk of hypertension. The influence of several of these factors on adult Samoan blood pressures has been examined in various surveys from the 1970s to the 1990s. These include studies of ecological modernization and migration (McGarvey and Baker 1979; McGarvey and Schendel 1986), obesity and salt intake (McGarvey and Schendel 1986; McGarvey 1992; Gershater and McGarvey 1995), and specific measures of psychosocial stress including lifestyle incongruity (Chin-Hong and McGarvey 1996) and coping with anger (Steele and McGarvey 1996).

Table 8.11. *Prevalence of hypertension^a in Samoa, 1979–2003*

Age group	1979–82		1991		2003	
	%	n/N	%	n/N	%	n/N
Males						
18–24	13.6	9/66	–	–	10.0	7/70
25–34	13.5	5/37	19.0	27/142	21.0	17/81
35–44	10.0	3/30	18.7	20/107	23.8	25/105
45–54	20.6	7/34	23.8	20/84	41.5	34/82
55–64	31.6	6/19	–	–	64.6	31/48
65–74	40.0	4/10	–	–	36.8	14/38
All ages	17.3	34/196	20.1^b	67/333	30.2	128/424
Females						
18–24	3.6	2/56	–	–	6.2	5/81
25–34	7.0	3/43	5.9	9/152	13.5	15/111
35–44	10.7	3/28	10.8	13/120	20.0	21/105
45–54	16.7	6/36	24.8	26/105	40.7	37/91
55–64	33.3	6/18	–	–	56.7	34/60
65–74	40.0	4/10	–	–	58.2	32/55
All ages	12.6	24/191	12.7^b	48/377	28.6	144/503

Notes:
^{a} Hypertension: either systolic BP ≥ 140 mmHg, diastolic BP ≥ 90 mmHg, or current use of anti-hypertensive medication;
^{b} Ages 25–54 years only.

In the years 2002 and 2003, awareness of hypertension was very low in adults below the age of 45 years in American Samoa, and in adults of all ages in Samoa. In American Samoa, more than 85% of young men and almost 80% of young women who were found to be hypertensive were unaware of it. Among American Samoan adults with a prior diagnosis of hypertension, between 62% and 85% of them, depending on sex and age group, used anti-hypertensive medication at the time of survey. In Samoa, 96% of men below the age of 45 years and 80% of men above 45 years of age who were found to have hypertension were unaware of it. Furthermore, in excess of 80% of young women and 58% of older women with hypertension were not aware of it. The use of anti-hypertensive medication in Samoa among those with a prior diagnosis of hypertension ranged from 33% in young men to 81% in older women. Hypertension awareness, diagnosis and adherence remain serious clinical and public health challenges in the contemporary Samoas.

Blood pressure of Samoan children has been less studied. However, data collected in the 1970s showed blood pressure to be related to age, body size and levels of body fatness (McGarvey and Schendel 1986). In a comparison of childhood blood pressure of 4 to 19-year-olds in 1979 and 1991 to 1993, age-adjusted systolic blood pressure for 10 to 19-year-olds was significantly higher in the period 1991 to 1993 than in 1979, the difference becoming insignificant after adjustment for BMI (Roberts *et al.* 2004). The proportion of those aged 10 to 19 years with elevated blood pressure ranged from 11% to 15% in the 1979 study sample, to approximately 25% in the period 1991 to 1993.

Type 2 diabetes

Along with increasing levels of body fatness, there has been an increase in the prevalence of type 2 diabetes in Samoans (Zimmet *et al.* 1981; Collins *et al.* 1994, 1995; McGarvey 2001). We report here on type 2 diabetes prevalence rates as determined in the 1990 and 2002 surveys in American Samoa and in the 1991 and 2003 surveys in Samoa (Table 8.12). The prevalence rate of diabetes among men aged 25 to 54 years in American Samoa in 1990 was 12.9%, increasing to 17.2% by the year 2002. Among American Samoan women of the same age range, the prevalence rate doubled from 8.1% in 1990 to 16.7% in 2002. Type 2 diabetes prevalence rates among adults aged 18 to 74 years in 2002 were 21.6% and 18.0% in men and women, respectively. In all cases, diabetes prevalence in Samoa was far lower than in American Samoa, but between 1991 and 2003, there were striking increases in prevalence rates in all sex and age groups. For example, 2.4% of men aged 25 to 54 years were diabetic in 1991; this increased to 6.0% in 2003. In women of the same age range, prevalence rose from 3.0% to 8.2% across the same period. In 2003, when a wider age range was sampled, diabetes prevalence rates in Samoa were 9.3% and 12.6% in men and women, respectively.

Four-year incidence rates of diabetes were calculated for the period 1990 to 1994 in American Samoa, and for the period 1991 to 1995 in Samoa (McGarvey 2001); perhaps predictably, incidence rates were much higher in American Samoa than in Samoa. In Samoa, 4% of the older (≥45 years) men and women developed type 2 diabetes over this four-year period. However, in American Samoa, 9% of younger (<45 years) men and women, 16% of the older men and 8% of older women developed diabetes across a similar four-year period. These increasing levels of type 2 diabetes

Table 8.12. *Prevalence of type 2 diabetes[a] including participants taking medications, 1990–2003*

| | American Samoa | | | | Samoa | | | |
| | 1990 | | 2002 | | 1991 | | 2003 | |
Age group	%	n/N	%	n/N	%	n/N	%	n/N
Males								
18–24	–	–	5.8	3/52	–	–	0.0	0/68
25–34	4.1	3/74	7.0	5/71	1.4	2/141	0.0	0/75
35–44	9.9	10/101	21.1	16/76	1.9	2/107	2.1	2/96
45–54	27.3	18/66	25.5	13/51	4.8	4/84	16.3	13/80
55–64	–	–	44.7	21/47	–	–	26.1	12/46
65–74	–	–	50.0	11/22	–	–	29.4	10/34
All ages	12.9[b]	31/241	21.6	69/319	2.4[b]	8/332	9.3	37/399
Females								
18–24	–	–	3.0	2/66	–	–	1.3	1/78
25–34	6.0	7/117	5.6	5/90	7.6	1/146	5.8	6/104
35–44	9.3	12/129	17.7	20/113	2.1	2/114	3.0	3/99
45–54	9.2	7/76	29.9	20/67	11.8	8/102	17.0	15/88
55–64	–	–	41.9	18/43	–	–	31.7	19/60
65–74	–	–	29.6	8/27	–	–	31.5	17/54
All ages	8.1[b]	26/322	18.0	73/406	3.0[b]	11/362	12.6	61/483

Notes:
[a] Type 2 diabetes: either fasting serum glucose ≥126 mg/dl, or current use of diabetes medications;
[b] Ages 25–54 years only.

across both Samoan polities are undeniably tied to increasing rates of obesity on the islands. Without intervention, the prevalence of diabetes will likely continue to increase long before it levels off (Collins *et al.* 1995).

Despite the rapidly increasing prevalence of diabetes among Samoans over the last two decades, there are many individuals unaware of their diabetes status. In American Samoa, 40% of younger men and women who were diagnosed with type 2 diabetes in 2002 did not have a prior diagnosis; 17% to 18% of older men and women with type 2 diabetes were previously undiagnosed. More than 80% of all American Samoan men and older women with prior diagnosis of type 2 diabetes used diabetes medication. The same was true for 65% of younger women there. In Samoa, about half of the older men and women who were identified as being diabetic in the surveys had not been diagnosed previously. In Samoa, 85% of older adults diagnosed with diabetes took medication for their condition. Thus, while few people are diagnosed as diabetic in the Samoas, those who are diagnosed receive and adhere to their treatment plans.

Children are also likely to be susceptible to type 2 diabetes. It is difficult to undertake primary prevention of type 2 diabetes in childhood, as it involves familial changes in diet and physical activity and alterations in societal environments that structure children's nutritional behaviours. Considerable applied and translational research (research which focuses on intervention) is needed to understand the best ways to foster healthy life habits to avoid obesity and the diseases associated with it, in both children and adults.

Dietary intake patterns

Both archaeological and cultural evidence suggests that the traditional Samoan diet consisted of starchy vegetables such as yam, taro, banana, and breadfruit, along with resources from the reef, coconut, and on rare occasions, domesticated pig (Hanna *et al.* 1986; WHO 2003). This traditional diet may have been an optimal one, since Samoans at the time of early contact did not display any signs of nutritional disorders. Furthermore, the initial introduction of a more modernized diet did not cause changes in stature at the population level, as it often does when introduced to communities with diets that are nutritionally insufficient in some way (Hanna *et al.* 1986).

Since the early twentieth century, modernization has changed the Samoan diet (Hanna *et al.* 1986). Samoans from both nations who have more apparent income show an increased reliance on externally produced, processed and purchased foods (Galanis *et al.* 1999). These foods are often higher in protein, cholesterol and sodium than native foods, and their consumption has been associated with increased risk of cardiovascular disease, type 2 diabetes and various types of cancer (WHO 1990). Contemporary Samoan diets are high in energy, total fat and saturated fat, while being low or lacking in dietary fibre (Galanis *et al.* 1999).

Since the early 1900s, there has been a large influx of imported foods to the Samoan islands. Between 1961 and 2000, there was in excess of a fourfold increase in food imports to Samoa (Food and Agriculture Organization 2002). It has become customary to give cabin crackers, *pisupo* (canned corn beef) and canned tuna as gifts at traditional *fa'alavelave* celebrations, along with the food types which were brought on the boats of the first Polynesians, such as taro, yam and banana. Walking into many stores on the island of

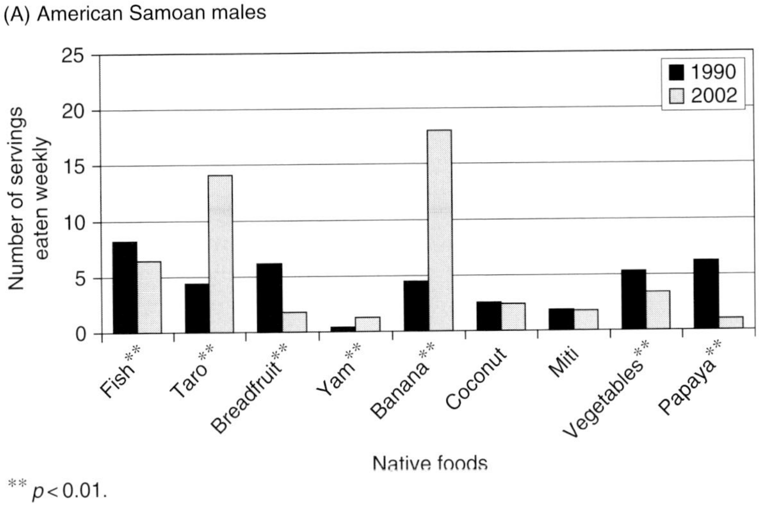
(A) American Samoan males
1990
2002
Number of servings eaten weekly
25
20
15
10
5
0
Fish**
Taro**
Breadfruit**
Yam**
Banana**
Coconut
Miti
Vegetables**
Papaya**
Native foods
** p<0.01.

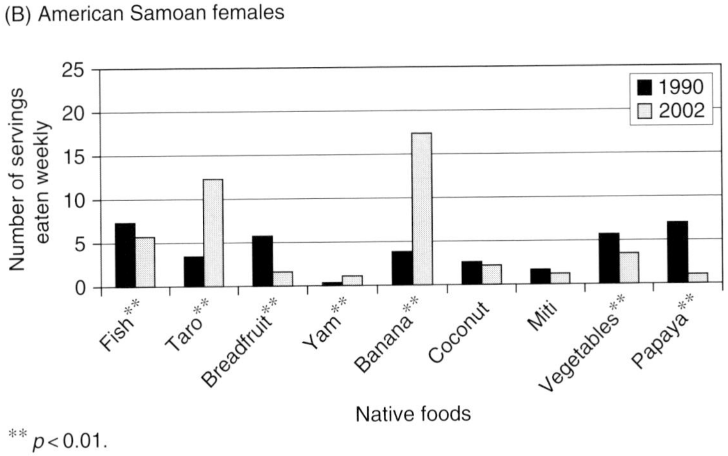
(B) American Samoan females
1990
2002
Number of servings eaten weekly
25
20
15
10
5
0
Fish**
Taro**
Breadfruit**
Yam**
Banana**
Coconut
Miti
Vegetables**
Papaya**
Native foods
** p<0.01.

Fig. 8.2. Intake of native foods.

Tutuila, American Samoa, the shelves are lined with canned beef, pasta, prepackaged noodles, candy, soda, chips, frozen turkey tails, with a few costly but low quality vegetables at the very back of the store.

Figures 8.2 and 8.3 give comparisons of self-reported dietary intakes of adults aged 25 to 74 years during 1990 to 1991 and 2002 to 2003, respectively. Weekly numbers of servings were calculated by multiplying the reported number of times a food was eaten per week by the portion size eaten. Reported serving sizes, however, might be underestimates of actual portions

(C) Samoan males living in Samoa

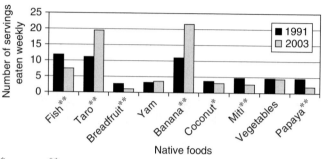

$^* p < 0.05;$ $^{**} p < 0.01.$

(D) Samoan females living in Samoa

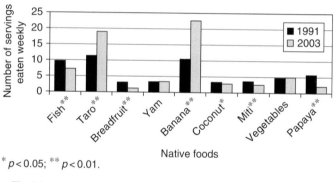

$^* p < 0.05;$ $^{**} p < 0.01.$

Fig. 8.2. (cont.)

consumed at meals. Across both sexes and nations, there were significant changes in the consumption of native foods including fish, starchy fruits and vegetables, as well as papaya (Fig. 8.2A–D). Among men and women in American Samoa, there were significant changes in the consumption of green vegetables, while in Samoa there were changes in the consumption of coconuts and *miti* (coconut cream). Consumption of fish, which has been an important source of protein since the earliest establishment of Polynesian culture, rapidly decreased between 1990 to 1991 and 2002 to 2003 in both nations. It is likely that fish is being replaced by other sources of protein.

There has been a significant decrease over time in the consumption of papaya across both nations and sexes, and a decrease in consumption of green vegetables in American Samoa (Fig. 8.2). Among American Samoan women, papaya was eaten about once a day in 1990, compared to once a week in 2002. In American Samoa in 2002, both men and women ate only

(A) American Samoan males

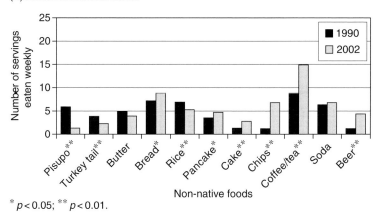

* $p < 0.05$; ** $p < 0.01$.

(B) American Samoan females

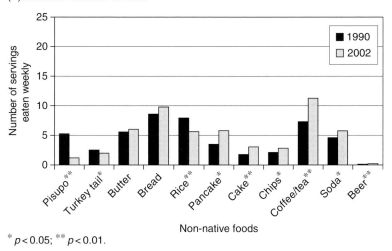

* $p < 0.05$; ** $p < 0.01$.

Fig. 8.3. Intake of non-native foods.

about 3.5 servings of green vegetables per week. These decreases have important dietary consequences, papaya and green vegetables being good sources of vitamins, antioxidants and dietary fibre. Furthermore, they have high water content, and provide satiety while contributing few calories to the diet (Guthrie 1975; Rolls *et al.* 2004). Diets composed primarily of locally grown foods have been shown to prevent and reduce obesity at population level (WHO 2003).

In both nations, there were also significant changes in the consumption of native starchy fruits and vegetables (Fig. 8.2). The apparent decrease in

(C) Samoan males living in Samoa

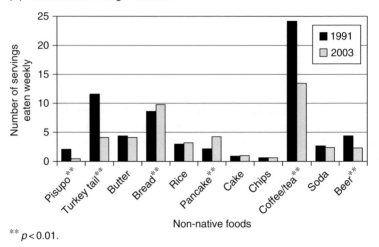

** $p < 0.01$.

(D) Samoan females living in Samoa

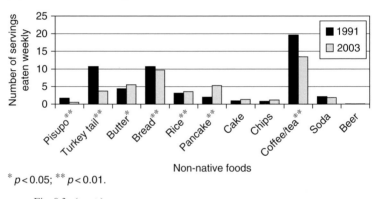

* $p < 0.05$; ** $p < 0.01$.

Fig. 8.3. (cont.)

breadfruit consumption in both nations is most likely explained by seasonal variations in breadfruit availability, since fieldwork in the 1990s and 2000s took place during different seasons. Other researchers have found seasonal variation in consumption of breadfruit in the Samoas (Gage 1986). There was a large increase in the consumption of taro and starchy banana between the 1990s and 2000s, the quantity of starchy vegetables being consumed in the latter period being impressive. American Samoan men and women in the 2000s ate roughly 4.5 starchy fruits (breadfruit) and vegetables (yam, taro and banana) per day, while their Samoan counterparts consumed about 6.5 of these starchy vegetables daily. Often these starchy staples are eaten with coconut cream, which contributes a substantial amount of fat and energy to the diet.

Patterns of consumption of non-native foods have also changed (Fig. 8.3), some of them positive in relation to health. In both nations, there has been a decrease in the consumption of *pisupo* (canned corn beef) and turkey tails, both of which are high in fat and sodium. More generally, however, there has been a marked increase in the consumption of potato chips, pancakes, coffee and tea (Fig. 8.3A–D). American Samoan adults have increased their consumption of pancake, chips, cake, coffee and tea, and beer. Women increased their consumption of soda, while men increased their bread consumption. In addition to the large amounts of traditional starchy vegetables eaten, men and women in American Samoa also eat between 2.5 and 3 servings of non-native carbohydrates (pancake, rice and bread) per day. Most men and women eat cake or chips with soda most days of the week. All of these changes point to the increase in tastes for imported energy-dense, nutrient-poor foods. In Samoa, cake, chips, soda and beer are consumed at much lower frequencies than in American Samoa. This probably reflects differences in levels of modernization, access to processed foods, costs of foods compared to income levels and dietary preferences. Samoans are more involved in farming than their American Samoan counterparts in Samoa, have greater access to traditional foods and may not need to supplement their diet with store-bought foods; alternatively, they may not have much income with which to buy it. Despite this, there has been an increase over time in the consumption of pancake in both men and women, rice among men, and butter and bread among women, suggesting that the diet of Samoans living in Samoa is also shifting to incorporate more non-native foods. In an earlier report on the impact of food aid after the December 1991 cyclone in Samoa, an increase in rice and pancake consumption was shown to be connected to the provision of rice and flour (Galanis *et al.* 1995). Thus, the secular changes in the types of carbohydrates used by Samoans may have been caused by specific events, as well as general processes of modernization.

As the Samoan islands have become more modernized, dietary habits have changed across generations (Figs. 8.4, 8.5 and 8.6). In the years 2002 and 2003, Samoan children and adolescents ate more non-native, energy-dense foods and fewer papaya and green vegetables than older adults. Across both nations and sexes, Samoan youth aged 6 to 17 years ate more cake and chips than adults aged 45 years and above. In Samoa, children and adolescents ate cake or chips more than three times per week. In American Samoa, the same age group ate more than one serving of cake or chips daily. Children and adolescents, with the

(A) American Samoan males

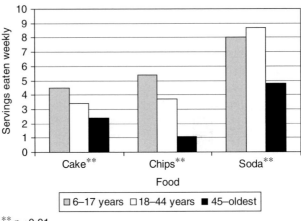

** $p < 0.01$.

(B) American Samoan females

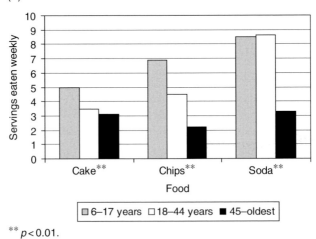

** $p < 0.01$.

Fig. 8.4. Intake of cake, potato chips and soda by age.

exception of males from Samoa, drank significantly more soda than adults aged 45 years and above. In American Samoa, boys drank eight servings of soda per week, while girls drank 8.5 servings. On average, young American Samoans consumed over 1,200 calories per week from soda alone. American Samoan boys aged 15 to 17 years consumed an average of 12.4 servings of soda weekly, this contributing almost 1,900 calories to their weekly diet. Likewise, American Samoan girls aged 15 to

(C) Samoan males living in Samoa

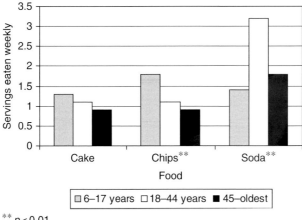

** $p < 0.01$.

(D) Samoan females living in Samoa

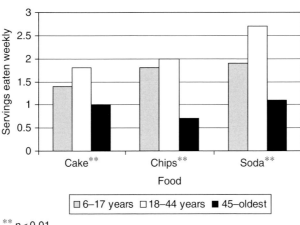

** $p < 0.01$.

Fig. 8.4. (cont.)

17 years consumed 13.5 servings weekly, more than 2,000 calories worth of soda.

While the consumption of non-native, energy-dense and nutrient-poor foods has increased among children and adolescents relative to their grandparents, consumption of papaya and green vegetables has decreased. All children and adolescents ate fewer papaya and green vegetables than adults aged 45 years and above. In Samoa, children in the youngest age group ate fewer than two servings of papaya weekly, while in American

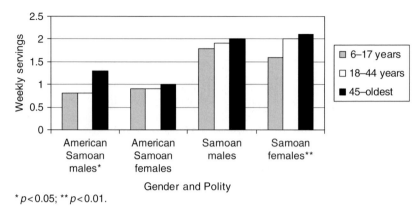

*p<0.05; **p<0.01.

Fig. 8.5. Intake of papaya by location, sex and age, 2002 and 2003.

Samoa they ate less than one serving weekly. Children and adolescents from both nations ate significantly fewer green vegetables than adults aged 45 years and above. Children and adolescents from Samoa ate fewer than three servings of green vegetables weekly, while their American Samoan counterparts ate less than two servings weekly. Samoan youth from both nations ate fewer servings of papaya and green vegetables than cake, chips, and soda combined. American Samoan boys ate on average only 2.6 servings of papaya and green vegetables per week while consuming 17.9 servings of cake, chips, and soda combined. American Samoan girls also ate less than three servings of papaya and green vegetables weekly, while eating more than 20 servings of cake, chips and soda. These large increases in the consumption of non-native, energy-dense foods and simultaneous decreases in the consumption of fruits and green vegetables are likely to have important and substantial consequences for the health of Samoan children and adolescents. A dietary shift towards more processed, energy-dense foods is usually linked to a marked increase in caloric intake along with a decrease in intakes of most other nutrients.

The dietary patterns that children have adopted are likely to have contributed to an increase in obesity levels. Studies in the Pacific suggest that consumption of fat from imported foods increases the risk of obesity and type 2 diabetes by 2.2- and 2.4-fold respectively (WHO 2003). Health risk is enhanced because consumption of imported foods has often been in addition to, rather than instead of, traditional diets. Some Polynesian groups have continued to consume large quantities of root crops (over 3 kg per day) in addition to consuming imported foods (WHO 2003). This, in association with decreased energy expenditure from physical activity, has created chronic positive energy balance, and has contributed to the levels of obesity among Samoans now.

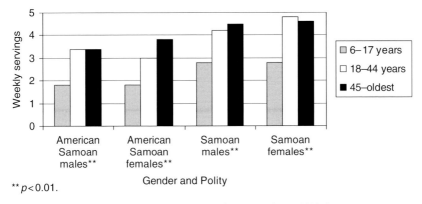

Fig. 8.6. Intake of vegetables by location, sex and age, 2002–3.

Physical activity and health

There have been dramatic shifts in lifestyle, physical activity patterns and energy expenditure among Samoans over the past 50 years. Studies of Samoans living in Hawaii, Samoa and American Samoa as well as other populations have shown that individuals with more modern lifestyles have decreased physical fitness (Greksa and Baker 1982; Lukaski 1977; Shephard 1978; Epstein *et al.* 1981). In the Samoan archipelago, professional sedentary jobs have increased in number as society has modernized. These require much lower levels of energy expenditure than traditional subsistence activities. For example, a 65 kg Samoan man on average expends around 6 to 7 kcal/min while participating in traditional subsistence agriculture, whereas office employees only expend between 1.7 and 1.9 kcal/min (Greksa and Baker 1982). Studies of Samoans living in Samoa in 1982 showed that aerobic fitness was dependent upon area of residence and occupation, in addition to leisure activities and levels of body fatness (Pelletier 1988).

Levels of physical activity from farming and sports participation varied greatly by sex and age group in the years 2002 and 2003 (Table 8.13). In American Samoa, only men attained the recommended minimum amount of physical activity of 2.5 hours of moderate activity per week (Centers for Disease Control 2005), women attaining only very low levels of physical activity. Younger men from Samoa participated in more hours of physical activity than any other sex and age group, largely because more of them were involved in subsistence farming. There was a large decline in physical activity with age among Samoans living in Samoa. Although both younger and older men performed more than 2.5 hours of moderate physical

Table 8.13. *Average number of hours per week spent participating in farm work or sports in American Samoa, 2002, and in Samoa, 2003*

| | | American Samoa | | Samoa | |
		Males	Females	Males	Females
18–44	N	203	270	257	297
	X (SD)	4.8 (6.6)	2.2 (3.8)	11.3 (17.6)	3.3 (7.1)
45–oldest	N	130	151	181	223
	X (SD)	2.7 (4.8)	1.0 (2.3)	5.1 (8.9)	1.4 (3.8)

activity per week, women aged 45 years and above participated in less than 1.5 hours of moderate physical activity per week. Since the lifestyles of women tend to rapidly become more sedentary with modernization than males, the concurrently higher average BMI levels among women than men may be due to primarily greater declines in their physical activity, rather than greater food intakes.

In Samoa in 2003, 75.1% of men and 30.4% of women aged 18 years and over reported participating in farm work at least weekly. By contrast, far fewer adults in American Samoa participated in farm work, less than half of the men and less than one quarter of women having done so in 2002. Of all measured forms of physical activity, farm work had the strongest, most consistent association with lower levels of BMI. American Samoan men of all ages and American Samoan women aged 45 years and above who reported participating in farm work had significantly lower BMIs (Table 8.14). Similarly in Samoa, all men and women aged 18 to 44 years who participated in farm work had significantly lower BMIs (Keighley *et al.* 2006).

Participation in farm work has at least two likely impacts on lifestyle. The first is that it entails a consistent amount of physical activity. Women living in American Samoa and Samoa who participate in farm work spend on average 3.6 (±4.5) hours per week on the farm, while American Samoan men spend 4.2 (±4.6) hours and Samoan men spend 7.4 (±9.1) hours in the same work, respectively. All of these averages exceed the recommended minimum of 2.5 hours of moderate activity per week (Centers for Disease Control 2005). The second impact of farm work is that people working on a farm have greater ready access to fruits and vegetables and therefore may have diets with more of these foods in them. Regardless of the direct mechanisms by which farm work may influence BMI, there is a strong association between the two, across both sexes and

Table 8.14. *Association between farm work and BMI in adults aged 18–74 years*

	Age		American Samoa					Samoa				
			X	se	N	F	p	X	se	N	F	p
Males	18–44	No farm work	34.7	0.72	103			29.5	0.70	41		
		Farm work	32.0	0.72	101	6.886	0.009	27.6	0.31	214	6.701	0.010
	45–oldest	No farm work	35.9	0.93	57			32.5	1.10	26		
		Farm work	32.9	0.89	62	5.354	0.022	31.8	0.47	143	0.293	0.589
Females	18–44	No farm work	36.9	0.56	213			32.7	0.42	206		
		Farm work	35.9	1.07	58	0.691	0.407	30.4	0.64	91	9.178	0.003
	45–oldest	No farm work	37.9	0.75	104			34.5	0.66	138		
		Farm work	34.6	1.32	34	4.412	0.038	34.4	0.97	66	0.010	0.921

Notes:
X: mean; se: standard error; F: F ratio.

both Samoas, despite varying social roles and the degree of industrialization in each of them. Although participation in farm work may also be indicative of differing socioeconomic status, this suggests that in the face of modernization, maintenance of at least some traditional lifestyle activities may help to maintain lower BMI and improve the overall health of the people.

Children's participation in farm work, sport and school gym class varied across the two Samoas. In Samoa in 2003, 29.5% of boys and 18.5% of girls participated in farm work (Table 8.15), about twice as many as in American Samoa in 2002. In both polities, girls on average spent less than half an hour per week doing farm work, while boys in Samoa and American Samoa worked for three quarters of an hour and over an hour, respectively. Across both polities and sexes, children and adolescents on average participated in over 2.5 hours of sport per week, in addition to

Table 8.15. *Number of times per week children aged 6–17 years participated in farm work, sports and gym class, by children in American Samoa, 2002, and Samoa, 2003*

		Farm work (hours)	Sports (hours)	Gym class (days)
American Samoa				
Males	N	283	280	235
	X (SD)	0.7 (2.6)	3.9 (5.2)	2.2 (1.8)
Females	N	261	260	205
	X (SD)	0.2 (0.7)	4.2 (12.0)	2.3 (1.8)
Samoa				
Males	N	233	233	223
	X (SD)	1.2 (3.3)	4.8 (9.6)	1.7 (1.9)
Females	N	250	250	240
	X (SD)	0.4 (1.1)	2.8 (4.8)	1.5 (1.8)

participating in gym class. American Samoan girls reported spending more time playing sport than American Samoan boys, while in Samoa, this trend was reversed, with girls spending about half as much time as boys playing sport. In both locales, children reported participating in gym class less than three days per week.

While there are no significant relationships between BMI or percent BF and participation in gym class, there are direct correlations between measures of adiposity in children and participation in sports and farm work. Percent BF was calculated using bioelectrical impedance measurements and equations designed for Polynesian populations (Swinburn *et al.* 1999; Rush *et al.* 2003). Participation in sports is significantly associated with a lower percent BF among girls aged 6 to 17 years living in Samoa in 2003 ($p = 0.006$). In 2002, American Samoan boys aged 6 to 17 years who participated in farm work had a significantly lower percent BF ($p = 0.01$) than those who did not. Among girls aged six to eight years living in Samoa, farm work was associated with an increase in BMI and BF. This may be due to confounding effects; girls doing farm work may do less physically active jobs than boys, so farm work might not have as large an impact on energy balance in girls, especially in younger age groups. Overall, there is no significant decrease in BMI with physical activity. However, there are some problems with the use of BMI as a measure of adiposity, especially in children (Tennefors and Forsum 2004). Furthermore, many children participating in physical activity may be considerably stronger than children participating in agriculture, and therefore have a higher BMI due to muscle mass regardless of BF content.

In addition to directly affecting physical activity levels, participation in farm work influences diet. Among American Samoan girls and Samoan boys, participation in farm work was associated with eating 1.3 and 1.5 servings of salad, green vegetables and fruits in the previous day, while their peers who did not participate in farm work ate on average 0.9 and 1.3 servings of these foods, respectively. As among farm-working adults, children who grow up having worked on a farm have greater access to locally grown fruits and vegetables, and may consume them more frequently. Thus participation in this traditional activity may help children to maintain healthier body weights and eat diets richer in traditional foods.

Conclusion

Samoans, along with other indigenous groups across the world, have suffered from sharp increases in obesity and chronic disease as their lifestyles have become more modernized. Between 1976 and 2003, the prevalence of obesity more than doubled among adult men living in American Samoa and more than quadrupled among men living in Samoa. Among women in Samoa, the prevalence of obesity more than doubled, and among American Samoan women, the already-elevated rates of obesity increased to 71% by 2002. Levels of overweight and obesity among adults in American Samoa in the 1970s match the levels in Samoa some 30 years later, and levels of overweight and obesity in children and adolescents from Hawaii in the 1970s presaged overweight in youth living in American Samoa in 2002. If current trends continue, we speculate that they may also be indicative of future levels of overweight and obesity among youth living in Samoa.

Obesity rates among Samoans are likely to continue their increase before levelling off, and may shift to younger age groups, given the rapidly increasing prevalence of obesity among children, especially in American Samoa. Associated with such a trend would be an earlier onset of chronic disease, and decreases in quality of life and life expectancy. However, although the rise in obesity among Samoans is consistent with rapid and ubiquitous exposure to obesogenic nutritional environments at all levels of Samoan society, not all Samoans are fated to become obese or overweight. David Tua, mentioned at the opening of this chapter, is just one example of the many Samoans who have become world-class athletes and who have stayed fit and healthy in a modernized setting. The data presented here show that participation in physical activities, such as farm work, along with consumption of a healthy diet can help Samoans maintain lower body weights despite general modernization of society. In the face of varying degrees of modernization, the maintenance of some traditional activities may have protective health effects.

The data presented in this chapter demonstrate an immediate need for health intervention in the Samoan islands. These should be aimed particularly at children, adolescents and young women, where the prevalence of obesity is rising most rapidly. They should promote the maintenance of traditional activities that already help many people stay healthy, as well as the consumption of locally grown fruits and vegetables that are both accessible and affordable. In addition, we have shown that while those who are diagnosed with NCDs such as hypertension and type 2 diabetes often comply with clinicians and take medications, the majority of cases are not diagnosed at all.

Acknowledgements

This work would not have been possible without the generosity of the Samoan people. Their eagerness to participate in research has allowed us to gain important insights into the trends in health occurring among them. We hope that these results will be useful in helping to develop effective means of delaying and preventing chronic disease among Samoans and others around the world.

References

Ahlburg, D. A. (1995). Migration, remittances, and the distribution of income: evidence from the Pacific. *Asian Pacific Migration Journal* **4**, 157–67.

Baker, P. and Crews, D. (1986). Mortality patterns and some biological predictors. In *The Changing Samoans: Behavior and Health in Transition*, ed. P. T. Baker, J. M. Hanna and T. S. Baker. New York: Oxford University Press, pp. 275–96.

Baker, P. T., Hanna, J. M. and Baker, T. S. (1986). *The Changing Samoans: Behavior and Health in Transition*. New York: Oxford University Press.

Beaglehole, J. (1961). *The Journals of Captain Cook on His Voyages of Discovery*. Cambridge: Cambridge University Press.

Bindon, J. R. and Baker, P. T. (1985). Modernization, migration and obesity among Samoan adults. *Annals of Human Biology* **12**, 67–76.

Bindon, J. R. and Zansky, S. M. (1986). Growth patterns of height and weight among three groups of Samoan preadolescents. *Annals of Human Biology* **13**, 171–8.

Brewis, A. A. and McGarvey, S. T. (2000). Body image, body size, and Samoan ecological and individual modernization. *Ecology of Food and Nutrition* **39**, 105–20.

Brewis, A. A., McGarvey, S. T., Jones, J. and Swinburn, B. A. (1998). Perceptions of body size in Pacific Islanders. *International Journal of Obesity and Related Metabolic Disorders* **22**, 185–9.

Centers for Disease Control (2005). *Physical activity for everyone: Recommendations: How active do adults need to be to gain some benefit.* online: http://www.cdc.gov/nccdphp/dnpa/physical/recommendations/adults.htm

Chin-Hong, P. V. and McGarvey, S. T. (1996). Lifestyle incongruity and adult blood pressure in Western Samoa. *Psychosomatic Medicine* **58**, 131–7.

Cole, T. J., Bellizzi, M. C., Flegal, K. M. and Dietz, W. H. (2000). Establishing a standard definition for childhood overweight and obesity worldwide: international survey. *British Medical Journal* **320**, 1240–3.

Collins, V. R., Dowse, G. K., Toelupe, P. M., *et al.* (1994). Increasing prevalence of NIDDM in the Pacific island population of Western Samoa over a 13-year period. *Diabetes Care* **17**, 288–96.

Collins, V. R., Dowse, G. K., Plehwe, W. E., *et al.* (1995). High prevalence of diabetic retinopathy and nephropathy in Polynesians of Western Samoa. *Diabetes Care* **18**, 1140–9.

Coustan, D. R., Carpenter, M. W., O'Sullivan, P. S. and Carr, S. R. (1993). Gestational diabetes: predictors of subsequent disordered glucose metabolism. *American Journal of Obstetrics and Gynecology* **168**, 1139–44.

Damcott, C. M., Sack, P. and Shuldiner, A. R. (2003). The genetics of obesity. *Endocrinology and Metabolism Clinics of North America* **32**, 761–86.

Department of Health (1986). *Annual Report of the Department of Health 1986.* Apia, Samoa: Health Planning, Information and Research Division.

(1993–94). *Annual Report of the Department of Health 1993–94.* Apia, Samoa: Health Planning, Information and Research Division.

Epstein, Y., Keren, G., Udassin, R. and Shapiro, Y. (1981). Way of life as a determinant of physical fitness. *European Journal of Applied Physiology* **47**, 1–5.

Filiga, V. and Levin, M. J. (1988). *Population Profile of American Samoa (1980 Census).* Pago Pago, American Samoa: Economic Development and Planning Office, American Samoa Government.

Food and Agriculture Organization (2002). *FAOSTAT Database*, http://apps.fao.org/default.htm (accessed June–July 2002).

Freedman, D. S., Dietz, W. H., Srinivasan, S. R. and Berengson, G. S. (1999). The relation of overweight to cardiovascular risk factors among children and adolescents: the Bogalusa heart study. *Pediatrics* **103**, 1175–82.

Gage, T. B. (1986). Environment and exploitation. In *The Changing Samoans: Behavior and Health in Transition*, ed. P. T. Baker, J. M. Hanna and T. S. Baker. New York: Oxford University Press, pp. 19–38.

Galanis, D. J., Chin-Hong, P. V., McGarvey, S. T., Messer, E. and Parkinson, D. (1995). Dietary intake changes associated with post-cyclone food aid in Western Samoa. *Ecology of Food and Nutrition* **34**, 137–47.

Galanis, D. J., McGarvey, S. T., Quested, C., Sio, B. and Afele-Fa'amuli, S. A. (1999). Dietary Intake of modernizing Samoans: implications for risk of cardiovascular disease. *Journal of the American Diabetic Association* **99**, 184–90.

Gershater, E. and McGarvey, S. T. (1995). Fourteen year changes in adiposity and blood pressure in Samoa adults. *American Journal of Human Biology* **7**, 597–606.

Gilson, R. P. (1970). *Samoa 1830 to 1900: The Politics of a Multi-cultural Community*. New York: Oxford University Press.

Government of Samoa (2001). *Census of Population and Housing 2001*. Samoa: Department of Statistics.

Green, R. C., Davidson, J. M. and Bernice Pauahi Bishop Museum (1969). *Archeology in Western Samoa*. Auckland, New Zealand: Auckland Institute and Museum.

Greksa, L. and Baker, P. (1982). Aerobic capacity of modernizing Samoan men. *Human Biology* **54**, 777–88.

Guo, S. S., Roche, A. F., Chumlea, W. C., Gardner, J. D. and Siervogel, R. M. (1994). The predictive value of childhood body mass index values for over-weight at age 35 y. *American Journal of Clinical Nutrition* **59**, 810–19.

Guthrie, H. A. (1975). *Introductory Nutrition*. Saint Louis: The CV Mosby Co.

Hanna, J. M., Pelletier, D. L. and Brown, V. J. (1986). The diet and nutrition of contemporary Samoans. In *The Changing Samoans: Behavior and Health in Transition*, ed. P. T. Baker, J. M. Hanna and T. S. Baker. New York: Oxford University Press, pp. 275–96.

Holmes, L. (1992). *Samoan Village: Now and Then*. Fort Worth, Texas: Harcourt Brace College.

Horgen, K. B., and Brownell, K. D. (2002). Confronting the toxic environment: environmental, public health actions in a world crisis. In *Obesity: Theory and Therapy* (2nd edn), ed. T. A. Wadden and A. J. Stunkard. New York: Guilford, pp. 95–106.

Jones, D. C. (2004). Body image among adolescent girls and boys: a longitudinal study. *Developmental Psychology* **40**, 823–35.

Keighley, E. D., McGarvey, S. T., Turituri, P. and Viali, S. (2006). Farming and adiposity in Samoan adults. *American Journal of Human Biology* **18**, 112–21.

Kirch, P. V. (2000). *On the Road of the Winds: An Archaeological History of the Pacific Islands before European Contact*. Berkeley: University of California Press.

Lukaski, H. C. (1977). Some observations on the coronary risk of male Samoan migrants on Oahu, Hawaii. M.A. thesis, The Pennsylvania State University, University Park, PA.

McArthur, N. A. (1956). *The Populations of the Pacific*. Part III: *American Samoa* and Part IV: *Western Samoa and the Tokelau Islands*. Canberra: Australian National University, Department of Demography.

   (1967). *Island Populations of the Pacific*. Canberra: Australian National University Press.

McGarvey, S. T. (1991). Obesity in Samoans and a perspective on its etiology in Polynesians. *American Journal of Clinical Nutrition* **53**, 1586S–94S.

(1992). Biocultural predictors of age increases in adult blood pressure among Samoans. *American Journal of Human Biology* **4**, 27–36.

(1994). The thrifty gene concept and adiposity studies in biological anthropology. *Journal of the Polynesian Society* **103**, 29–42.

(2001). Cardiovascular disease (CVD) risk factors in Samoa and American Samoa, 1990–95. *Pacific Health Dialog* **8**, 157–62.

McGarvey, S. T. and Baker, P. (1979). The effects of modernization on Samoan blood pressure. *Human Biology* **51**, 461–79.

McGarvey, S. T. and Schendel, D. E. (1986). Blood pressure in Samoans. In *The Changing Samoans: Behavior and Health in Transition*, ed. P. T. Baker, J. M. Hanna and T. S. Baker. New York: Oxford University Press, pp. 351–93.

McGarvey, S. T., Bindon, J., Crews, D. and Schendel, D. (1989). Modernization and adiposity: causes and consequences. In *Human Population Biology: A Trans-disciplinary Science*, ed. M. A. Little and J. D. Haas. New York: Academic Press, pp. 263–79.

McGarvey, S. T., Levinson, P. D., Bausserman, L, Galanis, D. L. and Hownick, C. A. (1993). Population change in adult-obesity and blood lipids in American Samoa from 1976–1978 to 1990. *American Journal of Human Biology* **5**, 17–30.

McGarvey, S. T., Bausserman, L., Viali, S. and Tufa, J. (2005). Prevalence of the metabolic syndrome in Samoans. *American Journal of Physical Anthropology* **40**(Supplement), 14–15.

McKinley, N. M. (1999). Women and objectified body consciousness: mothers' and daughters' body experience in cultural, developmental, and familial context. *Developmental Psychology* **35**, 760–9.

Meleisea (1987). *Lagaga: A Short History of Western Samoa*. University of the South Pacific.

Neumark-Sztainer, D. (1999). Factors influencing food choices of adolescents: findings from focus-group discussions with adolescents. *Journal of the American Dietetic Association* **99**, 929–37.

O'Meara, T. (1990). *Samoa Planters: Tradition and Economic Development in the Polynesia*. Fort Worth, Texas: Holt, Rinehart and Winston.

Pelletier, D. (1987). The relationship of energy intake and expenditure to body fatness in western Samoan men. *Ecology of Food and Nutrition* **19**, 185–99.

(1988). The effects of occupation, leisure activities and body composition on aerobic fitness in western Samoan men. *Human Biology* **60**, 889–99.

Philbrick, N. (2003). *Sea of Glory: America's Voyage of Discovery*. New York: Viking Press.

Popkin, B. M. (2004). The nutrition transition: an overview of world patterns of change. *Nutrition Reviews* **62**, S140–3.

Popkin, B. M. and Gordon-Larsen, P. (2004). The nutrition transition: worldwide obesity dynamics and their determinants. *International Journal of Obesity and Related Metabolic Disorders* **28**(Supplement 3), S2–9.

Roberts, S. T., McGarvey, S. T., Viali, S. and Quested, C. (2004). Youth blood pressure levels in Samoa in 1979 and 1991–93. *American Journal of Human Biology* **16**, 158–67.

Rolls, B. J., Ello-Martin, J. A. and Tohill, B. C. (2004). What can intervention studies tell us about the relationship between fruit and vegetable consumption and weight management? *Nutrition Reviews* **62**, 1–17.

Rush, E. C., Puniani, K., Valencia, M. E., Davies, P. S. and Plank, L. D. (2003). Estimation of body fatness from body mass index and bioelectrical impedance: comparison of New Zealand European, Maori and Pacific Island Children. *European Journal of Clinical Nutrition* **57**, 1394–401.

Shephard, R. J. (1978). *Human Physiological Work Capacity.* Cambridge: Cambridge University Press.

Snyder, E. E., Walts, B., Perusse, L., *et al.* (2004). The human obesity gene map: the 2003 update. *Obesity Research* **12**, 369–439.

Speakman, J. R. (2004). Obesity: the integrated roles of environment and genetics. *Journal of Nutrition* **134**(Supplement), 2090S–105S.

Steele, M. S. and McGarvey, S. T. (1996). Expression of anger by Samoan adults. *Psychological Reports* **79**, 1339–48.

Swinburn, B. A., Ley, S. J., Carmichael, H. E. and Plank, L. D. (1999). Body size and composition of Polynesians. *International Journal of Obesity and Related Metabolic Disorders* **23**, 1178–83.

Tennefors, C. and Forsum, E. (2004). Assessment of body fatness in young children using the skinfold technique and BMI vs body water dilution. *European Journal of Clinical Nutrition* **58**, 541–7.

Tsai, H. J., Sun, G., Smelser, D., *et al.* (2004). Distribution of genome-wide linkage disequilibrium based on microsatellite loci in the Samoan population. *Human Genomics* **1**, 327–34.

Turner, G. (1861). *Samoa: Nineteen Years in Polynesia.* Reprinted 1986. Apia, Samoa: Commercial Printers.

    (1884). *Samoa: A Hundred Years Ago and Long Before.* London: Macmillan.

Vohr, B. and McGarvey, S. T. (1997). Growth patterns for large-for-gestational-age and appropriate-for-gestational-age infants of gestational diabetic mothers and control mothers at age 1 year. *Diabetes Care* **20**, 1066–72.

Vohr, B. R., McGarvey, S. T. and Garcia-Coll, C. (1995). Effects of maternal gestational diabetes and adiposity on neonatal adiposity and blood pressure. *Diabetes Care* **18**, 467–75.

Vohr, B. R., McGarvey, S. T. and Tucker, R. (1999). Effects of maternal gestational diabetes on offspring adiposity at 4 to 7 years of age. *Diabetes Care* **22**, 1284–91.

World Health Organization (WHO) (1990). *Diet, Nutrition, and the Prevention of Chronic Diseases: Report of a WHO Study Group.* Geneva, Switzerland. Technical Report Series No. 797. Geneva: World Health Organization, pp. 1–203.

    (2002). *Workshop on Obesity Prevention and Control Strategies in the Pacific: Apia, Samoa 26–29 September 2000.* Manila, Philippines: Regional Office for the Western Pacific (July).

(2003). *Diet, Food Supply and Obesity in the Pacific*. Manila, Philippines: Regional Office for the Western Pacific.

Zimmet, P. (2000). Globalization, coca-colonization and the chronic disease epidemic: can the Doomsday scenario be averted? *Journal of Internal Medicine* **247**, 301–10.

Zimmet, P., Faaiuso, S., Ainuu, J., *et al.* (1981). The prevalence of diabetes in the rural and urban Polynesian population of Western Samoa. *Diabetes* **30**, 45–51.

9   *Health patterns of Pacific Islanders and Asians in the United States*

W. PARKER FRISBIE, ROBERT A. HUMMER,
T. ELIZABETH DURDEN AND YOUNGTAE CHO

Introduction

Any analysis of the health of the Asian and Pacific Islander (API) populations in the United States immediately confronts the obstacle that, as late as
the year 2000, 'Asian and Native Hawaiian/Pacific Islanders have to a large
degree been "invisible" in public health debates', and that there are large
gaps in our understanding of the 'factors that influence their health and
quality of life' (Srinivasan and Guillermo 2000: 1731).[1] A major reason for
such gaps is that the comparatively small size of many API subpopulations
results in data sets with an insufficient number of cases to allow stable and
reliable statistical estimates. Assessing API health in the United States
requires drawing on a number of different data sources across which information is not always fully consistent. With the partial exception of
Hawaiians, information on specific Pacific Islander populations is especially
scarce, so that it is often necessary to assess the health of this group as if it
were an undifferentiated whole. In fact, much more often than is desirable,
Pacific Islanders are subsumed within a combined Asian *and* Pacific Islander
category. Nevertheless, by employing information from a number of different sources, including our own analyses, we believe a useful assessment of the
physical health of the API population in the United States can be achieved.[2]

---

[1] In US statistics, the Asian population is defined as 'those having origins in any of the original
peoples of the Far East, Southeast Asia or the Indian subcontinent,' and ' "Pacific Islander"
refers to those having origins in any of the original peoples of Hawaii, Guam, Samoa, or
other Pacific Islands' (Reeves and Bennett 2003: 1). Although these populations are listed under
the general rubric of 'race groups', as has long been the case in the social sciences, race
is conceived as a social, *not* a biological or genetic, category.

[2] We are aware of a literature on the emotional and mental health of APIs (e.g., Lin and
Cheung 1999; Chow 2002). However, we have elected to focus on the physical health of APIs in
as much depth as space limitations permit, as opposed to a more cursory discussion of both
physical and mental health.

Health Change in the Asia-Pacific Region: Biocultural and Epidemiological Approaches,
ed. Ryutaro Ohtsuka and Stanley J. Ulijaszek. Published by Cambridge University Press.
© Cambridge University Press 2007.

Demographic and socioeconomic characteristics

The API population is quite heterogeneous in demographic, socioeconomic status (SES) and cultural terms, as well as in length of settlement (Srinivasan and Guillermo 2000; Barnes and Bennett 2002; Ro 2002; Reeves and Bennett 2003). Given such diversity, variation in health outcomes is to be expected across specific API groups, which, in turn, means that it will be instructive to distinguish and compare as many distinct API groups as the data permit. A brief consideration of basic demographic and socioeconomic characteristics provides essential context for the assessment of health that follows it.

Demographic characteristics

In the year 2000, the US census enumerated 11.9 million Asian individuals, or about 4.2% of the total US population of 281.4 million. Of these, 10.2 million (or 3.6% of the total) chose one, and only one, Asian category, with the other 1.7 million (0.6% of the total) self-reporting Asian in combination with some other race group (Barnes and Bennett 2002). The number of Pacific Islanders (including Native Hawaiians) enumerated was 874,000 (or about 0.3% of the total for the entire US population). In this case, the tendency to self-identify in multiracial terms was greater. In fact, about 399,000 persons self-identified as being only Pacific Islander, while about 476,000 reported as being Pacific Islander along with some other race (Grieco 2001).[3] Spread over more than 40 respondent-reported subgroups, it is not surprising that the numbers for many specific API populations are very small – often totalling less than a thousand and sometimes less than a hundred – even allowing for multiracial combined responses. Thus, detailed comparisons of health characteristics can usually be achieved only for the larger groups, along with a residual 'all other API' category.

Table 9.1 gives the latest census counts for the six largest Asian populations (plus a residual 'Other Asian' category) and for Native Hawaiians (plus a residual 'Other Pacific Islander' category). For the most part, these

---

[3] These data pertain only to the population residing in the 50 states (and the District of Columbia). Excluded are residents of the islands of Guam, American Samoa, Commonwealth of the Northern Marianas and the U.S. Virgin Islands. Also, the total listed for those who selected an API identity, plus some other racial identity, will (obviously) be larger than the total for those who reported only one API identity.

Table 9.1. *The largest Asian and Pacific Islander populations in the United States in 2000*

Specific group	One API group only		Specific API group in any combination[a]	
	Frequency	%	Frequency	%
Chinese[b]	2,314,537	23.10	2,734,841	22.98
Philippino	1,850,314	18.47	2,364,815	19.87
Asian Indian	1,678,795	16.76	1,899,599	15.96
Vietnamese	1,122,528	11.20	1,223,736	10.28
Korean	1,076,872	10.75	1,228,427	10.32
Japanese	796,700	7.95	1,148,932	9.66
Other Asian	1,179,659	11.77	1,298,478	10.91
Total Asian	*10,019,405*	*100.00*	*11,898,828*	*100.00*
Native Hawaiians	140,652	36.10	401,162	45.88
Other Pacific Islander	248,960	63.90	473,252	54.12
Total Pacific Islander	*389,612*	*100.00*	*874,414*	*100.00*

Notes:
[a] Based on responses, not respondents;
[b] Excludes Taiwanese.
Source: Barnes and Bennett (2002).

are the groups on which the few extant analytic studies (as opposed to basic descriptive studies) of health status and health care have been conducted. Some surveys for local areas, especially cities, may provide insight into some of the smaller groups, but one cannot safely generalize from local area data to the US population as a whole.

The six largest groups shown in Table 9.1 make up almost 90% of the total Asian population, and Hawaiians, while they do not constitute a majority of Pacific Islanders in the United States, are by far the largest single group in this category. Perhaps because Hawaiians comprise such a large proportion of all Pacific Islanders, data specific to Hawaiians can sometimes be located, even when data for certain specific Asian groups that are larger in size may not be available.

Socioeconomic characteristics

Taken as an undifferentiated whole, the API population is characterized as having considerably higher levels of education than any other race group in the United States. Recent census-based reports show that almost 44% of APIs aged 25 years and older have a college degree, compared to only 26.1%, 16.5% and 10.6% of whites, blacks and Hispanics in descending order. The annual median family income for APIs also tops the list at approximately $56,000, a figure $5,000 above the white median. Median family income for blacks and Hispanics is lower, at about $32,000 for both groups (United States Census Bureau 2001). Health, of course, varies positively with SES, although the effects are reciprocal (Morris 1996).

Poverty in the API population is similar to that of the overall population, but higher compared to the non-Hispanic white population (National Center for Health Statistics 2003). In 1990, 12.2% of all APIs in the United States were below the poverty line as compared to 8.8% of non-Hispanic whites. By 2001, poverty rates stood at 10.2% and 7.8% for the same groups, respectively. Hispanics, and particularly blacks, are characterized by poverty rates which are much higher than the US average, but have also experienced the most rapid declines in poverty.

Adult health

Discussion of adult health outcomes begins with examination of mortality among the age group at which a vast majority of deaths occur in the United States: individuals aged 65 years and above. The National Center for Health Statistics (NCHS) constructs official mortality rates based on Vital Statistics (numerator) and Census (denominator) data. Although highly informative, there are some well-known limitations in regard to the quality and reliability of official death rates by race/ethnicity, especially among the elderly. To illustrate, Rosenberg *et al.* (1999) found that for population groups other than non-Hispanic whites and blacks, 'levels of mortality are biased from mis-reporting in the numerator and under-coverage in the denominator of the death rates'. Their findings suggest that officially reported death rates for APIs might be under-estimated by 11%. Even with correction, their refined estimates of age-adjusted death rates across the life course suggest that API death rates still remain the lowest in the country, in comparison to non-Hispanic whites, non-Hispanic blacks, Hispanics and Native Americans.

Table 9.2 presents official death rates per 100,000 population, by race/ethnicity and sex, for five-year age groups among the elderly population

Table 9.2. *Death rates per 100,000 by race/ethnicity for the elderly*
population of the United States, official mortality data, 1999

Age group	Non-Hispanic black	Hispanic origin	Asian/Pacific Islander	Non-Hispanic white
Females				
65–69	2,231	1,126	863	1,515
70–74	3,516	1,662	1,404	2,372
75–79	5,123	2,591	2,273	3,803
80–84	7,714	4,301	4,262	6,492
85+	14,474	8,839	8,397	15,285
Males				
65–69	3,567	1,841	1,358	2,433
70–74	5,237	2,705	2,395	3,781
75–79	7,455	3,913	3,828	5,712
80–84	10,546	5,697	5,958	9,287
85+	16,321	9,842	11,343	17,539

Source: Hoyert *et al.* (2001).

in the United States in 1999. Reported mortality rates for the API
population are much lower than for non-Hispanic whites and other
race/ethnic groups at 65 to 69 years of age, and this difference is
typically maintained across subsequent age groups. For example, the
reported rate for API women at 65 to 69 years of age (862.7 per 100,000)
is 43% lower than that for non-Hispanic white women (1,515.0 per
100,000); above the age of 85 years, the advantage for API women compared to whites is 45%. While levels of mortality are higher among men
than women for each race/ethnic and age group, the relative disparities by
race/ethnicity vary little by sex. Elderly API women and men have significantly lower mortality rates than non-Hispanic whites and blacks. API
rates also compare favourably with Hispanic rates for most age and sex
groups.

Recent research has evaluated and re-estimated Asian American older
adult mortality data for various sources of bias and provided perhaps the
most accurate assessments to date. Importantly, in the revised estimates,
Asian American older adult mortality remained lower than that for whites,
although the advantage is not as great as indicated by the official data
(Lauderdale and Kestenbaum 2002). This was the case for both men and
women. The mortality probabilities across Asian subpopulations were
quite similar to one another, although Vietnamese men had a lower

probability of death than did other Asian men, and South Asian Indian women had a higher probability than other Asian women.

Another detailed analysis of API mortality among persons of all ages (Hoyert and Kung 1997) used official mortality data from seven states (California, Hawaii, Illinois, New Jersey, New York, Texas and Washington) in 1992. Hoyert and Kung (1997) acknowledge a number of factors which might create bias in these calculations, as well as small numbers of deaths for some specific groups. Nevertheless, this study showed that age-adjusted death rates were highest for Samoan and Native Hawaiian populations and lowest for the South Asian Indian, Korean and Japanese populations. API life expectancy at birth (LEB), without any adjustments for data quality, was calculated to be 80.3 years, and ranged from slightly higher than 85 years for South Asian Indians and Guamanians to about 68 years for Hawaiians and Samoans. The Japanese, Chinese, Philippino and Korean subpopulations were also shown to have LEBs of over 80 years, and the Vietnamese LEB was estimated to be 78.8 years. In that particular year, the LEB of non-Hispanic whites was much lower, at 75.1 years.

Table 9.3 presents age-standardized, cause-specific mortality rates (per 100,000 population) by race/ethnicity for the leading causes of death among the elderly population by sex. As with overall mortality rates, caution in interpretation is warranted, particularly for APIs and Hispanics. API women (Panel A, Table 9.3) stand out as having the lowest rate for several of the causes, exhibiting especially low rates of respiratory diseases, influenza/pneumonia, Alzheimer's disease, diabetes mellitus and septicaemia. The age-standardized rate of respiratory disease mortality for API women, at 92.4 per 100,000, is roughly 69% lower than among non-Hispanic white women (294.9 per 100,000). Older API women also display the lowest rate of mortality due to heart disease, which is the leading cause of death in the United States. Cancer rates for APIs are also only slightly above those of Hispanics, and far below those of non-Hispanic whites and non-Hispanic blacks. As with women, the older adult API male population is characterized by the lowest overall rate of death and lowest rates for several morbid conditions, including diabetes, accidents and Alzheimer's disease. Perhaps most impressively, death rates due to heart disease and cancer are approximately 30% to 50% lower than those for non-Hispanic whites and blacks at these ages. Indeed, API death rates are lower than for non-Hispanic whites and blacks for each of the leading causes of death listed in Table 9.3 for which comparisons are possible. Even if the death rates for APIs are adjusted upward by 10% to 15%, the overall conclusion of a distinct API advantage in comparison to whites and blacks would not change.

Table 9.3. *Race/ethnic disparities in cause-specific mortality rates (per 100,000 population, standardized to the age distribution of non-Hispanic whites) for the top ten causes of death among the US elderly population, 1999*

Underlying cause of death	Non-Hispanic black	Hispanic origin	Asian/Pacific Islander	Non-Hispanic white
Panel A: Females				
Diseases of the heart	2,030.8	1,129.1	972.2	1,715.2
Malignant neoplasms	1,068.7	545.2	569.6	955.4
Cerebrovascular diseases	556.5	268.3	363.6	481.1
Chronic lower respiratory diseases	149.2	113.0	92.4	294.9
Influenza and pneumonia	157.3	107.9	97.2	173.3
Alzheimer's disease	102.3	70.9	*	167.4
Diabetes mellitus	315.2	215.1	130.1	129.0
Accidents	*	45.8	48.5	85.3
Nephritis, nephrotic syndrome and nephrosis	171.8	60.0	62.2	72.8
Septicemia	152.7	48.1	22.5	67.5
Residual (all other causes)	1,079.6	577.4	541.0	903.0
Total	5,784.0	3,180.8	2,899.3	5,044.8
Panel B: Males				
Diseases of the heart	2,137.3	1,242.0	1,173.7	1,929.9
Malignant neoplasms	1,860.4	845.1	867.0	1,428.9
Chronic lower respiratory diseases	311.9	176.7	197.2	403.5
Cerebrovascular diseases	499.9	262.9	353.9	380.4

Influenza and pneumonia	183.1	111.1	128.0	172.2
Diabetes mellitus	255.7	198.1	116.4	142.4
Accidents	127.6	77.8	72.0	112.2
Alzheimer's disease	*	*	*	99.8
Nephritis, nephrotic syndrome and nephrosis	187.6	74.6	72.9	92.6
Septicemia	157.4	49.1	46.1	65.6
Residual (all other causes)	1,102.4	652.8	582.0	902.9
Total	6,823.3	3,690.2	3,609.2	5,730.3

Note:

The top ten causes for non-Hispanic whites are listed.

* indicates that this particular cause of death was not listed in the top 10 for this particular race/ethnic group and is therefore included in the residual category.

Source: Anderson (2001).

Adult health outcomes and behaviours

A very general comparative overview of one global indicator of adult health status, self-reported health (ranked from poor to excellent) for six broad race/ethnic categories in the United States around 1999, is given in Fig. 9.1 (Pleis and Coles 2003). Self-reports of health have repeatedly been shown to be both useful indicators of actual health status and predictors of mortality risk (Idler and Benyamini 1997; Rogers *et al.* 2000; Frisbie *et al.* 2001; Benjamins *et al.* 2004). Over 70% of Asians report very good or excellent health, a proportion higher than that for any other group, followed by whites (66%) and then Pacific Islanders (including Native Hawaiians) and Hispanics. The situation is somewhat different with respect to health that is only fair or poor. Asians have the lowest proportion (<10%) in this category, but Pacific Islanders have far and away the largest proportion (over 25%) with fair-to-poor health, followed by blacks and American Indians.

We also investigated several health indicators using pooled data from the 1997 to 1998 National Health Interview Surveys (NHIS). In Table 9.4,

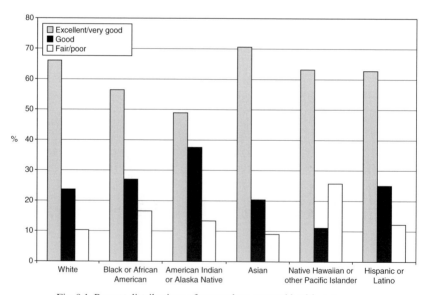

Fig. 9.1. Percent distributions of respondent-assessed health status among persons 18 years of age and over, by race/ethnicity: United States, 1999. (Source: Pleis and Coles 2003.)

Table 9.4. *Age-standardized percentage distributions of health indicators by race/ethnicity and sex, US, 1997–98*

	White		Black		Chinese		Philippino		Asian Indian		Other APIs	
	Male	Female	Male	Female	Male	Female	Male	Female	Male	Female	Male	Female
Body mass index												
Normal	36.4	53.9	33.6	33.2	67.0	82.4	48.3	60.8	68.2	59.9	59.9	73.7
Overweight	43.8	25.1	40.8	31.4	26.8	13.1	44.6	31.8	23.2	30.3	31.0	17.3
Obese	19.9	21.0	25.7	35.3	6.2	4.5	7.2	7.5	8.7	9.7	9.1	8.9
Functional limitations												
Limited	30.2	40.2	30.6	42.9	17.5	30.3	12.2	33.7	16.8	25.2	22.4	29.3
Unlimited	69.8	59.8	69.4	57.1	82.5	69.7	87.8	66.3	83.2	74.8	77.6	70.7
Smoking status												
Current smoker	26.9	23.5	30.7	21.3	14.3	9.8	16.7	8.1	11.7	0.2	23.7	15.7
Former smoker	30.1	21.4	19.9	13.1	21.9	5.9	26.6	9.7	18.1	2.7	30.4	8.9
Never smoker	43.0	55.1	49.4	65.5	63.9	84.3	56.6	82.2	70.2	97.2	46.0	75.4
Alcohol drinking status												
Current drinker	72.4	61.7	57.3	39.6	53.2	27.8	72.7	30.0	57.1	11.4	59.1	38.3
Former drinker	16.0	15.8	21.3	19.9	15.5	11.4	9.5	11.9	7.0	4.8	13.5	9.2
Never drinker	11.6	22.6	21.4	40.5	31.3	60.8	17.8	58.1	35.9	83.8	27.5	52.5
Frequency of moderate exercise 20+ min												
At least once/week	50.9	49.9	37.8	32.9	44.6	35.2	61.1	44.1	47.7	52.7	54.3	44.3
None or unable	49.1	50.1	62.2	67.1	55.4	64.8	39.0	56.0	52.3	47.3	45.7	55.7
Have usual place to visit when sick												
Yes	82.7	91.4	80.3	90.5	79.1	89.4	82.3	87.4	71.3	79.0	77.4	83.3
No	17.3	8.6	19.8	9.5	20.9	10.6	17.7	12.6	28.7	21.0	22.6	16.7
Unweighted N	20,175	25,341	3491	5675	182	181	126	207	148	113	372	469

Note:
Age standardization based on the non-Hispanic white age structure.
Source: National Health Interview Survey, Adult Sample Files (1997–8 pooled).

the health of adults in Chinese, Philippino and South Asian Indian subgroups is compared with Other APIs, non-Hispanic whites and non-Hispanic blacks. The data are standardized to the age structure of non-Hispanic whites.

Across most indicators, the API subgroups exhibit more favourable distributions than non-Hispanic whites and blacks. Levels of obesity (body mass index $> = 30 \, kg/m^2$), as determined by self-reported weight and height, are 21% for white females and over 35% for black females. Levels are much lower, and never reach 10%, for any of the API subgroups, regardless of sex. For example, just 4.5% of Chinese women and 6.2% of Chinese men are obese. Again, however, there is considerable heterogeneity across API subgroups; for example, the 8.9% level of obesity for women in the Other API category is about double that of Chinese women. A recent literature review of obesity among API populations found the highest rates of obesity and related risks to be among Native Hawaiians and Samoans (Davis *et al.* 2004).

The percentage of individuals who report a physical limitation is also lower among each of the API subgroups compared to whites and blacks. While these levels vary considerably by sex for each race/ethnic group, with women reporting much higher levels of limitation than men, there are consistently lower levels among API groups than among blacks and whites. However, recent work has shown much higher disability rates among some South East Asian (Laotian, Cambodian and Hmong) subgroups and Pacific Islanders than among the larger Asian subgroups in the United States (Cho and Hummer 2001). Such differences are probably best reflected in the current data when comparing Other API men (22.4% with disability) to Philippino men (12.2% with disability). However, the groups with the highest percentages of disability reported by Cho and Hummer (2001) are also of smaller size than groups with low levels of disability and functional limitations (Chinese, Philippino, South Asian Indian and Japanese). Thus, again, the overall prevalence of functional limitation shown for APIs is favourable because the relatively large size of the healthiest groups overwhelms the relatively small size of the less healthy groups in national-level, large-scale analyses.

Turning to health behaviour, the API groups in general show more favourable distributions than whites and blacks, but again there is some notable variation within the API population. This is most clear in the case of cigarette smoking. While 26.9% of white males and 30.7% of black males are current smokers, only 11.7% of South Asian Indian men are current smokers. Furthermore, only 0.2% of Asian Indian women report that they are current smokers, by far the lowest percentage exhibited by any of the groups. Other APIs again display the least favourable distribution, with 23.7% of men and 15.7% of women reporting current smoking, respectively.

The alcohol-use and exercise distributions less clearly favour the API populations. Philippino males, for example, exhibit the highest percentage of current drinking of any API group (72.7%); however, Philippino, Chinese and South Asian Indian women show the lowest percentages of current alcohol use. The Other API group is somewhat similar in current drinking patterns to blacks, both of which are lower than the percentage for whites. Almost all groups of adults shown here report very high levels of physical inactivity. Among APIs, Chinese adults are characterized by the highest level of inactivity (55.4% of men and 64.8% of women) among the API groups, with the other groups (including whites) hovering around 50%. While these percentages are more favourable than among blacks, of whom 62.2% of men and 67.1% of women do not exercise even once a week, levels of inactivity are high throughout.

The last indicator in Table 9.4 shows whether or not individuals have a usual place to go for medical attention. The API subgroups display less favourable distributions than whites and blacks, with South Asian Indians (28.7% of men and 21.0% of women) least likely to report having a usual place to visit when sick. Koreans have been shown to have less access to a usual source of health care than whites and a majority of API subgroups (Frisbie *et al.* 2001). Thus, health status indicators and health access indicators do not always coincide.

Measures of health and mortality, health behaviour and health access vary by place of birth and duration of residence for a number of populations in the United States, including APIs (Hummer *et al.* 1999; Cho and Hummer 2001; Frisbie *et al.* 2001; Singh and Siahpush 2002; Singh and Miller 2004). Most generally, foreign-born individuals exhibit advantages compared to native-born individuals of the same race/ethnic group, although such advantages seem to erode with increased duration in the United States. Earlier analyses (Cho and Hummer 2001; Frisbie *et al.* 2001) found such patterns both for the overall API population and across a large number of specific API groups. Although some scholars attribute this erosion to negative acculturation, it is worth noting that an alternative explanation is that immigrants tend to be less adequately served by the formal health care system as indicated by data on both utilization of services and degree of coverage by health insurance (LeClere *et al.* 1994; Thamer *et al.* 1997).

In a study of specific API subgroups, Frisbie *et al.* (2001) showed that age-adjusted levels of self-reported health, activity limitations and annual bed days were generally least favourable among Pacific Islander and Vietnamese individuals, and most favourable among Japanese, Chinese, Philippino, South Asian Indian and Korean individuals. When nativity, duration of residence, and social and demographic risk factor controls were added to the models, Japanese individuals continued to have the most positive health outcomes, with Chinese individuals also generally doing very well. In contrast, Pacific Islanders were the only group examined that recorded significantly less favourable outcomes in comparison to Japanese individuals across all three health indicators.

Infant and child health

Infant mortality

Inasmuch as infant mortality has often been employed as 'a synoptic indicator of the health and social condition of a population' (Gortmaker and Wise 1997: 147), considerable attention is given to infant mortality of the API population relative to other race/ethnic groups in the United States. Table 9.5 presents a description of infant mortality per 1,000 live births based on linked-period birth/infant death files in the United States for all reporting states and for 11 states with larger API populations. While geographic coverage is lost, the data for the 11 states allow separate consideration of the Vietnamese, South Asian Indian and Korean populations. Rates are shown for all infant deaths (IMR), neonatal mortality (NMR; deaths of infants under 28 days of age) and post-neonatal mortality (PNMR; deaths of infants from the 28th day through the first year of life).

With the exception of Hawaiians, both panels of Table 9.5 demonstrate that current IMRs for specific API groups are fairly close to the rates for whites. Rates for Hawaiians are approximately double those for most Asian groups, are moderately above the rates for American Indians and are exceeded only by the rates for blacks. Hummer *et al.* (1999) employed the linked birth/infant death cohort files for 1989 to 1991 to analyse race/ethnic differences for four Asian groups (Chinese, Japanese, Philippinos and a residual category) in comparison to non-Hispanic whites, non-Hispanic blacks and several Hispanic populations. In logistic regression

Table 9.5. *Infant mortality rates* by race: United States, 2000*

Maternal race/ethnicity	All reporting states			Maternal race/ethnicity	11 states with larger API populations		
	IMR	NMR	PNMR		IMR	NMR	PNMR
All races	6.9	4.6	2.3	All races	6.2	4.1	2.1
White	5.7	3.8	1.9	White	5.3	3.5	1.8
Black	13.5	9.1	4.3	Black	12.3	8.2	4.1
Asian/Pacific Islander	4.9	3.4	1.4	Asian/Pacific Islander	4.9	3.5	1.4
Chinese	3.5	2.5	1.0	Chinese	3.4	2.5	0.8
Japanese	4.5	2.6	**	Japanese	4.6	**	**
Philippino	5.7	4.1	1.6	Philippino	5.6	4.0	1.6
Hawaiian	9.0	6.2	**	Vietnamese	4.4	2.9	1.5
Other API	4.8	3.4	1.4	South Asian Indian	4.5	3.5	0.9
American Indian[‡]	8.3	4.4	3.9	Korean	4.5	2.8	**
				Hawaiian	8.4	5.9	**
				Samoan	**	**	**
				Guamanian	**	**	**
				Other API	5.9	4.3	1.7
				American Indian[‡]	7.0	3.3	3.6

Notes:

The 11 states in the right hand panel are: California, Hawaii, Illinois, Minnesota, Missouri, New York, New Jersey, Texas, Virginia, Washington and West Virginia.

* All rates per 1,000 live births. IMR = infant mortality rate; NMR = neonatal mortality rate; PNMR = postneonatal mortality rate;

** Not reported when based on fewer than 20 deaths;

[‡] Includes Eskimos and Aleuts.

Source: Matthews *et al.* (2002).

models with race/ethnicity as a covariate, and net of controls for a wide range of risk factors, the odds of infant death for three of the Asian populations were statistically equivalent to that of the non-Hispanic white population. The risk estimated for the fourth group, Japanese Americans, was significantly lower than that for non-Hispanic whites. With birth outcomes (birth weight and gestational age) controlled for, the risk for every Asian group was significantly lower than the risk for non-Hispanic whites. For every Asian group, the odds of infant death were lower among infants born to immigrant mothers compared to their US-born counterparts, although a significant effect of maternal place of birth occurred only among Philippinos.

Birth outcomes

Birth outcomes (birth weight and gestational age) warrant discussion in a comparative analysis of API infant mortality and child health for several reasons. Low birth weight and short gestation have consistently been found to be the strongest predictors of the risk of infant mortality (Cramer 1987; Kiely and Susser 1992; Frisbie *et al.* 1996). Furthermore, compromised birth outcomes are associated with greater risk of congenital anomalies, neurodevelopmental and behavioural disorders, lower educational achievement and poorer family functioning (McCormick 1985; Kline *et al.* 1989; Hack *et al.* 1995). Thus, it is of interest that some API groups have relatively high percentages of low weight births. In 1999, 8.3% of Philippino births were of low weight, higher than that for any other group for which data were available except for blacks and Puerto Ricans, who recorded low birth weight rates of 13.1% and 9.0%, respectively. The risk of low birth weight was also higher for Japanese (7.9%) and Hawaiian (7.7%) infants than for whites (6.6%). Moreover, for every group, except blacks, the rates of pre-term and low weight births increased across the 1990s (Frisbie and Song 2003).

Such findings make the relatively low infant mortality observed among most of the larger API populations seem slightly anomalous. An explanation may be found in research indicating that use of the conventional measure of prematurity (<37 weeks gestation) and low birth weight ($<2,500$ g) may fail to adequately distinguish high risk infants across different race/ethnic populations. Wilcox and Russell demonstrated that more meaningful comparisons of the infant mortality rate across populations can be achieved 'by plotting each weight-specific mortality curve relative to its own birthweight distribution' (1986: 188). The most profound implications of this measurement issue can be summarized as follows:

> Given that the general norms for preterm, post-term, and fetal growth measures may be largely derived from White populations, more information is needed to assess whether or not the ongoing use of these one-size-fits-all standards may result, for some ethnic groups, in invalid risk assessments and the misidentification of infants in need of intervention services.
>
> (Alexander *et al.* 1999: 77)

Child health and health care

The death rates of children are lower for APIs as a whole than for any other major race/ethnic category (Health Resources and Services Administration 2001). Taking the two extremes for illustrative purposes, the death rate per 100,000 for children aged one to four years in 1980 stood at approximately 100 for blacks and a little over 40 for API children. Risk of death for young children has declined substantially over time, so that by 1998, the rate for young black children was down to 62 per 100,000 and that for API children had dropped to 19 per 100,000. Death rates among children aged one to four years for the entire US population fell from a little over 60 to 35 per 100,000 over the same period, and API rates were similar to (and often slightly lower than) the rates for non-Hispanic whites. The race/ethnic ranking over time has remained the same for children aged 5 to 14 years, but death rates are much lower than those for young children, and the trajectory of decline is much less steep (Health Resources and Services Administration 2001).

The low API death rates are consistent with an analysis of health behaviours reported in the National Youth Risk Behaviour Survey, which is representative of all US children enrolled in the 9th through 12th grades (Grunbaum *et al.* 2000). Results show that API students were less likely to have consumed alcohol or marijuana, had sexual intercourse, carried a weapon, or fought than white, black or Hispanic students. Cigarette smoking was much less common among API students compared to their white and Hispanic counterparts, but API students were somewhat more likely to smoke than were black students. Nevertheless, drawing on unpublished data, Grunbaum *et al.* (2000) also point to differences across small areas, noting, for example, that cigarette smoking by male students was reported as 34% among Philippinos in San Diego, 13% among Chinese, but 28% among Philippinos in San Francisco, and 26% among

Hawaiians/Part Hawaiians in Hawaii. Male student reports of sexual intercourse ranged from 11% of Chinese people in San Francisco to 49% of Philippinos in San Diego (Grunbaum *et al.* 2000).

Table 9.6 shows the percentage distributions for two indicators of child health, global health status and selected categories of number of school days missed due to illness, based on data from the NHIS in the year 2000. Note that all percentages for Pacific Islander children (and in a few instances, percentages for other groups) must be used with caution because small sample sizes did not meet conventional standards of precision and reliability. The pattern that emerges for the global measures of child health in the first panel of Table 9.6 is quite straightforward. Asians and whites have the highest proportions (each about 56%) reporting excellent health, while the proportion reporting excellent health is around 43% to 44% for every other group except American Indians (36.8%). The relative proportion of children whose health is reported as being good or worse is between 15% and 17% for the first two groups mentioned, ranging narrowly from 24% to 28% for the others. Well over 40% of both Asians and Pacific Islanders report no school days missed due to illness, followed by blacks and Hispanics (both at 36% to 37%). The frequency distribution of school days missed due to illness for Pacific Islander children is bimodal; these children experienced the highest proportion of zero absences due to illness, but also the largest percentage that missed 6 to 10 days of school for that reason. It is likely that this unusual pattern results at least in part from instability of estimates due to small sample size, but given their relatively poor health indicated in the first panel of the table, one should not discount the possibility that morbidity is an obstacle to higher levels of attendance among some Pacific Islander children.

A recent study (Yu *et al.* 2004) employed the 1997 to 2000 NHIS to examine differentials in both health status and health service access for Chinese, Philippino, South Asian Indian and Other API children in the United States, in comparison with non-Hispanic white children. Among the child health status measures were the following: respondent (i.e., parent) assessed health; congenital diseases; missing school due to illness; and learning disabilities. Health services access and use indicators included whether the family had health insurance and a regular place to access health care, and whether seeking health care was delayed because of cost.[4] Logistic regression models showed that Asian children had more

---

[4] Information on health care access is limited for all groups in the sense that the data for the nation as a whole do not allow an assessment of either quality of care received or obstacles that may be encountered in attempts to acquire adequate treatment.

Table 9.6. *Percent distributions on two child-health indicators by race/ethnicity: United States, 2000*

All children under 18 years

Race/ethnic group[a]	Respondent-reported health			
	Excellent	Very good	Good	Fair/poor
White	56.9	28.4	13.2	1.6
Black	44.5	29.3	22.6	3.6
American Indian	36.8	35.3	26.6	1.2
Asian	56.3	27.2	16.1	0.4[b]
Pacific Islander	43.2[b]	30.1[b]	26.7[b]	*[b]
All Hispanic	43.7	32.0	22.1	2.1
Mexican	42.6	32.3	23.4	1.8

All children 5–17 years

Race/ethnic group	School days missed due to illness (selected categories)		
	None	1–2 days	6–10 days
White	24.7	29.6	11.2
Black	37.0	24.9	8.6
American Indian	16.1	20.1	16.2[b]
Asian	42.4	29.2	4.9
Pacific Islander	46.6[b]	18.6[b]	28.2[b]
All Hispanic	36.1	23.0	11.6
Mexican	36.5	23.5	11.0

Notes:

[a] Data for whites, blacks, American Indians, Asians and Pacific Islanders who selected one, and only one, race. Hispanics (including Mexicans) may be of any race or combination of races;

[b] Use with caution. Estimates do not meet standard of reliability or precision (standard error >30%);

*Quantity zero.

Source: Blackwell *et al.* (2003).

favourable health status, but less access to, and utilization of, health care, as compared to non-Hispanic white children. These same patterns were seen in our data for adults shown above – notably in comparisons of immigrants with those born in the United States.

Maternal health and health care

The lack of knowledge regarding the health of API women is of particular concern, especially in the larger context of less attention having been paid to primary and preventive health care among women than among men more generally (Ro 2002: 518). However, there is some information available on maternal health and health care that pertains to specific morbid conditions and medical procedures for several API populations. Table 9.7 shows comparisons of maternal health and health care markers for 11 race/ethnic categories, including several specific API populations, for women who gave birth in the year 2000 (Martin *et al.* 2002). The data, drawn from US birth certificates and presented as rates per 1,000 live births, are divided into three categories: (1) maternal medical risks; (2) complications of labour and delivery; and (3) obstetric procedures. Asian and Pacific Islander women have the lowest rates of anaemia overall. It is notable, however, that while Chinese women are at very low risk of anaemia (11.7 per 1,000 live births), the rate for Hawaiians is exceptionally high (49.9 per 1,000), and second only to the rate for American Indians (52.4 per 1,000). Much the same is true in regard to pregnancy-induced hypertension, although high and similar rates are observed among whites, blacks, Hawaiians and especially American Indians. At odds with the notion of 'healthy Asians and Pacific Islanders' is the extremely high rate of Type 2 diabetes for several API groups. With the exception of the Japanese, this condition is much more prevalent in every API population than it is for either whites or for all races taken together. Type 2 diabetes has long been recognized as a significant problem among American Indians (United States Department of Health and Human Services 2000). This is reflected in the rate of 51.0 per 1,000 live births for American Indian women who gave birth in 2000, but incidence among Philippinos is essentially identical (50.8 per 1,000 live births), and the rate for Chinese women (48.8 per 1,000) is also comparatively quite high. In contrast, the incidence of uterine bleeding is similar across all groups included in Table 9.7, except for the moderately lower rates for Hispanic women.

The prevalence of morbid conditions among women giving birth depends not only on risk of contracting the condition, but also to some extent on the quantity and quality of medical care. It is plausible that for

Table 9.7. *Maternal health characteristics by race/ethnicity: United States, 2000*

						Rates per 1,000 live births						
	All Races	White	Black	Amer. Indian[a]	All API	Chinese	Japanese	Hawaiian	Philippino	Other API	All Hisp.	Mexican
Maternal medical risks												
Anemia	23.9	21.2	37.5	52.4	18.7	11.7	17.7	49.9	16.2	20.0	24.5	23.3
Diabetes	29.3	28.3	27.4	51.0	46.9	48.8	29.8	37.4	50.8	47.0	28.3	27.3
Hypertension[b]	38.8	39.3	41.7	48.2	20.5	11.9	16.5	40.4	30.4	19.7	27.9	27.0
Uterine bleeding[c]	6.6	6.8	5.8	7.6	6.0	4.8	6.7	7.8	5.8	6.2	4.2	3.9
Complications of labor and delivery												
Meconium, moderate/heavy	53.9	50.2	72.2	55.0	55.9	51.8	47.3	64.7	63.5	55.2	58.9	58.0
Prem. rupture membranes	24.6	23.7	29.3	36.2	22.7	21.5	24.3	21.5	22.6	23.1	17.8	15.6
Dysfunctional labor	28.2	28.2	25.9	36.4	33.4	45.7	37.0	27.1	34.5	29.7	24.2	20.6
Breech/malpresentation	38.8	40.6	30.7	37.6	34.9	35.2	37.4	38.4	33.5	34.9	30.4	28.5
Cephalopelvic disproportion	17.2	17.8	12.8	16.1	21.6	22.4	18.7	14.7	25.8	20.9	13.2	13.2
Fetal distress[c]	39.2	27.5	48.5	38.4	36.5	32.4	27.3	26.9	34.2	39.6	30.4	27.8
Obstetric procedures												
Amniocentesis	24.0	25.1	15.5	15.2	34.8	60.1	75.7	25.4	31.8	25.8	11.2	7.9
Electronic Fetal monitoring	842.0	843.2	850.7	831.2	797.6	787.6	758.4	731.6	779.7	811.8	798.2	785.6
Induction of labor	198.8	209.7	163.9	200.8	133.1	122.8	144.5	136.5	117.0	139.5	131.6	124.2
Ultrasound	670.0	682.8	610.7	620.4	660.9	701.5	691.1	578.4	655.9	652.6	583.5	562.4
Stimulation of labor	179.5	182.0	164.9	165.1	187.3	196.1	179.5	125.8	162.6	195.4	166.4	160.1

Notes:

Persons in the two Hispanic groups may be of any race. Each of the other groups may include Hispanics.

[a] Includes Eskimos and Aleuts;

[b] Pregnancy induced;

[c] Texas does not report data on this item.

Source: Martin *et al.* (2002).

groups such as Hawaiians, in which health insurance coverage is comparatively low, a number of ailments may be under-diagnosed. Such a bias, if it exists, can be expected to be most serious among race/ethnic groups characterized by low socioeconomic status and among immigrants not fully integrated into the formal health care system.

A consideration of complications of labour and delivery (second panel, Table 9.7) shows that women of Mexican origin tend to have low rates for most categories of complications. The presence of moderate to heavy meconium (intestinal discharges of the newborn) is most common among black women (72.2 per 1,000). By contrast, Philippino women have the highest rate of cephalopelvic disproportion (25.8 per 1,000), twice the rate observed among blacks. Occurrence of dysfunctional labour is greatest among Chinese mothers (45.7 per 1,000), followed rather distantly by Japanese and American Indian women (37.0 and 36.4 per 1,000 respectively). The variation in the distribution of breech births and other malpresentations would be small except for the high rate for whites and the low rate for Hispanics and blacks. Fetal distress is disproportionately high among black women, with the lowest rates occurring among whites, Hawaiians, Japanese and women of Mexican origin (roughly 27 per 1,000 for each of the 4 groups).

Ultrasound (sonogram) estimates of fetal development and electronic fetal monitoring are common obstetrics procedures that are employed in the vast majority of pregnancies and deliveries so that timely medical intervention can be initiated if so indicated. It is somewhat surprising and disturbing that rates of electronic fetal monitoring are lower among API women in general, and lowest for Hawaiians, who represent the API population at greatest risk of infant mortality. However, Chinese American women are the only group for which ultrasound use was greater than 700 per 1,000 live births.

Application of other procedures occurs primarily when a problem in pregnancy is detected or suspected. Thus, amniocentesis and induced or stimulated labour can be viewed as having both positive and negative implications, positive in the sense that due medical vigilance is being exercised, and negative in the sense that some heightened risk is associated with the pregnancy. The relatively high amniocentesis rates among Chinese and Japanese women might be viewed as troubling, in that this procedure is a diagnostic tool that tests for the possible existence of congenital anomalies and other morbid conditions. On the other hand, this may indicate a greater desire among Asian women to take the fullest possible advantage of medical diagnostics associated with pregnancy. For example, research indicates that 'the number of requests for amniocentesis, after a diagnosis of minor fetal ultrasound abnormality [is] higher among Asian women than among other ethnic or racial groups' (De Santo *et al.* 2003: 118S).

Prenatal care

The long-held conclusion that adequate prenatal care (PNC) is of major benefit for the prevention of low weight births, and therefore a key to reducing infant mortality (Institute of Medicine 1988), has been challenged, based on evidence that the apparent beneficial effect stems primarily from selectivity bias (Alexander and Kotelchuck 2001). Regardless of its influence on birth weight, PNC represents a package of health-related services highly relevant to pregnant women (Shiono and Behrman 1995; Alexander *et al.* 1999). Adequate PNC (that is, care started early in the pregnancy, with regular visits to a provider thereafter) is an important mechanism through which the woman and medical personnel can become aware of existing maternal morbidities and/or problems in fetal development well before the onset of labour. If receipt of PNC is an indication of the degree of integration into the formal system of health care, then this, in turn, may have important implications for access to high quality medical care both before and after childbirth.

Information on PNC was accessible for a few specific API populations for the 1990s from the United States Census Bureau (2001). The API population as a whole fared well with respect to PNC, the same being true for most specific groups for which data are available. For example, in 1990, the number of API women who began to receive PNC in the first trimester of pregnancy ranged from 65.8% for Hawaiians to 87.0% for Japanese women in the United States, with Chinese and Philippino women receiving intermediate levels of PNC. By 1999, PNC coverage had improved for all race/ethnic groups. In the latter year, 79.6% of Hawaiian women received early care, and Japanese women were again the most likely to begin PNC during the first trimester. At both points in time, each specific API group for which we have data (with the exception of Hawaiians) was about as likely, and sometimes more likely, to receive early care as compared to whites.

A more rigorous assessment of race/ethnic disparities in PNC can be achieved via another measure, Kotelchuck's Adequacy of Prenatal Care Utilization (APNCU) index (Kotelchuck 1994), recently adopted by the US National Center for Health Statistics (NCHS). The APNCU is superior on several grounds; among other things, it takes into account not only the month that PNC was begun, but also the number of visits, adjusted for gestational age (Alexander and Kotelchuck 1996, 2001). We identified only one analysis that employs the APNCU in API comparisons. Hummer *et al.* (1999), using information from the 1989 to 1991 NCHS linked birth/infant death files, compared distributions on this index for both US-born and immigrant mothers for a number of specific race/ethnic groups. The proportion of women receiving inadequate care was 10.8% for US-born non-Hispanic white women and 12.8% for their foreign-born counterparts. Except for Cuban immigrants (9.3% inadequate), only the Chinese and Japanese populations had proportions of women experiencing inadequate care below 10% (6.2% and 7.9% for US-born Chinese and Japanese women, respectively, and 9.6% for Japanese immigrants). The proportion of Philippino and Other Asian women with inadequate PNC ranged from approximately 15% to over 20%, depending on maternal nativity, but the latter degree of disadvantage was eclipsed by the 37.9% of immigrants from Mexico for whom care, as measured by the APNCU, was inadequate.

Conclusions

This chapter gives a summary evaluation of the physical health, health care and mortality status of API populations living in the United States. Three general conclusions are drawn. First, when the API population is examined as an undifferentiated whole, its health status most often ranks at the top of race/ethnic groups in the country. In this sense, the 'myth of healthy Asian Americans and Pacific Islanders' (Chen and Hawks 1995) is not a myth at all. Looking across health outcomes, health behaviours, and different age and sex groups, the API population as a whole usually exhibits the most favourable health outcomes of any race/ethnic group in the United States. Second, however, there is substantial heterogeneity in health across API subgroups, as reflected in nearly all the outcomes we examined. Most of the larger Asian American subgroups (Chinese, Philippino, South Asian Indian, Korean and Japanese) exhibit fairly healthy profiles. The Vietnamese subgroup also fares well in most comparisons with other API subgroups and non-Hispanic whites and blacks. On the other hand, Native Hawaiians and Samoans, when data are available to examine, fare worst among all API subgroups, and also do not compare favourably with the non-Hispanic white population. Very limited data on some other smaller Asian American subgroups (Laotians, Cambodians and Hmong) also show poorer health compared to the larger Asian subgroups. Such findings reinforce the notion that the healthy profile of the API population as a whole is not generalizable to all of the API subgroups.

The larger size of the healthiest subgroups, then, is largely what drives the overall healthy profile of APIs. Third, health care use and access for APIs seem to be generally less favourable when compared to non-Hispanic whites. Whether this pattern is due to less need, greater use of alternative care, barriers due to the higher percentage of immigrants among many API groups or other factors is not assessed here. It is a concern, however, because less access to and use of care could lead to less favourable health outcomes in the future.

We would be remiss in closing if we did not reiterate the need for more data on specific API subpopulations. For example, even the NHIS is quite limited in terms of its ability to examine more than just a few specific API subgroups. The need for more and better data is perhaps most acute for some of the relatively small API subpopulations who do not seem to be faring as well (including Samoans, Hawaiians, Hmong, Laotians and Cambodians) as other and larger race/ethnic groups in the United States.

Acknowledgement

The authors gratefully acknowledge the support for this analysis provided by the National Institute of Child Health and Human Development: Grants RO1-HD41147 and RO1-HD49754.

References

Alexander, G. R. and Kotelchuck, M. (1996). Quantifying the adequacy of prenatal care: a comparison of indices. *Public Health Reports* **111**, 408–18.

(2001). Assessing the role and effectiveness of prenatal care: history, challenges, and directions for future research. *Public Health Reports* **116**, 306–16.

Alexander, G. R., Tompkins, M. E., Allen, M. C. and Hulsey, T. C. (1999). Trends and racial differences in birth weight and related survival. *Maternal and Child Health Journal* **3**, 71–9.

Anderson, R. N. (2001). Leading causes for 1999. *National Vital Statistics Reports* **49**, 1–87. Hyattsville, Maryland: National Center for Health Statistics.

Barnes, J. S. and Bennett, C. (2002). The Asian Population: 2000. *Census 2000 Brief*. Washington, DC: United States Census Bureau. http://www.census.gov/population/www/cen2000/briefs.html

Benjamins, M., Hummer, R. A., Eberstein, I. and Nam, C. (2004). Self-reported health and cause-specific adult mortality. *Social Science and Medicine* **97**, 1297–306.

Blackwell, D. L., Vickerie, J. L. and Wondimu, E. A. (2003). *Summary Health Statistics for U.S. Children. National Health Interview Survey, 2000. National Center for Health Statistics.* **10**(213).

Chen, M. S., Jr and Hawks, B. L. (1995). A debunking of the myth of healthy Asian Americans and Pacific Islanders. *American Journal of Health Promotion* **9**, 261–8.

Cho, Y. and Hummer, R. A. (2001). Disability status differentials across fifteen Asian and Pacific Islander groups and the effect of nativity and duration of residence in the U.S. *Social Biology* **48**, 171–95.

Chow, J. (2002). Asian American and Pacific Islander mental health and substance abuse agencies: organizational characteristics and service gaps. *Administration and Policy in Mental Health* **30**, 79–86.

Cramer, J. (1987). Social factors and infant mortality: identifying high-risk groups and proximate causes. *Demography* **24**, 299–322.

Davis, J., Busch, J., Hammatt, Z., *et al.* (2004). The relationship between ethnicity and obesity in Asian and Pacific Islander populations: a literature review. *Ethnicity and Disease* **14**, 111–19.

De Santo, D., Culhane, J., Guaschino, S., McCollum, K. and Weiner, S. (2003). Diagnosis of intracardiac echogenic focus and request of amniocentesis among Asian pregnant women. *Obstetrics and Gynecology* **101**(Supplement 1), 118S.

Frisbie, W. P. and Song, S. (2003). Hispanic pregnancy outcomes: differentials over time and current risk factor effects. *Policy Studies Journal* **31**, 237.

Frisbie, W. P., Forbes, D. and Pullum, S. G. (1996). Compromised birth outcomes and infant mortality among racial and ethnic groups. *Demography* **33**, 469–81.

Frisbie, W. P., Cho, Y. and Hummer, R. A. (2001). Immigration and the health of Asian and Pacific Islander adults in the United States. *American Journal of Epidemiology* **153**, 372–80.

Gortmaker, S. L. and Wise, P. H. (1997). The first injustice: socioeconomic disparities, health services technology and infant mortality. *Annual Review of Sociology* **23**, 147–70.

Grieco, E. M. (2001). The Native Hawaiian and Other Pacific Islander population. *Census 2000 Brief.* Washington, DC: US Census Bureau.

Grunbaum, J. A., Lowry, R., Kann, L. and Pateman, B. (2000). Prevalence of health risk behaviors among Asian American and Pacific Islander high school students. *Journal of Adolescent Health* **27**, 322–30.

Hack, M., Klein, N. K. and Taylor, H. G. (1995). Long-term developmental outcomes of low birth weight infants. *The Future of Children* **5**, 176–96.

Health Resources and Services Administration (2001). Indicators of children's well-being. *America's Children 2001.* Vienna, VA. http://childstats.gov.

Hoyert, D. L. and Kung, H. C. (1997). Asian or Pacific Islander mortality, Selected States, 1992. *Monthly Vital Statistics Report* **46**(Supplement 1). Hyattsville, Maryland: National Center for Health Statistics.

Hoyert, D. L., Arias, E., Smith, B. L., Murphy, S. L. and Kochanek, K. D. (2001). Deaths: final data for 1999. *National Vital Statistics Reports* **48**, 1–113. Hyattsville, Maryland: National Center for Health Statistics.

Hummer, R. A., Biegler, M., De Turk, P. B., *et al.* (1999). Race/ethnicity, nativity, and infant mortality in the United States. *Social Forces* **77**, 1083–118.

Idler, E. and Benyamini, Y. (1997). Self-rated health and mortality: a review of 27 community studies. *Journal of Health and Social Behavior* **39**, 21–37.

Institute of Medicine (1988). *Reaching Mothers, Reaching Infants*. Washington, DC: National Academy Press.

Kiely, J. L. and Susser, M. (1992). Preterm birth, intrauterine growth retardation, and perinatal mortality. *American Journal of Public Health* **82**, 343–5.

Kline, J., Stein, Z. and Susser, M. (1989). *Conception to Birth: Epidemiology of Prenatal Development*. New York: Oxford University Press.

Kotelchuck, M. (1994). An evaluation of the Kessner Adequacy of Prenatal Care Index and a proposed adequacy of prenatal care utilization index. *American Journal of Public Health* **84**, 1414–20.

Lauderdale, D. S. and Kestenbaum, B. (2002). Mortality rates of elderly American populations based on medicare and social security data. *Demography* **39**, 529–40.

LeClere, F. B., Jensen, L. and Biddlecom, A. E. (1994). Health care utilization, family context, and adaptation among immigrants to the United States. *Journal of Health and Social Behavior* **35**, 370–84.

Lin, K. and Cheung, F. (1999). Mental health issues for Asian Americans. *Psychiatric Services* **50**, 774–80.

Martin, J. A., Hamilton, B. E., Ventura, S. J., Menacker, F. and Park, M. M. (2002). Births: final data for 2000. *National Vital Statistics Reports* **50**(5). Hyattsville, Maryland: National Center for Health Statistics.

Matthews, T. J., Menacker, F. and MacDorman, M. F. (2002). Infant mortality statistics from the 2000 period linked birth/infant death data set. *National Vitals Statistics Reports*, Col. 50, No. 12, Tables A and C. Hyattsville, Maryland: National Center for Health Statistics.

McCormick, M. (1985). The contribution of low birth weight to infant mortality and childhood morbidity. *The New England Journal of Medicine* **312**, 82–90.

Morris, N. M. (1996). The influence of socioeconomic position on health – and vice versa. *American Journal of Public Health* **86**, 1649–50.

National Center for Health Statistics (2003). *Health, United States, 2003*. Hyattsville, Maryland: National Center for Health Statistics.

Pleis, J. R. and Coles, R. (2003). Summary Health Statistics for U.S. Adults: National Health Interview Survey, 1999. *Vital and Health Statistics*, **10**(2). Washington, DC: US Government Printing Office.

Reeves, T. and Bennett, C. (2003). The Asian and Pacific Islander population in the United States: March 2002. *Current Population Reports*, P20–540. Washington, DC: US Census Bureau.

Ro, M. (2002). Moving forward: addressing the health of Asian American and Pacific Islander women. *American Journal of Public Health* **92**, 516–19.

Rogers, R. G., Hummer, R. A. and Nam, C. B. (2000). *Living and Dying in the USA: Behavior, Health, and Social Differences in Adult Mortality*. San Diego: Academic Press.

Rosenberg, H. M., Mauer, J. D., Sorlie, P. D., *et al.* (1999). Quality of death rates by race and Hispanic origin: a summary of current research, 1999. *Vital and*

Health Statistics **2**(128), 1–13. Hyattsville, Maryland: National Center for Health Statistics.

Shiono, P. H. and Behrman, R. E. (1995). Low birth weight: analysis and recommendations. *The Future of Children: Low Birth Weight* **5**, 4–18.

Singh, G. K. and Miller, B. A. (2004). Health, life expectancy, and mortality patterns among immigrant populations in the United States. *Canadian Journal of Public Health* **95**, 114–21.

Singh, G. K. and Siahpush, M. (2002). Ethnic-immigrant differentials in health behaviors, morbidity, and cause-specific mortality in the United States: an analysis of two national data bases. *Human Biology* **74**, 83–109.

Srinivasan, S. and Guillermo, T. (2000). Toward improved health: disaggregating Asian American and Native Hawaiian/Pacific Islander data. *American Journal of Public Health* **90**, 1731–4.

Thamer, M., Richard, C., Casebeer, A. W. and Ray, N. F. (1997). Health insurance coverage among foreign-born U.S. residents: the impact of race, ethnicity, and length of residence. *American Journal of Public Health* **87**, 96–102.

United States Census Bureau (2001). *Statistical Abstract of the United States: 2001.* Washington, DC: United States Government Printing Office.

United States Department of Health and Human Services (2000). *Healthy People 2010: Understanding and Improving Health*, 2nd edn. Washington, DC: United States Government Printing Office. http://www.healthypeople.gov/.

Wilcox, A. J. and Russell, I. T. (1986). Birthweight and perinatal mortality: III. Towards a new method of analysis. *International Journal of Epidemiology* **15**, 188–96.

Yu, S. M., Huang, Z. J. and Singh, G. K. (2004). Health status and health services utilization among U.S. Chinese, Asian Indian, Filipino, and Other Asian/Pacific Islander children. *Pediatrics* **113**, 101.

10 Impacts of modernization and transnationalism on nutritional health of Cook Islanders

STANLEY J. ULIJASZEK

Introduction

As elsewhere in the Pacific, the population of the Cook Islands is characterized by the rapid increase of obesity, hypertension and type 2 diabetes, as well as profound out-migration across the past 40 years or so. Cook Islander migrations to New Zealand, and subsequently Australia, began in the 1950s, but have proceeded at such a rate that Cook Islander migrants now outnumber indigenes on the Cook Islands by about two to one (Ulijaszek 2005). The effects of economic modernization on blood pressure, body fatness and type 2 diabetes have been largely attributed to commonly measured risk factors, including dietary change associated with increased penetration of the world food system, and reduced physical activity associated with increased mechanization of life. Highly palatable and energy-dense foods are available, affordable and widely consumed in the Cook Islands (Ulijaszek 2002), and explanations invoking dietary change (Ulijaszek 2001a, 2002) and reductions in physical activity (Evans and Prior 1969; Ulijaszek 2001b) have been put forward for the high prevalence rates of obesity there.

In this chapter, trends in blood pressure, body size and diabetes status across recent decades are described for adult Cook Islanders living on Rarotonga, the most economically developed of the Cook Islands. Relationships between their blood pressure, body mass index (BMI) and fasting blood glucose are also described. These are then related to their diet, physical activity and different modernization factors in multiple regression models. In addition, blood pressure, BMI, blood pressure and fasting blood glucose of Cook Islanders living in Melbourne, Australia, are compared with their counterparts on Rarotonga.

Health Change in the Asia-Pacific Region: Biocultural and Epidemiological Approaches, ed. Ryutaro Ohtsuka and Stanley J. Ulijaszek. Published by Cambridge University Press.

Cook Islander populations

The population of the Cook Islands increased steadily across the first half of the twentieth century after a population collapse in the mid nineteenth century (Beaglehole 1957) soon after colonization by the British. There has been little increase in population between 1950 and the year 2000 (National Statistical Office 2001), despite there having been a dramatic decline in crude death rate and infant mortality rate since 1962 (Table 10.1). Migrants to New Zealand were few prior to 1950, with the number of Cook Islanders in that country being 103 in 1936, and around a thousand in 1951 (Beaglehole 1957). An estimate of emigrants from the Cook Islands, based on known resident population and population in excess of replacement level calculated from total fertility rates of 7.3 in 1955 and 3.3 in 1991 (assuming a linear decline across this period), suggests that by the year 2000, emigrants are likely to have outnumbered residents by at least two to one (Fig. 10.1). The emigrant values exclude those born in New Zealand and Australia, and are therefore smaller than the numbers given for those resident in those two countries in Fig. 10.2.

The vast majority of Cook Islander migrants live in New Zealand and Australia; however, migration to the latter country became numerically important only from the 1980s onward. Figure 10.2 shows the Cook Islander populations of New Zealand and Australia between 1986 and 2001; these are aggregate numbers of Cook Islanders both born in those countries and who are immigrants (New Zealand Census 1996; Australian Bureau of Statistics 2002). It is likely that remittances from relatives in these countries have provided much of the economic basis for the purchase of imported foods by the general Cook Islander population since the 1950s. Various accounts obtained from elderly subjects during fieldwork recalled to me in 1996 indicate that quite significant sums are remitted from

Table 10.1. *Demographic indicators, Cook Islands, 1962–2000*

	Crude death rate (/1000 population)	Infant mortality rate (/1000 population)	Proportional mortality indicator (deaths over the age of 50 years/all deaths)
1962	8.3	48	39
1982	6.5	21	68
2000	7.9	19	68
2004	7.3	17	

Source: Katoh (1988); World Bank (2001); Cook Islands Statistics Office (2006c).

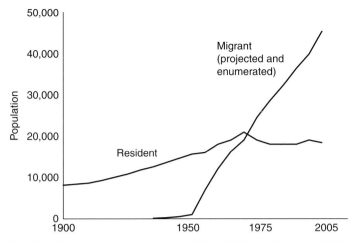

Fig. 10.1. The Cook Islander population, 1900–2005 (from Beaglehole 1957; National Statistical Office 1996, 2001; Cook Islands Statistics Office 2006c; Statistics New Zealand 2006).

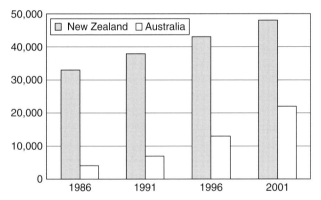

Fig. 10.2. Cook Islanders in New Zealand and Australia, 1986–2001 (Australian Bureau of Statistics 2002; Statistics New Zealand 2006).

adult offspring living overseas, predominantly to their parents. Indicators of economic growth, such as gross national product, show the Cook Islands to have among the fastest growing economies of the Pacific Island nations, against a high baseline in 1985 (Fig. 10.3). Food prices also dropped relative to the cost of all other consumer goods between the mid 1970s and mid 1990s. Figure 10.4 shows the consumer price index of food as a percentage of all consumer groups combined (food, housing, household operation, apparel, transport, tobacco and alcohol, miscellaneous).

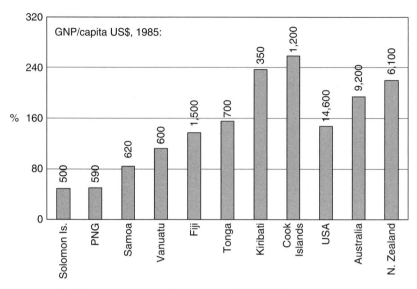

Fig. 10.3. Economic growth per capita, 1985–2000, Pacific Island nations,
United States, Australia and New Zealand (World Bank 2001).

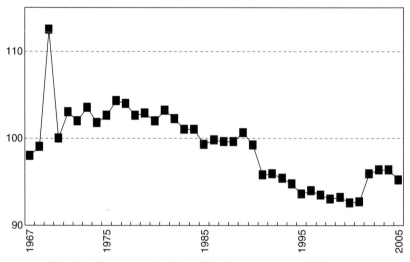

Fig. 10.4. Consumer price index of food as a percentage of all consumer groups
combined (food, housing, household operation, apparel, transport, tobacco and
alcohol, miscellaneous) (calculated from National Statistical Office, Cook
Islands 1996; Cook Islands Statistics Office 2006a,b).

The vast majority of Cook Islander migrants to Australia and New Zealand have settled predominantly in one of three cities: Auckland, Sydney and Melbourne. The population structure of Cook Islanders in New Zealand is the youngest of all the Pacific Islander communities resident in that country, which are in turn much younger than the total New Zealand population (Cook *et al.* 1999). Furthermore, adult Cook Islanders in Melbourne are much younger than the general adult population on Rarotonga. There has been scant study of nutritional health of Cook Islanders in New Zealand, and none at all in Australia. In 1998, I carried out a small survey among Cook Islanders resident in the Brunswick and Clayton districts of Melbourne to obtain the first measures of nutritional health in this growing migrant population.

Dietary change among Cook Islanders, 1952–2000

The diet of Cook Islanders in the Cook Islands has undergone considerable change across the past half century. Food availability and consumption is described from food frequency data collected on Rarotonga in 1952 by Fry (1957a,b), as part of fieldwork there in 1996 (Ulijaszek 2002) and among Cook Islanders in Melbourne, Australia, in 1997. This is compared with national food availability data for other nations in the Asia-Pacific region, and on the Pacific Rim across the period 1961 to 2002, using analyses from the Food and Agriculture Organisation (FAO 2006) FAOSTAT food balance dataset.

Early dietary surveys on Rarotonga show that a largely traditional Cook Islander diet (which is similar to diets elsewhere in the Pacific, including Tonga and Samoa) was consumed in 1952, but that this was already supplemented with significant amounts of imported foods (Fry 1957a,b). The typical rural diet of Cook Islanders on Rarotonga in 1952 largely comprised of the root crop taro (*Colocasia* species), banana, breadfruit, fresh fish and coconut, with significant amounts of bread and tea or coffee with sugar also being consumed (Fry 1957a,b). Table 10.2 gives a comparison of food consumption frequency data collected on Rarotonga in 1952 and 1996, and in Melbourne in 1997. Data collection strategies were different in the two Rarotonga surveys; the 1952 data shows the percentage of families having eaten different foods across a seven- to ten-day period, while the 1996 data gives the proportion of women having eaten different foods in the previous 24 hours, obtained using a short self-administered 16-item food-frequency questionnaire (FFQ) among 379 Cook Islanders aged 22 to 86 years. The short FFQ was chosen as a minimally invasive method because subjects were

Table 10.2. *Diet of Cook Islanders, from food frequency questionnaire, Rarotonga and Melbourne*

	Rarotonga			Melbourne
	% families 1952	% all adults 1996	% younger adults[a] 1996	% younger adults[a] 1997–98
Traditional staples (tubers)	100	91	89	61
Coconut	100	39	26	31
Fresh fish	95	37	34	23
Fruits and vegetables	62	100	69	75
Bread and rice	100	94	87	100
Tinned meat	83	72	85	93

Note:
[a] Proportion of adults aged 18–40 years consuming different food categories during the previous day.

resistant to the longer questionnaire piloted prior to the study. The short questionnaire was drawn up and pilot-tested as a consequence. The 16 items used in the FFQ were taro, cassava, banana, coconut, fresh fish, tinned fish, prawns, green vegetables, other vegetables, fruit of all types, beer, alcoholic liquor or wine, fresh meat, tinned meat, biscuits and bread. The 1997 Melbourne survey uses the same methods as the 1996 Rarotonga survey.

On Rarotonga, in both 1952 and 1996, bread and rice were eaten as often as traditional staples, although greater quantities of traditional staples were more likely to have been consumed than imported wheat and wheat products, and rice. The frequency of consumption of coconut and fresh fish was apparently much lower in 1996 than in 1952. Since fish and coconut cream are eaten in combination as part of the traditional diet of Cook Islanders (Fry 1957a; Ulijaszek 2002), and availability of coconut oil showed considerable decline across the period 1972 to 1987 (Ulijaszek 2003), it is likely that the difference in fish consumption frequency is a reflection of true dietary change. In 1952, fresh fish consumption was 1,733 g/family/day, while consumption of coconut cream (not whole coconut) was 1,393 g/family/day, both items being eaten daily by the vast majority of families (Fry 1957b). In 1996, various informants, especially older ones, lamented the decline in subsistence fishing to the author, again suggesting a decline in the consumption of fish caught using traditional methods. The pattern of food consumption in Melbourne is quite similar to that of the younger residents of Rarotonga, suggesting that modernization of the diet on both the capital Cook Island and among migrants in Australia may have proceeded in tandem.

Table 10.3. *Daily per capita (grams) availability, 1961*

	1961				2002			
	Fat				Fat			
	Vegetable	Animal	Total	Meat	Vegetable	Animal	Total	Meat
Cook Islands	174	5	179	8	78	49	127	131
Fiji	159	10	169	35	51	38	89	107
French Polynesia	31	12	43	74	60	62	122	279
Kiribati	36	2	38	44	84	21	105	93
New Caledonia	39	10	49	126	67	48	115	179
Papua New Guinea[a]	33	3	36	46	53	4	57	61
Vanuatu	21	8	29	32	58	29	87	91
New Zealand	2	59	61	286	33	82	115	292
Australia	9	43	52	285	58	73	131	298
United States	33	30	63	243	85	72	157	340
Japan	10	3	13	21	50	35	85	120

Note:
[a] No data for 2002, so data for 2000 given.
Source: From FAO (2006)

National food balance data (FAO 2006) shows the Cook Islands diet in 1961 to have contained more fat and less meat than that of New Zealand, the United States and Japan (FAO 2006) (Table 10.3). Dietary fat availability to Cook Islanders was similar to that of Fiji, but much higher than French Polynesia, Kiribati, New Caledonia, Papua New Guinea and Vanuatu. The Cook Islands had the lowest meat availability among all Pacific Island nations for which there is data. However, all Pacific Island nations showed an increase in meat availability between 1961 and the early 2000s. Dietary fat availability increased in French Polynesia, Kiribati, New Caledonia, Papua New Guinea and Vanuatu across the same period, but decreased in both Fiji and the Cook Islands.

Blood pressure, body fatness and type 2 diabetes

Blood pressure

In the Cook Islands, studies in 1953 and 1964 showed adults living on the remote island of Pukapuka to have low mean blood pressure (Murphy 1955; Prior *et al.* 1968), while adults on the modernizing island of Rarotonga in 1964 were shown to have an increase in blood pressure with age, and to be taller and heavier, with higher blood pressure than subjects on Pukapuka in the same year (Prior *et al.* 1968). A comparison of mean blood pressure of 425 adults on Rarotonga in 1996 (Ulijaszek and Koziel 2003) with the 1964 data showed no difference across this period for males, and a decline for females. Blood pressure of males was elevated at both times and across all age groups in both 1964 and 1996, but did not show any increase with increasing age in 1996. Blood pressure of females was not elevated in the youngest age group in both 1964 and 1996. However, the elevation with age in both systolic and diastolic blood pressure is much less marked in 1996 than in 1964 (Fig. 10.5). This difference is most likely due to more extensive use of anti-hypertensive medication among those diagnosed as having high blood pressure since 1964.

Factors associated with blood pressure among adults on Rarotonga in 1996 were examined in more detail, using forward stepwise multiple regressions with systolic and diastolic blood pressure as dependent variables, respectively. Independent variables used in these analyses were body size (BMI), diabetes status (fasting blood glucose), physical activity level (averaged across three days), diet and modernization factors (education, occupation, remittances from overseas, number of close relatives in New Zealand and Australia) (Ulijaszek 2001a,b,c). For males, systolic blood pressure in 1996 is positively associated with BMI, remittances from relatives overseas, age and fasting blood glucose, but negatively associated with use of anti-hypertension medication ($r^2 = 0.25$; $F = 6.2$; $p < 0.001$). For females, it is positively associated with remittances from relatives overseas, age and BMI ($r^2 = 0.15$; $F = 11.3$; $p < 0.001$). Diastolic blood pressure of males is positively associated only with remittances from relatives overseas ($r^2 = 0.06$; $F = 6.3$; $p < 0.05$). For females, it is positively associated with BMI, age and number of close relatives living in New Zealand, but negatively associated with use of anti-hypertensive medication ($r^2 = 0.13$; $F = 7.2$; $p < 0.001$). Expected interrelationships of blood pressure with increasing age, overweight, obesity and diabetes are confirmed for this population. However, this analysis also demonstrates the importance of remittances and transnational connections for elevated blood pressure, and supports the view that anti-hypertension medication is responsible for reducing high blood pressure levels in the face of multiple factors that would otherwise contribute to its elevation in this population.

A comparison of blood pressure, anthropometry and diabetes status of Cook Islanders resident in Melbourne, Australia, with those of their counterparts living on Rarotonga, measured in 1996, of the same age group, is given in Table 10.4. Cook Islanders aged 18 to 45 years in Melbourne had

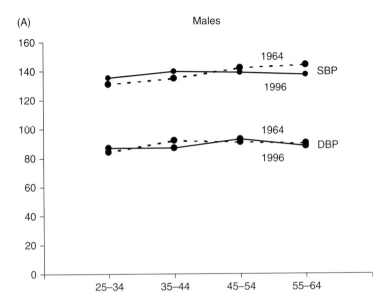

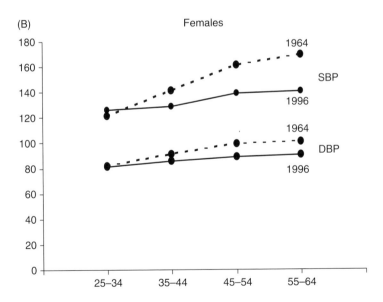

SBP – Systolic blood pressure
DBP – Diastolic blood pressure

Fig. 10.5. Blood pressure of Cook Islanders on Rarotonga, 1964 and 1996 (from Ulijaszek and Koziel 2003).

Table 10.4. *Anthropometry and blood pressure of Cook Islander males and females aged 18–45 years on Rarotonga and in Melbourne, Australia*
(A) Males

Sample	Rarotonga		Melbourne	
	16		23	
	Mean	SD	Mean	SD
Age (years)	36.1	7.0	37.3	7.6
Stature (cm)	174.6	7.2	170.8	5.4
Weight (kg)	101.2	13.3	92.8	18.8
BMI	33.3	4.5	31.7	5.8
Systolic BP	134.6	16.3	135.0	16.6
Diastolic BP	86.4	17.6	91.5	13.1
Fasting blood glucose (mg/dl)	6.4	4.0	6.0	3.1

Note:
All not statistically significant.

(B) Females

Sample	Rarotonga; Blood pressure and/or diabetes medication		Rarotonga; no blood pressure and/or diabetes medication		Melbourne		One way ANOVA
	10		43		42		
	Mean	SD	Mean	SD	Mean	SD	F
Age (years)	35.7	6.1	35.5	7.0	35.6	6.6	0.3
Stature (cm)	160.0	7.4	163.7	7.9	162.5	7.7	0.6
Weight (kg)	101.5*	18.1	88.8	16.7	89.2	19.5	1.7
BMI	39.8**	7.6	33.1	5.6	33.5	8.5	5.5**
Systolic BP	133.4	17.8	125.1	15.9	127.3	18.3	2.0
Diastolic BP	86.7	12.3	82.4	12.2	84.8	13.5	1.3
Fasting blood glucose (mg/dl)	7.0	5.6	5.8	4.1	6.2	5.2	1.8

Note:
*$p < 0.05$; **$p < 0.01$.

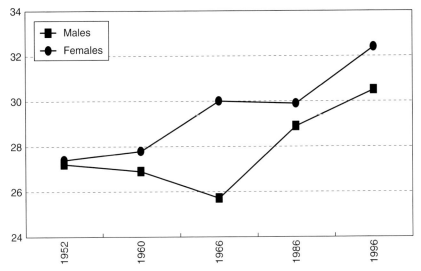

Fig. 10.6. Mean body mass index of Cook Islander adults, 1950s to the 1990s (data from Fry 1957b; Hunter 1962; Evans and Prior 1969; Katoh *et al.* 1990; Ulijaszek 2001a).

similar blood pressures to those on Rarotonga who did not take blood pressure medication at the time of survey.

Body size

The body size of adult Cook Islanders living on the main island of Rarotonga has increased since at least 1952 (Ulijaszek 2003) (Fig. 10.6). While males and females had similar BMIs in 1952 (27.2 kg/m^2, males; 27.4 kg/m^2, females), this increased by 5.0 kg/m^2 for females and 3.3 kg/m^2 for males by 1996. There has also been an associated secular trend towards increased stature of adults at least since 1966, largely because of improved nutritional status in childhood which has resulted in increased height as well as weight (Ulijaszek 2001a). The rates of secular trend in stature (cm/decade) and weight (kg/decade) of women between 1952 and 1996 are given in Table 10.5. In all cases, the rate of increase was higher between 1966 and 1996 than between 1952 and 1966, being much greater for weight than for stature. This suggests that there were much greater increases in the rates of overweight and obesity in the later period than the former, concomitant with greater rates of stature increase. Table 10.6 shows the proportion of adults that would be classified as obese by the World

Table 10.5. *Rates of secular increase in adult body size in Cook Islander women aged 20–40 years*

	Stature (cm/decade)		Weight (kg/decade)	
Period	20–29 years	30–39 years	20–29 years	30–39 years
1952–66	0.5	0.5	0.6	3.2
1966–96	0.63	1.0	7.3	5.1

Table 10.6. *Proportion of the sample classified as obese (body mass index (BMI) $\geq 30\,kg/m^2$), by age group*

	Sample size				Proportion with BMI > 30 (%)			
	Males		Females		Males		Females	
Age group (years)	1966	1996	1966	1996	1966	1996	1966	1996
20–29	83	5	73	18	20	100	30	56
30–39	44	10	30	27	11	70	50	63
40–49	39	27	39	48	13	59	56	58
50–59	39	40	32	85	10	53	53	56
60+	33	65	32	99	15	37	43	53
Total					14	52	44	57

Source: Adapted from Ulijaszek (2001a).

Health Organization (2000) criterion of BMI $\geq 30\,kg/m^2$. There has been a significant increase in obesity among both males and females (Ulijaszek 2001a). The proportion of obese males had increased from 14% in 1966 to 52% in 1996. Among females, obesity prevalence rates increased from 44% in 1966 to 57% in 1996. The secular trend towards increased body size and obesity on Rarotonga has been shown to be associated with economic modernization and level of education (Ulijaszek 2001b).

Factors associated with body size among adults on Rarotonga in 1996 were examined in detail, again using forward stepwise multiple regressions, with BMI as the dependent variable. Independent variables used in this analysis were blood pressure (systolic and diastolic), diabetes status (fasting blood glucose), physical activity level (averaged across three days), diet (different food and food categories) and modernization factors (education, occupation, remittances from overseas, number of close relatives in New Zealand and Australia). BMI was positively associated with systolic blood pressure but negatively associated with age and weekday physical activity level for males ($r^2 = 0.27$; $F = 8.4$; $p < 0.001$); it is positively associated with systolic blood pressure but negatively associated with age for females ($r^2 = 0.15$; $F = 16.9$; $p < 0.001$). The expected relationships between blood pressure, increasing age and BMI are confirmed for both males and females. This analysis also demonstrates an inverse relationship between weekday physical activity and BMI for males only. The variable 'occupation' reflects physical activity in the workplace, as well as income. Since it showed no relationship with BMI in this analysis, it suggests that physical activity during non-work time is more important for the control of BMI than is physical activity at work. Cook Islander females in Melbourne aged 18 to 45 years have similar mean stature, weight and BMI to women born on Rarotonga who did not take anti-hypertension medication (Table 10.4); the latter are heavier and with higher BMI than the former. Males in Melbourne have similar stature, weight and BMI to their counterparts on Rarotonga.

Type 2 diabetes

With respect to type 2 diabetes, 27% of females and 26% of males on Rarotonga had fasting blood glucose greater than 8 mg/l, a level suggestive of diabetes (World Health Organization 1985). Factors associated with type 2 diabetes were examined in more detail by using binary logistic regression analysis with diabetes status (fasting blood glucose less than or greater than 8 mg/l) as the dependent variable. The independent variables used were physical activity level, diet, the same modernization factors employed in the earlier analyses, BMI and blood pressure. For the males, type 2 diabetes was associated with the use of anti-diabetes medication (Wald statistic 8.1, $p < 0.01$) and systolic blood pressure (Wald statistic 4.0, $p < 0.05$) only. Among females, type 2 diabetes was associated with the use of anti-diabetes medication (Wald statistic 15.7, $p < 0.001$) and BMI (Wald statistic 3.8, $p = 0.05$). Both males and females in Melbourne had fasting blood glucose which was similar to their counterparts on Rarotonga, whether or not they took anti-diabetes medication.

Conclusions

Cook Islanders have experienced increased levels of obesity since at least the 1950s, attaining one of the highest prevalences of obesity in the world by the 1990s. The earliest surveys of blood pressure, in the 1960s, showed that increases had already taken place on the most modernized island, Rarotonga, but not elsewhere in the Cook Islands. Unlike body fatness and obesity, however, blood pressure stabilized among males, albeit at high levels, and declined for females, mostly because of the availability and use of anti-hypertension medication by a large proportion of the adult population, in the face of strong continuing modernizing influences. Type 2 diabetes prevalence was also high in 1996, but was kept in check largely by anti-diabetes medication. The importance of transnational connections (in the form of remittances and perhaps social influences of close relatives living in New Zealand) for elevated blood pressure is also confirmed, as is the importance of physical activity for lower BMI among males but not females. Diet does not emerge as a factor contributing to the prevalence of obesity, hypertension and diabetes in the Cook Islands, perhaps because transnational factors are stronger influences for all aspects of health-related behaviour, and not for dietary patterns alone. Perhaps the most important modernising influence has been that of medication for the control of diabetes and hypertension. No such medications are currently available for the control of obesity; if there were, Cook Islanders would be likely to embrace them wholeheartedly.

References

Australian Bureau of Statistics (2002). *Intercensal data, 1996.* <http://www.abs.gov.au> (accessed July–September 2002).

Beaglehole, E. (1957). *Social Change in the South Pacific. Rarotonga and Aitutaki.* London: George Allen and Unwin Ltd.

Cook, L., Didham, R. and Khawaja, M. (1999). *On the Demography of Pacific People in New Zealand.* Wellington: Statistics New Zealand.

Cook Islands Statistics Office (2006a). *Social Statistics.* http://www.stats.gov.ck
(2006b). *Consumer Price Index.* http://www.stats.gov.ck
(2006c). *2001 Census of Population and Dwellings.* http://www.stats.gov.ck

Evans, J. G. and Prior, I. A. M. (1969). Indices of obesity derived from height and weight in two Polynesian populations. *British Journal of Preventive and Social Medicine* **23**, 56–9.

Food and Agriculture Organisation (2006). *FAOSTAT Database,* <http://apps.fao.org/default.htm> (accessed January 2006).

Fry, P. C. (1957a). Dietary survey on Rarotonga, Cook Islands. I. General description, methods and food habits. *American Journal of Clinical Nutrition* **5**, 42–50.
(1957b). Dietary survey on Rarotonga, Cook Islands. II. Food consumption in two villages. *American Journal of Clinical Nutrition* **5**, 260–73.

Hunter, J. D. (1962). Diet, body build, blood pressure and serum cholesterol levels in coconut eating Polynesians. *Federation Proceedings* **21**, 36–43.

Katoh, K. (1988). Statistical and epidemiological research for health in the Cook Islands. In *People of the Cook Islands – Past and Present*, ed. K. Katayama and A. Tagaya. Osaka: City University Medical School, pp. 257–301.

Katoh, K., Yamauchi, T. and Hiraiwa, K. (1990). Blood pressure, obesity and urine cation excretion in two populations of the Cook Islands. *Tohoku Journal of Experimental Medicine* **160**, 117–28.

Murphy, W. (1955). Some observations on blood pressure in the humid tropics. *New Zealand Medical Journal* **54**, 64–73.

National Statistical Office (1996). *Cook Islands Annual Statistical Bulletin*. Rarotonga: National Statistical Office.

(2001). *Cook Islands Annual Statistical Bulletin*. Rarotonga: National Statistical Office.

Prior, I. A. M., Evans, J. G., Harvey, H. P. B., Davidson, F. and Lindsey, M. (1968). Sodium intake and blood pressure in two Polynesian populations. *New England Journal of Medicine* **279**, 515–20.

Statistics New Zealand (2006). *2001 Census of Population and Dwellings – Pacific Peoples Tables*. www.stats.govt.nz/census/pacific people table.htm.

Ulijaszek, S. J. (2001a). Increasing body size and obesity among Cook Islanders between 1966 and 1996. *Annals of Human Biology* **28**, 363–73.

(2001b). Socioeconomic status, body size and physical activity of adults on Rarotonga, the Cook Islands. *Annals of Human Biology* **28**, 554–63.

(2001c). Body mass index and physical activity levels of adults on Rarotonga, the Cook Islands. *International Journal of Food Science and Nutrition* **52**, 453–61.

(2002). Modernization and the diet of adults on Rarotonga, the Cook Islands. *Ecology of Food and Nutrition* **41**, 203–28.

(2003). Trends in body size, diet and food availability in the Cook Islands in the second half of the twentieth century. *Economics and Human Biology*, **1**, 123–37.

(2005). Modernisation, migration, and nutritional health of Pacific Island populations. *Environmental Sciences* **12**, 167–76.

Ulijaszek, S. J. and Koziel, S. (2003). Associations between blood pressure and economic modernization among adults on Rarotonga, the Cook Islands. *Anthropological Review* **66**, 65–75.

World Bank (2001). *World Tables Dataset*. Washington: World Bank.

World Health Organization (1985). *Diabetes Mellitus: Report of a WHO Study Group*. Technical Report Series 727. Geneva: World Health Organization.

(2000). *Obesity: Preventing and Managing the Global Endemic*. Technical report series No. 894. Geneva: World Health Organization.

11 *Mortality decline in the Pacific: economic development and other explanations*

ALISTAIR WOODWARD AND TONY BLAKELY

Introduction

The history of human mortality in the Pacific is one of peaks, troughs and contrasts. When James Cook first sailed to this part of the world in 1769, he commented on the vigour and good health of the peoples he met. Indeed, the evidence we now have suggests that Pacific Islanders of the time probably lived, on average, at least as long as their contemporaries in England. But mortality rates climbed steeply among indigenous populations following colonization, while European settlers in Australia and New Zealand were for a time the longest-lived people anywhere in the world. In the recent period, mortality rates have continued to fall in Australia and New Zealand, but again, the experience of indigenous and non-indigenous populations has not been the same. In this chapter, we describe the major patterns of mortality in the Pacific and review possible causes, concentrating especially on the role of economic development. We will focus most closely on recent changes in mortality in New Zealand, including the experience of Maori and Pacific peoples, and contrasts in mortality patterns between populations in New Zealand and Australia.

The longest-lived people in the world in 1840, when reliable national death statistics were first available, were Swedish women, whose life expectancy at birth (LEB) was about 45 years. Some 160 years later, maximum LEB is almost double that value (in 2002, Japanese women could expect to live on average for 85 years) (Oeppen and Vaupel 2002). This extraordinary increase in life expectancy (and the reduction in mortality that has driven it) defies explanation. Particularly striking is the fact that the improvement shows no sign of slowing down. Since the early observations in the 1840s, maximum LEB has increased by about two and a half extra years every decade in a remarkably regular fashion. The trend continues to the present day, calling into question conceptions of

Health Change in the Asia-Pacific Region: Biocultural and Epidemiological Approaches, ed. Ryutaro Ohtsuka and Stanley J. Ulijaszek. Published by Cambridge University Press.

biological limits on the survival of humans. The pattern of relentless improvement has not applied to individual countries, however. All having shown fluctuations, some have recently shown a flattening in mortality and LEB curves, and in a few the record has deteriorated (McMichael *et al.* 2004). Nations of sub-Saharan Africa and the former Soviet bloc in particular have experienced serious increases in mortality in the last two decades. However, the general tendency has been towards a decrease in mortality which has applied across all age groups, to both men and women, in high-income countries and in less developed parts of the world. Many explanations have been offered for this (Riley 2001). They include improved health care, prevention of epidemics and other mortality crises, better nutrition, improved living conditions (with provision of clean water, waste disposal and improved shelter) and education. While all these factors are likely to have played some part in mortality declines, none of them can account independently for the trends observed. For example, epidemics, famines and other extreme events cause only a small fraction of all deaths in the long run – the major force of mortality in most societies was, and remains, endemic disease. The reduction in mortality from infectious disease in the nineteenth century occurred prior to the development of effective medical interventions such as antibiotics (McKeown 1979). While nutrition is clearly related to mortality in populations with marginal intakes of dietary energy, protein and micro-nutrients, and heavy burdens of parasitic and infectious disease (Ulijaszek 1990), it is more difficult to identify a dietary explanation that accounts for the majority of continuing reductions in mortality in affluent countries. High levels of education have been commonly associated with improved population health, but are not a necessary prerequisite (Caldwell 1986).

Why is it important to tease out the reasons for mortality trends? Apart from the curiosity that we all have about human longevity, there are important policy issues at stake. These include the plans that need to be made for health and social services, since demands will depend on future mortality and morbidity patterns. No country can take health improvements for granted. Past gains may not be sustained as new cohorts age; for example, the full effects of child obesity may not become apparent until those individuals reach middle age and beyond. New threats may arise in the future, such as new forms of infectious diseases, consequences of civil disruption (as seen in the last 20 years in Eastern Europe), and environmental changes such as depletion of topsoil and water and increasing instability in the global climate. Consolidating historic gains, and preparing as best we can for new challenges, depends on understanding the determinants of population mortality.

There are many important causes of disability and reduced enjoyment of life that do not cause death, and so are not captured in mortality statistics. However, if the purpose is to trace changes over time in health status at the level of populations, and to compare groups, mortality serves reasonably well as a proxy for all lost health. There is much less reliable historic information on non-fatal health conditions than on causes of death. But what is available suggests that over most of the period described in this chapter, trends in the incidence of the major killing diseases have followed the trajectory of mortality. An important caveat is that in the last 20 to 30 years there have been significant improvements in survival for vascular disease and some cancers.

The relationships between economic development, reduced mortality and improved health have received considerable attention. In general terms, richer is undoubtedly better for mortality rates. Cross-sectional comparisons show that affluence, as measured in the standard fashion using indices such as gross domestic product (GDP), is associated with better health, particularly in low- and middle-income countries. In high-income countries, a relationship between wealth and mortality persists, but is much weaker. The same is true within countries; over time, economic development, with rising incomes and increasing levels of consumption, tends to accompany mortality decline. But which is the cause and which is the effect? The relationship probably runs both ways – affluence makes the conditions for good health, and a healthy population is more successful economically. Fogel (2004) has argued that increases in height and weight, resulting from better nutrition and disease control, account for about half the growth in national income in Britain since 1790.

But there are many inconsistencies in these patterns. For any given level of economic development, there is a wide spread in national mortality, especially at high incomes. Some countries have achieved spectacular reductions in mortality during long periods of very weak economic growth (Caldwell 1986). On the other hand, there have been times, for example in the United States and England in the middle of the nineteenth century, when rapid economic growth was accompanied by increasing mortality (Szreter 1997). Periods of rapid economic change tend to magnify inequalities in wealth and health, and improvements in average life expectancy may go alongside deteriorating conditions for minorities in the same population; later in this chapter we provide examples of this phenomenon in the Asia-Pacific region.

Common threads in the literature concerning economic influences on health include recognition that the relationship between economic development and health runs two ways, especially in low-income countries.

That is, poor health both causes poverty and is a consequence of it. And while some elements of economic development are health-enhancing but others are potentially health-damaging, distinguishing these and knowing how to alter the balance in favour of the positive are not straightforward. Other social changes probably need to take place in association with economic change in order that population health benefits are achieved (Sen 1999; Powles 2001). The Pacific region is potentially a fertile area in which to explore questions about the social and economic forces that most strongly influence mortality. The region is distinguished by diversity of cultural traditions; economic contrasts; large-scale migration; a very recent history of colonization and settlement; and throughout, very different experiences of indigenous and non-indigenous peoples.

Major trends in mortality in New Zealand

New Zealand is the southernmost Pacific state, and the most recently settled one. Polynesian peoples first came to these islands sometime between AD 900 and AD 1200, and for several centuries there was movement of goods and population between the new settlements and the source islands (principally the Cook, Marquesa and Society island groups) (King 2003). These ocean voyages, formidable in the distances covered and the navigational difficulties to be overcome, ceased for reasons unknown by about AD 1400. Contact with European explorers dates from the visit of Tasman in 1603, but there were no substantial encounters between the indigenous inhabitants (the Maori) and Europeans until the voyages of James Cook between 1769 and 1778. European settlement began in the early 1800s, and accelerated following the signing of the Treaty of Waitangi in 1840 between Maori chiefs and the British crown, which marked the establishment of New Zealand as a colony of Great Britain.

Estimates of mortality among Maori prior to European settlement are based principally on archeological evidence, especially of skeletal remains. The data are not strong but suggest that at the time of Cook's visits the health of Maori was at least as good as that of most European populations of the time. It is estimated that LEB was of the order of 32 years or more (Pool 1991), which exceeds the figures for England and France in the late 1700s. However mortality among Maori increased rapidly after European settlement took hold, along with declining fertility. As a consequence, the population reduced considerably in size, from about 86,000 in 1769 to about 50,000 in the early 1870s, and reaching its lowest level of about 42,000 by the end of the 1890s (Belich 1996).

Table 11.1. *Life expectancy at birth (years), New Zealand, Sweden and England and Wales, 1876, 1976 and 2000*

	About 1876	About 1976	About 2000
Males			
New Zealand			
Maori	25.3	63.4	69.0
Non-Maori	52.0	69.4	77.2
England and Wales	39.2	69.6	75.9
Sweden	45.3	72.1	77.6
Females			
New Zealand			
Maori	22.5	67.8	73.2
Non-Maori	54.2	75.9	81.9
England and Wales	42.5	75.8	80.6
Sweden	48.6	77.9	82.1

Sources: 1876 and 1976 data from Pool (1982). New Zealand 2000–02: www2.stats.govt.nz/domino/external/pasfull/pasfull.nsf/web/Hot+ Off+The+Press+New+Zealand+Life+Tables+2000-2002?open England and Wales 2001: http://www.statistics.gov.uk/CCI/nscl. asp?ID=7528 Sweden 2001: www.scb.se/templates/tableOrChart____25831.asp

In the late nineteenth century, life expectancy of Maori is estimated to have been less than half that of non-Maori New Zealanders (Table 11.1). But this was a period when the European newcomers thrived. It was a time of considerable immigration: the New Zealand non-Maori population grew much more rapidly in the first 50 years of the colony's life than comparable settlements in Australia, Canada or the United States (Belich 1996). Apart from the 'long depression' of the 1880s, this was a period of prosperity, based first on mining and forestry and then on agricultural industry, that capitalized on the country's temperate climate and remarkably fertile soils. The economic future of the colony was assured by the introduction of refrigerated shipping which took New Zealand's butter and meat to secure markets in the United Kingdom. Against this background, mortality rates for the settler group were among the lowest in the world. Indeed, from about 1860 to 1940, non-Maori New Zealand females had the highest life expectancy of any recorded worldwide (Oeppen and Vaupel 2002).

The selection processes that favoured relatively healthy migrants no doubt played a part in explaining the statistics for non-Maori New Zealanders, especially in the early years of this period. Migration, predominantly from England and Scotland, peaked in the early 1870s. But the British colonies of the same time in Australia did not match the high LEBs attained in New Zealand then. Other explanations for low mortality in New Zealand include the consistent and drought-free availability of plentiful and varied food supplies, especially of protein-rich foods, associated with abundant agriculture, the low density of settlement (even in the cities) and the early introduction of mass education and social services. New Zealand was a world leader in the introduction, before the turn of the century, of reforms such as free compulsory secular education for Maori and non-Maori peoples alike, old-age pensions, the vote for all adults (men and women, Maori and non-Maori) and industrial arbitration (Belich 1996). Rapid improvements in child survival in New Zealand around 1900 have been linked to changes in average family size. The birth rate for non-Maori New Zealanders almost halved between 1876 and 2001, falling considerably below that for Australia and England (Pool and Cheung 2002). This trend, caused largely by rising age at marriage, appeared first in the most prosperous parts of the country, and was later adopted elsewhere (Pool and Tiong 1991).

What role did economic factors play? The economy of New Zealand was strong by international standards from the 1870s onwards, and was distinguished by high levels of government spending on transport, housing and utilities. Public expenditure on infrastructure was 'almost seven times that spent in an average developed economy between 1865 and 1914. It was nearly three times the amount spent by South African governments and twelve times that spent by Canadian governments' (Belich 1996).

In this period, wages in New Zealand were higher than in Britain or in Australia, food was cheap and 'nearly all the people had a horse to ride to work' (Belich 1996). National policies at the turn of the century tended to be egalitarian. For example, legislation was passed to improve working conditions of most vulnerable employees, while land reforms divided up large estates. Alongside such social and institutional reforms, these relatively favourable economic conditions may have influenced health in a variety of ways. It is no surprise, for instance, if better educated mothers with smaller families and the means to provide healthy diets, backed up by health and welfare services such as Plunket (established in 1907) and the School Medical Service (which started in 1912), should have raised children more likely to survive to adulthood (Tennant 1991).

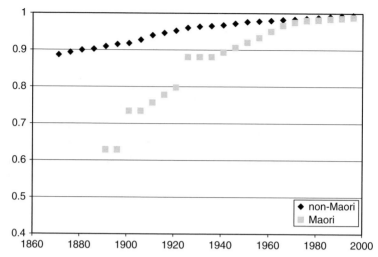

Fig. 11.1. Probability of survival from birth to one year of age, males, New Zealand non-Maori and Maori

The experiences of indigenous and non-indigenous populations in this period were linked by the massive transfer of land from Maori to non-Maori ownership. This was achieved in the first instance by confiscation of property from tribes that fought against the Crown in the New Zealand Wars between the 1840s and 1860s, and later by legislation that individual-ized Maori land titles and judicial processes (the Native Land Courts) to channel the transfer of ownership. For the European settlers, this provided farms, employment, relatively cheap and plentiful food supplies, access to lucrative export markets, and economic prosperity that rivalled any other country at the time (Hawke 1985). For Maori, loss of land was at the heart of the dispossession and dislocation that destroyed their economic, social and cultural base, and contributed in many ways to their heavy burden of disease and premature mortality. The effects included disruption of agri-culture, the break-up of social networks, crowding, deepening poverty and demoralization. Not all parts of the country were affected to the same extent – Pool (1991) has shown that Maori child-to-woman ratios in the later part of the 1800s were lowest in regions in which land alienation was taking place most rapidly.

The recovery of Maori in the twentieth century is illustrated by Fig. 11.1, which traces infant survival rates from 1900 onwards. The period of most rapid convergence of Maori and non-Maori statistics, 1900 to 1940, was a time of very mixed economic fortunes. Fluctuations in the overseas market for wool and meat meant that there were years of reduced incomes and high unemployment. The worldwide depression at the end of the 1920s affected New Zealand severely. But over this time there was a substantial improvement in the health of Maori, both in absolute terms, and relative to the remainder of the New Zealand population. Ian Pool, currently New Zealand's senior demographer, attributes the turnaround to (1) improved health and social services (including the coordinating and mobilizing activities of the new Department of Health); (2) leadership provided by senior Maori politicians and medical staff (Maui Pomare, Peter Buck and James Carroll among them); (3) the community-based health and development programmes that these men inspired; and (4) the diminishing force of epidemic disease due to increasing levels of natural immunity (Pool 1991; Pool and Cheung 2003).

Mortality trends from about 1950 onwards are shown in Fig. 11.2. The Maori life expectancy series has been corrected for substantial undercounting of Maori deaths in the 1980s and early 1990s (Blakely *et al.* 2005a). Such undercounting was less problematic in the late 1990s and early 2000s. For the 1950s to 1970s, any undercounting of Maori deaths is unknown and unquantifiable. But the overall picture, of a significant narrowing of the gap between Maori and non-Maori life expectancy in the first 30 years after the Second World War, followed by divergence in the late 1980s and 1990s, is robust. The equal improvements in Maori and non-Maori life expectancy in the early 2000s is promising for the future, but should be treated only as preliminary evidence at this point.

The 30-year period after the Second World War was not a time of social targeting nor was the period distinguished by a high level of awareness of the needs and entitlements of indigenous people. It preceded major institutional and policy changes such as the establishment of the Waitangi Tribunal to compensate Maori for breaches of the Treaty, substantial investment in health care by Maori for Maori and initiatives to promote the Maori language. Possible upstream reasons for the convergence of mortality include the improved access to health and social services that followed the large-scale move of Maori to towns and cities; relatively high levels of employment ensured by investment in public works; and population-wide interventions in health care, including immunization, that brought the greatest return to groups with the highest levels of disease. At a more biomedical level, much of the improving life expectancy of Maori in the 1950s to 1970s can be attributed to reduced infant mortality (Fig. 11.1) and waning health impacts of tuberculosis. It is also important to examine the non-Maori series and note that the peaking of ischaemic heart disease mortality in the 1960s and 1970s was responsible for slowing

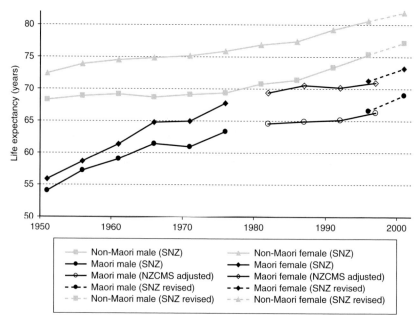

Fig. 11.2. Maori and non-Maori life expectancy trends, 1950–2002, using official Statistics New Zealand (SNZ) series and New Zealand Census-Mortality Study (NZCMS) estimates, adjusted for numerator–denominator bias

Notes:

- The non-Maori (SNZ) series from 1951 to 1996 is unadjusted for numerator–denominator bias. However, because non-Maori are the larger population, the point estimates are little affected by numerator–denominator bias.
- The Maori (SNZ) series from 1951 to 1976 is unadjusted for numerator–denominator bias. Thus, we cannot be certain of the accuracy of each point estimate. However, the large increase in Maori life expectancy during this period is not an artefact of numerator–denominator bias.
- The Maori (NZCMS adjusted) series from 1980–84 to 1996–99 is for the prioritized Maori ethnic group, and is adjusted for numerator–denominator bias.
- The Maori (SNZ revised) and non-Maori (SNZ revised) series are for the prioritized ethnic groups, and are adjusted for numerator–denominator bias in the late 1990s by SNZ using NZCMS adjusters. Preliminary comparisons by SNZ of census and mortality data from ethnicity record for 2001 suggests that these numerator–denominator differences are not significant enough to reliably adjust death numbers by age, sex and ethnicity (Statistics New Zealand 2004).

down the increase in the rate of life expectancy. The Maori were affected similarly by ischaemic heart disease across this period, but this did not curtail the overall trend of increased life expectancy.

From 1980 onwards, we have access to unbiased estimates of mortality by ethnicity, including Pacific Islanders in New Zealand, as well as Maori.

Table 11.2. *Life expectancy at birth by ethnicity and sex,*
adjusted for numerator–denominator bias, 1980–99

	1980–84	1985–89	1990–95	1996–99
Males				
Maori				
Prioritized	64.6	64.9	65.2	66.3
Sole	63.3	64.1	64.3	64.0
Pacific				
Prioritized	66.7	66.9	68.6	67.9
Sole	67.4	66.9	68.2	67.7
Non-Maori, non-Pacific	70.9	71.9	73.7	75.7
Females				
Maori				
Prioritized	69.4	70.5	70.2	71.0
Sole	68.0	69.6	69.2	68.7
Pacific				
Prioritized	74.2	73.2	75.0	74.2
Sole	74.8	72.9	74.6	73.9
Non-Maori, non-Pacific	77.2	77.9	79.4	80.8

Source: Data from Blakely *et al.* (2005b).

After the Second World War, there were just over 2,100 people from the Pacific Islands in New Zealand (about 0.1% of the national population). This number grew in the 1950s and 1960s to more than 65,000 at the 1976 census. Since then, high fertility rather than migration has been the major driver of Pacific Islander population growth in New Zealand, the 2001 census recording almost 232,000 Pacific people, 6.5% of the total population (Statistics New Zealand 2002). The category of 'Pacific' sweeps together many different ethnic and national groups, with distinct languages, cultures and traditions. In 2001, the largest group was Samoan (just under half of all Pacific Islanders living in New Zealand), followed by Cook Islanders (about 22%), Tongans (17%), Niueans (8%), and the remainder made up of Tokelauans, Fijians, and about 7,000 people from Tuvalu, Tahiti and Kiribas combined (Statistics New Zealand 2002). Routinely collected mortality statistics are not disaggregated by island group, if only because data would become too sparse for precise estimates.

It is evident from Table 11.2 that neither Maori nor Pacific people enjoyed much improvement in LEB in the 1980s and 1990s, while non-Maori, non-Pacific people enjoyed substantial gains. This relative stasis for Maori and Pacific people was due to some decreases in mortality from cardiovascular disease and injury (albeit lesser decreases in relative terms

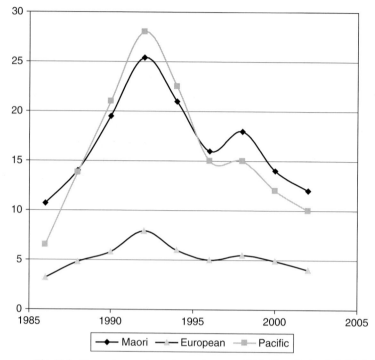

Fig. 11.3. Unemployment rate (annual average %) in New Zealand, 1986–2002,
by ethnicity (data from Ministry of Social Development 2004).

compared to non-Maori non-Pacific peoples) being offset by increases in
cancers and suicide deaths.

What determinants may have been responsible for the widening mortality inequalities during the 1980s and 1990s? We suggest that rapid social
change and macroeconomic reforms during this period penalized the most
vulnerable groups in New Zealand, and this included Maori and Pacific
peoples. One example is the different experience of the surge in unemployment that followed restructuring of the public sector (Fig. 11.3). Prior to
the 1980s, unemployment was low in New Zealand, below 5% for many
years, and this figure applied to all ethnic groups. But the changes of the
1980s had differential effects. At the peak, one in four Maori and Pacific
adults of working age were registered as unemployed, three times the
proportion of Europeans. Furthermore, Maori and Pacific unemployment
rates took longer to recover than non-Maori non-Pacific rates. This differential impact can be understood by noting that Maori and Pacific people
were more likely to be in occupations that were hit hardest by the structural reforms of the late 1980s and 1990s.

Similar patterns are seen in income trends. Real (inflation-adjusted) incomes of Maori and Pacific households dropped during this period and did not recover to the level they had reached in the early 1980s (Mowbray 2001). For example, the percentage of families with one or more European adults with a net-of-housing-cost income below 60% of the median income was 12.6% in 1987 and 1988, peaked at 23.3% in 1991 and 1992, and fell to 18.5% in 1997 and 1998. By contrast, the equivalent percentages for families with one or more Maori adults were 14.0%, 41.0% and 31.2%, respectively, and with one or more Pacific adults were 24.4%, 48.9% and 44.3% respectively (Te Puni Kokiri 2000). As with unemployment, the incomes of Maori and Pacific peoples were hit harder and for longer than non-Maori non-Pacific peoples. It has been suggested that there was also a significant drop in social assistance going to Maori communities in the early 1990s following the 'mainstreaming' of Maori social services (Blakely *et al.* 2005b).

How much of the differential life expectancy trends between Maori, Pacific and other groups can be explained by conventional measures of social and economic disadvantage? It is clear that income, occupation and education explain some, but not all, of the difference at any one time. For example, socioeconomic factors such as occupational class (Pearce *et al.* 1993), small area deprivation (Salmond and Crampton 2000), income and education (Blakely *et al.* 2002), when considered separately, each explain about a quarter to a third of ethnic disparities in mortality. Within categories of any given socioeconomic factor, Maori mortality rates are consistently higher than those of non-Maori peoples. A similar pattern is apparent for Pacific peoples, although interpretation is more difficult because of the highly skewed distribution of the Pacific population by socioeconomic status (40% of them reside in the most deprived 10% of small areas).

It is not enough merely to say that large proportions of the difference in mortality rates between ethnic groups can be explained by socioeconomic position. Rather, one has to note in addition that socioeconomic resources are unevenly distributed by ethnicity, and then ask why this is so. Both Maori and Pacific Islanders are greatly over-represented in the high deprivation areas and under-represented in the most affluent parts of the country. In 1991, 50% of Maori lived in the most deprived parts of the country compared to 15% of other ethnic groups (Salmond and Crampton 2000). Plainly there must be important prior causes in a sequence leading to material disadvantage and poor health. For Maori, the impacts of colonization that shifted economic resources from the indigenous to the colonizing population, with flow-on effects for generations thereafter, is a key explanation. For the more recently arrived Pacific population, it is relevant to consider New Zealand migration policies in the mid-twentieth century, which aimed to recruit such people to fill jobs of low socio-economic status in construction, forestry and service industries.

However, not all of the ethnic disparities in health can be explained by socioeconomic factors. Simultaneous adjustment for a range of socio-economic factors explains much, but far from the majority of ethnic disparities in health. Other factors must also be at play such as (1) access to, and quality of, health services that is not explained by socioeconomic position; (2) impacts of racism; (3) cultural, lifestyle and health behaviours not explained by socioeconomic position; and possibly (4) the health-damaging impacts of stress. Genetic and other biological differences between ethnic groups are unlikely to contribute much to ethnic disparities in health (Pearce *et al.* 2004) although gene–environment interactions, such as the ones involving diet in the predisposition to diabetes, should not be totally discounted.

The New Zealand experience is not an isolated one. For instance, the countries of the former Soviet Union also underwent a time of social upheaval, which was reflected in their mortality statistics. The social changes experienced by them were more drastic than those in New Zealand, and so were the responses in the health statistics. But a common pattern emerged whereby the effects of social disruption were strongest in the most disadvantaged sections of these populations. For instance, Plavinski *et al.* (2003) followed two cohorts of Russian men through the 1980s and 1990s. The increase in mortality during this period was concentrated amongst the least educated participants in the study. While there was no apparent change in mortality for men with university education, among those with less than high school education death rates rose by 75%.

A comparison of mortality trends between Australia and New Zealand

Further insights into mortality trends, and the role of economic change in their production, are provided by comparisons of New Zealand and its closest neighbour in the Pacific, Australia (Woodward *et al.* 2001). The two countries are very different in their physical characteristics, one being continental, geologically aged and arid, the other small, steep, wet and geothermally active. But they have much in common in other respects. Both sit on the southern edge of Asia, but their social systems and cultures are principally products of British settlement. The populations are largely

European in ancestry, with indigenous minorities, but with substantial recent migration from Asia and the Pacific. In 2001, LEB in New Zealand was 81.1 years for females and 76.3 years for males (Statistics New Zealand 2004). The corresponding figures for Australia were 82.1 years (females) and 76.6 years (males) (Australian Institute of Health and Welfare 2005). A difference of a year in life expectancy may seem trivial for an individual, but represents a substantial loss of years of life for a population. The magnitude of the effect can be illustrated by two examples: (1) in New Zealand, deaths due to unintentional injury reduce LEB by 1.1 years; (Ministry of Health 2000); and (2) if all smoking-related deaths were avoided, LEB would rise by 1.5 to 2 years (Bronnum-Hansen and Juel 2000).

New Zealand and Australia are, by global standards, both affluent countries, well past the 'shoulder' in graphs that compare average national wealth with health measures such as life expectancy. Cross-sectional comparisons show steep increases in LEB with increasing national wealth, to a point equivalent to a gross national product of roughly US$5,000 per person, and then a marked flattening of the curve. However, the line is not absolutely flat at higher levels of GDP. An analysis of data around the year of 1993 from all countries with a GDP over US$10,000 (adjusted for purchasing power parity) found a correlation between GDP per person and LEB of 0.51 (Lynch *et al.* 2000). Such an association explains about a third of the difference between life expectancy in New Zealand and Australia; in the mid 1990s GDP per person in Australia was approximately 25% greater than in New Zealand.

Mortality among Aboriginal people and Torres Strait islanders, the indigenous peoples of Australia, is reported variably between States and is thought to be underestimated overall. Even so, the poor health of this group stands out in almost every regard, and the gap between indigenous and non-indigenous life expectancies of almost 20 years is more than twice that in New Zealand. Aboriginal people and Torres Strait islanders make up a much smaller fraction of the national population than do Maori. If indigenous peoples were present in similar proportions, life expectancy in Australia would be more than 1.5 years below that of New Zealand.

Until the 1970s, LEB was greater in New Zealand than in Australia. In Fig. 11.4, LEB is charted for men in Australia and New Zealand, from 1961 to 2000. Data for women follow a similar pattern, although with rather less difference between the two countries. LEBs for the two countries followed similar unchanging trajectories until around 1970, when life expectancy began to improve, but at a faster rate in Australia than in New Zealand. This period of divergence lasted for about a decade, and from the

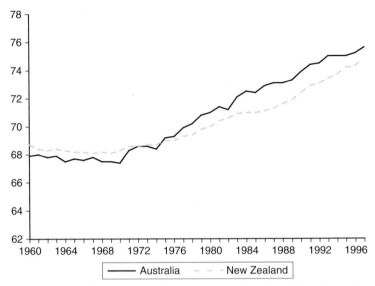

Fig. 11.4. Life expectancy at birth for males, Australia and New Zealand, 1961–2000 (data from Organization for Economic Cooperation and Development 2006. http://juno.sourceoecd.org/vl=1174864/cl=75/nw=1/rpsv/home.htm).

early 1980s onwards the two countries tracked more or less in parallel, with Australia enjoying between 2% and 3% higher LEB than New Zealand, with some convergence in the past five years or so.

This mortality 'crossover' cannot be explained by the poorer health of Maori and Aboriginal people, since time-series analyses show the same transition to have taken place for Maori and non-Maori alike. Indeed, during the period that LEB in New Zealand slipped behind that of Australia, Maori LEB was improving at its fastest rate ever. The pattern was driven by declining mortality in middle age and later life, since this is when most deaths occur; trends in infant mortality were very similar in the two countries. The changes observed in coronary mortality closely match the trends for all deaths: decline in coronary heart disease mortality occurred at about the same time in both countries, but the decline in deaths from heart disease was much sharper, initially, in Australia. Among the other major causes of mortality, deaths from stroke show a very similar mortality crossover to that observed for coronary heart disease. Trends in cancer deaths have a different pattern. Mortality from this cause has been consistently higher in New Zealand than Australia, especially among women, and this may be explained in large part by differences in smoking rates (Laugesen and Swinburn 2000).

Why did the course of decline differ in Australia and New Zealand? We focus here on differences in social and economic conditions that may be relevant, with brief reference to other factors. It is striking how closely New Zealand's economic performance matches its international rankings in population health. Immediately after the Second World War, New Zealand did as well as Australia, or better, on most economic indicators; until the 1960s, per capita GNP was higher in New Zealand than Australia. But at the end of that decade, New Zealand experienced a sharp fall in terms of trade and a loss of traditional markets for agricultural exports, preceding Britain's formal entry into the European Economic Community in 1972. The oil shocks of the early 1970s also had severe impacts on economic activity. Australia was subject to the same external forces, but enjoyed a greater diversity of raw resources and a much stronger manufacturing base. A similar pattern is also apparent when New Zealand is compared to per capita GDP of Organization for Economic Co-operation and Development (OECD) nations, where a slow decline in relative terms took place from the 1950s, the sharpest fall occurring in the 1960s and 1970s (Easton 1997).

There is no doubt that the New Zealand economy fared poorly by comparison with Australia at the time that mortality decline tipped to the advantage of the latter nation. But had conditions become so adverse that they could have caused these changes in death rates? In absolute terms, the differentials were modest, but it may be that the nature of the change was at least as important as the size of the decline in New Zealand. In the mid 1970s New Zealand began a period of rapid social change, swinging from the 'Think Big' extravagances of the Muldoon era, which involved large-scale public works such as hydroelectric development, to the abrupt winding down of public services by Roger Douglas and colleagues in the mid to late 1980s (King 2003). Australia moved in similar directions, but more slowly. There were incremental shifts in areas such as benefits, employment and public utilities, rather than relatively sudden changes of direction. As an example, the telecommunications industry was privatized in New Zealand by the early 1990s, but not until the end of that decade in Australia. Not only was the speed of change greater in New Zealand, but in some respects New Zealand had more to lose economically. For instance, there had been almost no unemployment in New Zealand since the 1940s. When employment services for job seekers were established in the 1970s, senior staff had to be recruited from Australia, as there were few locals with relevant experience.

The deterioration in the economic performance of New Zealand relative to Australia is reflected in the subsequent change in mortality, with a lag period of about five years. Eastern Europe in the 1990s showed that deteriorating social conditions can have an effect which is almost immediate on national mortality rates, and not just the causes of death such as injury, which are recognized to have short latencies (Zatonski *et al.* 1998). But it is an open question whether the more moderate changes that occurred in New Zealand could have had similar impacts.

Mortality patterns elsewhere in the Pacific

There are substantial uncertainties around even the most recent estimates of LEB in Pacific Island nations, due to under-enumeration of deaths, lack of accurate information on background populations and fluctuations in rates due to small numbers in some countries (Taylor and Bampton 2005). Although the estimates of life expectancy shown in Table 11.3 should be treated with caution, it is plain that mortality rates vary greatly, being highest in Papua New Guinea (one of the countries with the least complete statistical collections), intermediate in western Polynesia, and lowest in New Zealand and Australia. However, the greatest differences in LEB are found within countries. For example, the gap between the national figures for Papua New Guinea and Australia (the longest-lived nation in the Pacific) is less than the difference between Aboriginal and non-Aboriginal Australians. Within this group of nations, national income is

Table 11.3. *Gross national income for Pacific Island nations, 2003, and life expectancy at birth (males and females combined)*

	Gross national income (US$)	Life expectancy at birth (years)
Papua New Guinea	510	59.8
Marshall Islands	2710	62.7
Nauru	–	62.7
Kiribati	880	64.1
Solomon Islands	600	65.4
Fiji	2360	67.3
Vanuatu	1180	67.7
Samoa	1600	68.2
Niue		70.3
Tonga	1490	70.7
Cook Islands		71.6
New Zealand	15870	78.9
Australia	21650	80.4 (total population) 59.2 (Aboriginals)

Source: Data from World Health Report (2004); Australian Institute of Health and Welfare (2005).

very weakly correlated with life expectancy (excluding Australia and New Zealand, where the correlation coefficient is 0.22).

Conclusions

There are substantial variations in health and wealth between countries in the region, but the greatest differences in mortality are found between indigenous and non-indigenous populations. Social and economic deprivation undoubtedly contributes to the gap in life expectancy between Maori and non-Maori, and between Australian Aboriginal and non-Aboriginal people. But to attribute ethnic differences in mortality wholly to economic factors is unsatisfactory for two reasons. First, economic influences are not the whole story. Within categories of deprivation, rates for indigenous peoples are worse than those for non-indigenous peoples. Second, simply observing that wealth and advantage are unevenly distributed begs the question of why this should be so.

In the long run, mortality has declined considerably in the Pacific, adding almost 30 years to LEB in the last 130 years in New Zealand and Australia. But the trend has not been continuous. There has been a weak relationship between the surges in life expectancy and periods of economic advancement, apparent, for instance, in the way that statistics for New Zealand and Australia have tracked rather differently in the past 30 years. But the national averages do not tell the full story. In New Zealand in the second half of the nineteenth century, for example, and in the last decades of the twentieth century, the trend for Maori mortality diverged substantially from that for non-Maori. Others have shown that industrialization in Europe in the 1800s brought new health problems, particularly for those with few resources, at the same time that empires were built and fortunes amassed by others. We conclude that there is a rather similar picture in the Pacific region. The social consequences of periods of rapid social and economic change have been mixed, bringing benefits to some, but putting others in jeopardy, and overall, increasing inequalities in wealth and health.

References

Australian Institute of Health and Welfare (2005). *Mortality Data*. www.aihw.gov.au/mortality/data/life_expectancy.cfm

Belich, J. (1996). *Making Peoples. A History of the New Zealanders*. Auckland: Penguin Books.

Blakely, T., Kiro, C. and Woodward, A. (2002). Unlocking the numerator-denominator bias. II: adjustments to mortality rates by ethnicity and deprivation during 1991–94. *New Zealand Medical Journal* **115**, 43–8.

Blakely, T., Tobias, M., Robson, B., *et al.* (2005a). Widening ethnic disparities in New Zealand 1981–99. *Social Science and Medicine* **61**, 2233–51.

Blakely, T., Fawcett, J., Atkinson, J., Tobias, M. and Cheung, J. (2005b). *Decades of Disparity II: Socioeconomic Mortality Trends in New Zealand 1981–1999.* Wellington: Ministry of Health.

Bronnum-Hansen, H. and Juel, K. (2000). Estimating mortality due to cigarette smoking: two methods, same result. *Epidemiology* **11**, 422–6.

Caldwell, J. C. (1986). Routes to low mortality in poor countries. *Population and Development Review* **12**, 171–220.

Easton, B. H. (1997). *In Stormy Seas. The Post-war New Zealand Economy.* Dunedin: University of Otago Press.

Fogel, R. W. (2004). *The Escape from Hunger and Premature Death, 1700–2100: Europe, America, and the Third World.* Cambridge: Cambridge University Press.

Hawke, G. (1985). *The Making of New Zealand.* Cambridge: Cambridge University Press.

King, M. (2003). *History of New Zealand.* Auckland: Penguin.

Laugesen, M. and Swinburn, B. (2000). New Zealand's tobacco control programme 1985–1998. *Tobacco Control* **9**, 155–62.

Lynch, J. W., Smith, G. D., Kaplan, G. A. and House, J. S. (2000). Income inequality and mortality: importance to health of individual income, psychosocial environment or material conditions. *British Medical Journal* **320**, 1200–4.

McKeown, T. (1979). *The Role of Medicine; Dream, Mirage or Nemesis?* Oxford: Blackwell.

McMichael, A. J., McKee, M., Shkolnikov, V. and Valkonen, T. (2004). Mortality trends and setbacks: global convergence or divergence? *Lancet* **363**, 1155–9.

Ministry of Health (2000). *Our Health, Our Future. Haoura pakari, kioora roa.* Wellington, New Zealand: Ministry of Health.

Ministry of Social Development (2004). *The Social Report 2004.* Wellington, New Zealand: Ministry of Social Development.

Mowbray, M. (2001). *Distributions and Disparity: New Zealand Household Incomes.* Wellington, New Zealand: Ministry of Social Policy.

Oeppen, J. and Vaupel, J. W. (2002). Broken limits to life expectancy. *Science* **296**, 1029–31.

Pearce, N., Pomare, E., Marshall, S. and Borman, B. (1993). Mortality and social class in Maori and nonMaori New Zealand men: changes between 1975–77 and 1985–87. *New Zealand Medical Journal* **106**, 193–6.

Pearce, N., Foliaki, S., Sporle, A. and Cunningham, C. (2004). Genetics, race, ethnicity, and health. *British Medical Journal* **328**, 1070–2.

Plavinski, S. L., Plavinskaya, S. I. and Klimov, A. N. (2003). Social factors and increase in mortality in Russia in the 1990s: prospective cohort study. *British Medical Journal* **326**, 1240–2.

Pool, I. (1982). Is New Zealand a healthy country? The centenary of Dr Alfred Newman's affirmation 'that it is yet the healthiest on the face of the globe'. *New Zealand Population Review* **8**, 2–27.

(1991). *Te Iwi Maori.* Auckland: Auckland University Press.

Pool, I. and Cheung, J. (2002). Why were New Zealand levels of life-expectation so high at the dawn of the twentieth century? In *Paper presented to the Max Planck Institute*, Rostock.

(2003). A cohort history of mortality in New Zealand. *New Zealand Population Review* **29**, 107–38.

Pool, I. and Tiong, F. (1991). Sub-national differentials in the Pakeha fertility decline, 1876–1901. *New Zealand Population Review* **17**, 46–64.

Powles, J. (2001). Healthier progress: historical perspectives on the social and economic determinants of health. In *The Social Origins of Health and Well-Being*, ed. R. Eckersley and J. R. D. Dixon. Cambridge: Cambridge University Press, pp. 3–24.

Riley, J. C. (2001). *Rising Life Expectancy. A Global History*. Cambridge: Cambridge University Press.

Salmond, C. and Crampton, P. (2000). Deprivation and health. In *Social Inequalities in Health: New Zealand 1999*, ed. P. Howden-Chapman and M. Tobias. Wellington, New Zealand: Ministry of Health, pp. 9–64.

Sen, A. (1999). *Development as Freedom*. New York: Alfred A. Knopf.

Statistics New Zealand (2002). *Pacific Progress: A Report on the Economic Status of Pacific Peoples in New Zealand*. Wellington, New Zealand: Statistics New Zealand.

(2004). *New Zealand Life Tables 2000–2002*. Wellington, New Zealand: Statistics New Zealand.

Szreter, S. (1997). Economic growth, disruption, deprivation, disease and death: on the importance of the politics of public health for development. *Population Development Review* **23**, 693–728.

Taylor, R. and Bampton, D. (2005). Contemporary patterns of Pacific Island mortality. *International Journal of Epidemiology* **34**, 207–14.

Tennant, M. (1991). Missionaries of health: the School Medical Service during the inter-war period. In *A Healthy Country. Essays on the Social History of Medicine in New Zealand*, ed. L. Bryder. Wellington, New Zealand: Bridget Williams Books, pp. 128–48.

Te Puni Kokiri (2000). *Progress Towards Closing the Social and Economic Gaps between Maori and Non-Maori: A Report to the Minister of Maori Affairs*. Wellington, New Zealand: Te Puni Kokiri, Ministry of Maori Affairs.

Ulijaszek, S. J. (1990). Nutritional status and susceptibility to infectious disease. In *Diet and Disease*, ed. G. A. Harrison and J. C. Waterlow. Cambridge: Cambridge University Press, pp. 137–54.

Woodward, A., Mathers, C. and Tobias, M. (2001). Migrants, money and margarine: possible explanations for Australia–New Zealand mortality differences. In *The Social Origins of Health and Well-Being*, R. Eckersley, J. Dixon and R. M. Douglas. Cambridge: Cambridge Univesity Press, pp. 114–28.

World Health Organization (2004). *World Health Report 2004 – Changing History*. Geneva: World Health Organization.

Zatonski, W. A., McMichael, A. J. and Powles, J. W. (1998). Ecological study of reasons for sharp decline in mortality from ischaemic heart disease in Poland since 1991. *British Medical Journal* **316**, 1047–51.

12 Health changes in Papua New Guinea: from adaptation to double jeopardy?

ROBERT ATTENBOROUGH

Introduction

Are the health patterns of Papua New Guinea (PNG) changing? If so, in what ways are they doing so? Specifically, are they changing in line with theoretical models which invoke modernization, development and globalization to account for the decline of infections and the rise of lifestyle- and ageing-related health problems? Or does communicable disease – persistent, recrudescent or emergent – still predominate? In this review, I aim to discuss selected evidence on the health of Papua New Guineans, with a particular focus on examining past trends and on considering, so far as possible, future prospects – whether by projection from the past or on some other rationale. In the process, I shall consider the adequacy of conventional health modernization models to describe PNG's present situation.

To be modern is to have characteristics typical of recent times and the present day, by contrast with the more distant past. This sounds open-ended; but, in the health sciences as in the arts and social sciences, modernity is (explicitly or implicitly) expected to bring changes of a specific character. This follows usually from the concept that socioeconomic modernization, discussed by Ulijaszek (1995), is accompanied, anywhere in the world and whenever it may take place, by a demographic, health or epidemiological transition (Caldwell *et al.* 1990; Riley 2001). In such transitions, not only are death rates alleviated and life expectancies increased, but also one set of causes of ill-health (life-threatening and otherwise) declines in importance, to be replaced by a different set. In broad terms, infectious causes, often acting early in life and sometimes ecologically sensitive, decline, while non-infectious causes, usually late-acting and often related to diet or other aspects of lifestyle, become the leading health problems.

Thirty years ago, a sceptic could tenably have argued that the demographic transition was a unique phenomenon embedded in the culture and

Health Change in the Asia-Pacific Region: Biocultural and Epidemiological Approaches, ed. Ryutaro Ohtsuka and Stanley J. Ulijaszek. Published by Cambridge University Press.
© Cambridge University Press 2007.

historical experience of industrialization in Europe, and in settler nations predominantly populated from Europe (Teitelbaum 1975). The corollary of this was that attempts to extrapolate key features of transition theory to other parts of the world undergoing a later and faster economic development were inadequately grounded and perhaps wrong. Nowadays, such scepticism has largely evaporated. Debate continues as to both the theoretical standing of the central propositions and the details of the transition process in particular populations. But in broad outline it is now widely accepted that health and mortality transitions as just characterized (and most often transitions from high to low fertility as well) do regularly, albeit with variations, accompany socioeconomic modernization almost anywhere. Among the variations, however, improvements in morbidity and mortality are currently slowing down in most Pacific countries, even though developed-country levels have not yet been reached (McMurray 2004). Other chapters in this book provide more detailed evidence from the Asia-Pacific region, much of which essentially bears out the established expectations of declining infection and rising lifestyle-related health problems with modernization. How well do these expectations hold for PNG?

Papua New Guinea

The eastern portion of the large island of New Guinea, the Bismarck archipelago and many other smaller islands nearby together make up the modern nation of PNG. Now celebrating over 30 years of independence, PNG emerged in 1975 from a century or less of colonialism – effectively much less in the remotest areas – under Germany and Britain initially, and later Australia (Alpers and Attenborough 1992; Ulijaszek 1995). It retains strong economic and other links with Australia as well as other Asia-Pacific nations.

With a land area of about $463,000 \, km^2$ (Hanson *et al.* 2001) and a population of about 5.7 million, PNG is not very populous by Asian standards, but is easily the largest of the Pacific Community nations and territories, with 65% of their total population and one of their shorter population doubling times, currently 32 years (Secretariat of the Pacific Community 2004). It also has a larger population than New Zealand or Hawaii. In its famous diversity and complexity – linguistic and cultural, topographic and climatic – PNG bears comparison with any nation in the Asia-Pacific region. The circumstances as well as the timing of socioeconomic modernization in PNG are about as different as can be imagined from mid-eighteenth-century France, Sweden or Britain, where the earliest signs of a health transition can be traced (Riley 2001).

It would be a mistake to consider PNG's undocumented pre-colonial past as timeless or changeless. Two particularly salient transitions for which we have evidence are the development of taro and banana agriculture in the highlands, and the later advent of the South American sweet potato as an alternative basis for root crop agriculture (Denham *et al.* 2003; Ballard *et al.* 2005; Pawley *et al.* 2005). The possible implications of these for diet and lifestyle are discussed below. Nonetheless, the starting point for PNG's modern transition might be taken either at the earliest effective exertions of European colonial control (from the 1870s in some coastal regions to the 1950s in parts of the interior) or at the Second World War, during and following which major social and economic changes, with lasting impacts, took place in most parts of it. Since that war, and with gathering pace and increasing complexity since independence, modernization has brought to PNG astonishing changes within the lifetimes of individuals. Some people still alive now grew up outside the ambit of colonial control. Their parents may have witnessed wartime aerial dogfights without any prior awareness of foreign armies on New Guinea soil: many others, of course, were far more involved. Post-war development has proceeded in different ways and at different speeds in different regions, accentuating PNG's pre-existing socioeconomic diversity and resulting in what many have regarded as increasingly unbalanced development (Allen 1992; Hanson *et al.* 2001).

The starkest manifestation of that uneven development, arguably, is urbanization. Table 12.1 provides some urban settlement figures from the year 2000 census (National Statistical Office 2002). Port Moresby, the capital of PNG and of Central Province, has over a quarter of a million inhabitants. Other larger urban settlements (generally provincial capitals) range from nearly 80,000 people in Lae, through Arawa, Madang and Mount Hagen, down to approximately 20,000 each in Wewak, Popondetta, Goroka and Kokopo/Vunamami (which, following the September 1994 volcanic eruptions, has taken over from Rabaul as provincial capital). Nine of the other eleven provincial capitals, plus five non-capital centres, also have populations now over 5,000. Many of these centres have shown substantial recent population growth: Brinkhoff (2003) conveniently tabulates comparative figures for 1980, 1990 and the year 2000. The total urban population of PNG now stands at 686,000 (National Statistical Office 2003a), of whom all but 38,000 live in one of the 26 settlements listed in Table 12.1. This is a substantial urban population, though well short of the one million anticipated for the year 2000 by Allen (1992); and it is growing absolutely and as a proportion of the country's total population (from 10% in 1980 to 13% in the year 2000).

Table 12.1. *Populations of urban settlements, Papua New Guinea, Census of 9–15 July 2000*

Settlement	Province	Population
Port Moresby[a]	National Capital District	254,158
Lae[a]	Morobe	78,692
Arawa[a]	North Solomons (Bougainville Autonomous Region)	31,462
Madang[a]	Madang	28,547
Mount Hagen[a]	Western Highlands	27,877
Wewak[a]	East Sepik	20,257
Kokopo/Vunamami	East New Britain	19,933
Popondetta[a]	Northern (Oro)	19,866
Goroka[a]	Eastern Highlands	19,523
Mendi[a]	Southern Highlands	17,128
Kimbe[a]	West New Britain	14,184
Wau/Bulolo	Morobe	13,037
Daru[a]	Western	12,935
Kavieng[a]	New Ireland	11,274
Alotau[a]	Milne Bay	9,888
Vanimo[a]	West Sepik (Sandaun)	9,778
Tari	Southern Highlands	8,824
Kiunga	Western	8,295
Kundiawa[a]	Chimbu (Simbu)	8,087
Kainantu	Eastern Highlands	6,788
Lorengau[a]	Manus	5,874
Ialibu	Southern Highlands	5,479
Kerema[a]	Gulf	5,124
Wabag[a]	Enga	4,208
Rabaul[a]	East New Britain	3,907
Balimo	Western	3,143

Note:
[a] Denotes provincial capitals.
Source: 2000 census, National Statistical Office (2002).

Urban dwellers range from the more educated élite to the squatter settlements where often people have migrated from the villages to start a new life with their city *wantok*s (speakers of the same indigenous language). The larger of PNG's cities and towns have hospitals, university campuses and many of the same institutions and facilities as are found in urban centres worldwide.

The other 87% of the total population of PNG still live in rural areas, where, despite networks of government stations, aid posts, health centres, agricultural extension centres and schools, facilities are much more restricted. Agriculture, especially of root crops, and husbandry, especially of pigs, supplemented in some places by hunting and collecting, still supply rural people with their staple foods. Many use earnings from cash-cropping or plantation labour among other occupations to supplement what they can grow or acquire directly, or rely principally on these sources of income. Modern life in PNG contains many elements and contrasts which foreigners are apt to find arresting, even paradoxical, in their compression or bypassing of the earlier stages of technological modernity as experienced elsewhere. There is, for example, familiarity with air transport even in areas where there is no overland transport except on foot and in canoes; and for some, there is access to video and electronic technology even in areas where there are no cinemas or televisions and perhaps not even reticulated electricity.

Are modernization and globalization, at the scale and pace experienced in PNG, at risk of overwhelming people's capacity to cope, whether in urban or rural areas? Do the health deficits they bring outweigh the benefits? Do other extrinsic shocks, coming perhaps from the worlds of infectious microorganisms or of climate change, bring similar risks? Or do the assumptions behind such questions underestimate people's capacity to adapt, both biologically and socially, as the recent rise of interest in resilience (e.g., Barnett 2002) might lead us to think?

Perhaps this last question cannot be directly answered from conventional population health data. But a sampling of recent comparative statistics certainly offers little cause for complacency over PNG's present situation. At 93 per 1,000, PNG's rate of child mortality (under five years) is currently the 49th highest in the world, the 11th highest among non-African nations (United Nations Children's Fund 2004a: 105). Within the East Asia-Pacific region, only Cambodia and the Democratic People's Republic of Korea have seen slower progress than PNG towards the Millennium Development Goal of reducing under-five mortality by two-thirds between 1990 and 2015 (United Nations Children's Fund 2004b: 20). Among Pacific Community nations and territories, PNG has, if the data can be accepted, the highest infant mortality rate (IMR; death rate among those below one year of age: 64 per 1,000 live births) except for the Solomon Islands; a very high maternal mortality rate of 370 per 100,000 live births, only 45% of which are attended by trained personnel; the highest rate of HIV/AIDS diagnosis; and the lowest life expectancy (54 years for males, 55 for females) with the striking exception of Nauruan males (McMurray 2004; Secretariat of the Pacific Community 2004). Of the World Health Organization's 191 member states in 1997, PNG ranked 150th in overall health system attainment, a composite measure covering average health, health equality, system responsiveness, responsiveness equality and financial fairness; and 148th in overall health system performance, as measured

Table 12.2. *Overall health system attainment* and performance[†] rankings of 22 countries, selected from 191 World Health Organization member states, 1997*

Attainment		Performance	
Rank	Country	Rank	Country
1	Japan	1	France
6	France	10	Japan
9	UK	18	UK
12	Australia	32	Australia
15	USA	37	USA
26	New Zealand	41	New Zealand
54	Philippines	60	Philippines
78	Fiji	76	Sri Lanka
80	Sri Lanka	80	Solomon Islands
82	Samoa	92	Indonesia
85	Tonga	96	Fiji
88	Cook Islands	107	Cook Islands
100	Russian Federation	112	India
106	Indonesia	116	Tonga
108	Solomon Islands	119	Samoa
121	India	125	Brazil
125	Brazil	127	Vanuatu
132	China	130	Russian Federation
134	Vanuatu	144	China
150	**Papua New Guinea**	**148**	**Papua New Guinea**
175	Myanmar	190	Myanmar
191	Sierra Leone	191	Sierra Leone

Source: World Health Organization 2000, Annex Tables 9* and 10[†], which also present uncertainties around these rankings. See explanation and discussion in text.

by the efficiency with which health expenditure is translated into health outcomes (World Health Organization 2000). Table 12.2 presents selected international, including Asia-Pacific, comparisons on these rankings. Of course some of the comparison countries have *a priori* advantages; but whatever the reasons, PNG is not fortunately placed. While much remains uncertain, therefore, it seems clear that concerns as to PNG's health prospects raised by previous authors (e.g., Inaoka 1995 and n.d.; Ulijaszek 1995; Connell 1997; Naraqi *et al.* 2003) should remain a focus of attention.

In this chapter, then, my aim is to examine some (necessarily selected) evidence on health trends in PNG. I shall ask how far the changes evidenced, and those that one can project into the future, correspond to the picture, established from other countries' experience, of the health trends to be expected from modernization; and, on the other hand, how far other processes, such as those that underlie the persistence, emergence or re-emergence of infections, might blow the evolution of PNG's health trends off their expected course.

I shall begin at the beginning, with the earliest known human presence in New Guinea and the concept of human adaptation to the region's environment. In view of the paucity of available evidence, I shall be brief on the long unwritten pre-colonial history of human health in PNG. Even the literature from the earlier part of the colonial period, though informative in some ways (Attenborough and Alpers 1995), has severe limitations as a source of population health information. Frequently, as elsewhere in colonialism's history, it was marked by the power asymmetries and the now discarded beliefs about human difference characteristic of that period; by closer attention to the health of the colonizers than of the colonized; by conceptions of infectious disease processes and of epidemiological quantification then still nascent; and above all by biomedical problems of a variety and complexity disproportionate (even more than nowadays) to the research capacity available on the ground. In most contexts, therefore, my timescale will be little more than six decades into the past, often less, according to the availability since the Second World War of literature in which such problems are better controlled; and perhaps a decade or two into the future. At times I shall draw on national statistics, but for detail and time depth I shall sometimes opt instead for local studies, often better controlled than national ones can be, and sometimes (too seldom) repeated locally so as to allow trends to emerge, not obscured by extreme regional diversity.

It is not by chance that this chapter's title refers to health changes in the plural. There are of course many health changes at work – changes in different parts of the country, at different periods and on different timescales, and affecting different aspects of health. Some changes may be aptly described as quantitative trends, in that they are gradual shifts over a considerable time in the burden of mortality and morbidity in a consistent direction, often but not always the direction predicted by health transition theory and experience. Other changes have different characters. In addition to infections which may have reached New Guinea with the earliest people, PNG's medical history includes infections which arrived later than that but before European colonialism: Wigley (1977, 1991) has argued that this is true of tuberculosis, although also that it became a major health problem only under colonial conditions. There have also been introductions during the colonial period or after it – from the late nineteenth century in the case of some sexually transmitted infections for example (Attenborough and Alpers 1995; Hughes 1997), up to at least the late 1980s when PNG's first AIDS case was reported. There has been disappearance under colonialism followed by re-emergence, as in the case of yaws – apparently widely endemic at outsider contact and responsive to intervention (Gajdusek 1990), but later recrudescent (Garner *et al.* 1986; Manning and Ogle 2002). There have been introduced epidemic infections which never became recurrent or endemic, for example smallpox (Fenner *et al.* 1988; Allen 1989); and others which did, for example measles (Adels and Gajdusek 1963; Benjamin and Dramoi 2002). There have been major but finite epidemics under specific historical circumstances, such as the bacillary dysentery epidemic during the Second World War (Allen 1983; Burton 1983). Pigbel, until recently a major endemic cause of child death in the PNG highlands but responsive to vaccination campaigns, has been enormously reduced, although it has not disappeared entirely (Lawrence 1992; Poka and Duke 2003). And of course PNG's health changes are not confined to infectious disease. A range of non-infectious conditions, including but not solely the classical diseases of modernization, were essentially new in PNG in the later twentieth century: for example, asthma (Anderson and Woolcock 1992). Accidents, violence (including the Second World War) and sudden cataclysmic events have also taken a major toll, episodically in some cases. As an example of the last, the Sissano (Sandaun Province) tsunami of 17 July 1998, though less catastrophic than the Indian Ocean one of 26 December 2004, killed over 2,100 people and led to the hospitalization of at least 800 people, the permanent displacement of over 10,000 people and the abandonment of ten villages, all in one quite restricted area (Dengler and Preuss 2003).

In the space available here, I naturally cannot review PNG's health changes comprehensively, nor even cover adequately all the examples listed above. I shall review trends in overall mortality, as one indicator of health change; examine the rise of some diseases of modernization with attention to diabetes in particular; and finally consider, selectively, the question of infectious disease, old and new, persistent and emergent, with special focus on malaria and to a lesser extent HIV/AIDS. But first, I turn to early human adaptations in the New Guinea region.

Pre-colonial adaptation and health

By 40,000 years ago, people had reached a New Guinea then connected by land bridge to Australia and Tasmania (Groube *et al.* 1986), and not much later, via a further sea voyage, the Bismarck Archipelago (Specht 2005). By 30,000 years ago, the valleys of the central highlands were also reached

(Hope and Haberle 2005). For millennia, therefore, people have lived – for most of that time as hunter-gatherers – in the lands which now constitute PNG (Lilley 1992). With some riders, one must assume, then, that there has been sufficient time, on any plausible model, for them to adapt to specific critical features of Melanesia's humid tropical environment.

Riders to this statement must cover complications affecting both people's exposure to environmental pressures and those pressures themselves. On the one hand, some present-day Papua New Guineans will have some ancestors, including speakers of Austronesian languages, who arrived there more recently than the earliest peopling (Foley 1992; Kirk 1992; Spriggs 1997). Also, changes in ways of life (including agriculture), diet and rising population density will have exposed people to pressures additional to those intrinsic to the natural environment. On the other hand, the environment itself is, as already discussed, far from homogeneous; and it has changed significantly since earliest times (e.g., the infilling of the inland sea that once occupied the lower Sepik-Ramu basin: Swadling and Hope 1992, Swadling and Hide 2005).

No ecological change in the New Guinea region since the first peopling is likely to have been more important, or more devastating, for human adaptation and health than the arrival of malaria. All of the four species of the protozoon *Plasmodium* which infect humans occur in New Guinea. The global prehistory of *Plasmodium falciparum*, which as the cause of the most dangerous malaria will receive most attention here, remains under debate. While one school of thought proposes a 'Malaria's Eve' hypothesis whereby extant *P. falciparum* variants have evolved from a quite recent (3,000 to 8,000 years) common ancestor, another school argues for a much older one (100,000 to 200,000 years): see review by Hume *et al.* (2003). Recently, Joy *et al.* (2003) have presented evidence that while the origins of *P. falciparum* appear very ancient, much of its numerical expansion, including a second dispersion beyond Africa, appears to have occurred within the last 10,000 years. For PNG, their evidence suggests a rapid expansion event of the order of 13,000 to 19,000 years ago (but with a small sample and large uncertainties, so PNG data are hard to differentiate unequivocally from the African data). Thus *P. falciparum* may be a relatively recent – or if we follow Groube's (1993, 2000) speculation, very recent (within the last 1,000 to 2,000 years) – arrival in Melanesia. None of this necessarily holds true for the other three *Plasmodium* species which cause less severe malarias in humans; though in fact Clark and Kelly (1993) took seriously the possibility of malaria (species not stated) arriving with Austronesian-speakers in the fourth millennium before the present, and Leclerc *et al.* (2004) concluded that the worldwide expansion of *P. vivax* too was recent, perhaps within the last 10,000 years.

But despite indications of the relative recency of malaria, especially that caused by *P. falciparum*, it seems plausible that there has been sufficient time for substantial genetic adaptation to intense selection pressure, such as falciparum malaria at least does exert (Riley 1983). The sickle-cell gene, though not one of the suite of malaria adaptations indigenous to Melanesia (Lavu *et al.* 2002), supplies the main, albeit approximate, model for others. Wiesenfeld (1967) concluded that even 1,000–1,500 years would be enough for such adaptation; Livingstone (1989) found that present distributions were best approximated by 3,000–4,500 years of simulated malaria selection and diffusion; and Fix (2003) too was impressed by the potential rapidity of genetic change, with substantial change occurring even within 300 years. On balance then – and taking malaria as the classical exemplar of the environmental and biotic pressures at work – even if we cannot entirely exclude Groube's (1993) suggestion that accommodation to falciparum malaria is still under way, it seems reasonable to work with the proposition that by the beginning of the colonial era New Guineans had reached a high level of adaptive equilibrium. At any rate, a very effective functional adaptability has been well demonstrated (Lourie *et al.* 1992; Norgan 1997), even if specific genetic adaptations, other than to malaria, are hard to identify (Serjeantson *et al.* 1992). And if low population growth suggests some sort of adaptive equilibrium, that too is evidenced, at least in genealogical data covering recent centuries for one case, the Gidra of Western Province (Ohtsuka 1986).

A high level of adaptive equilibrium should not, of course, be equated with good population health and low mortality, as we would rate them on modern expectations. Indeed, it is one of New Guinea's conundrums that population density and cultural efflorescence such as those witnessed in the Sepik basin (Forge 1972, 1990) should have emerged and been sustained under conditions of holoendemic malaria, severe child growth deficit (Tyson 1987) and atrocious infant mortality (to cite an extreme example of the latter, Forge's estimate of 570 per 1,000 for the Abelam: Peters 1960). It is not particularly reassuring to reflect that, theoretically, some mortality régimes that strike us as appalling can be compensated by quite realistically attainable fertility régimes: for example, under definable conditions, a total fertility rate of 6.6 children is sufficient to offset a life expectancy as low as 20 years (Attenborough 2002).

The obstacle to assessing how good or bad population health was in New Guinea during the pre-colonial period (or any period not covered by systematic data collection) is self-evident. Our only direct evidence is that of palaeopathology, and for New Guinea this is limited by the paucity of studies as well as methodological constraints. In a skeletal sample of

42 individuals from Motupore Island near Port Moresby, dated at between AD 1200 and AD 1700, Webb (1995) diagnosed pathology of some kind in 22 individuals (both sexes, all but two of them adults), representing a higher prevalence of infection particularly than in his comparative Australian samples. The most common infectious disease diagnosis was treponematosis (11 individuals), with no definite signs of tuberculosis, in contrast to the nearby and roughly contemporary site of Nebira (Hope *et al.* 1983). Haematological disease was diagnosed in three cases, including one child considered to be a thalassaemia sufferer, suggestive of endemic malaria in the region by that time. Arthritis, of the cervical and lumbar vertebrae especially, was a very common diagnosis (16 individuals); trauma less so (2 individuals). Dental caries, attributed to a diet high in sago, was substantially more common than at Nebira. Greenburg-Hudd (1998), in a survey of protohistoric and historic dental palaeopathology museum specimens originating in PNG, found periodontal disease to be the most prevalent pathology: attrition was only moderate, tooth loss low to moderate and caries low, perhaps owing to the cariostatic properties of betel nut.

Beyond palaeopathology, we can only resort to indirect evidence and broad inferences. Caution is required if we are to assume that the long hunter-gatherer phase of New Guinea prehistory (and some of its history) resembles that of recent low-latitude hunter-gatherers, not in close interaction with agriculturalists, elsewhere in the world. But if we do make that assumption, it is likely (Froment 2001) that population density was low, and that the main causes of morbidity and mortality were certain parasites and infections (though others would have been absent or at low level, owing to the protections afforded by low population density and semi-isolation), plus accidental trauma, violence, snakebite and so forth, perhaps overt malnutrition on occasion, with no more than a minor role for degenerative disease except perhaps arthritis and poor dental health. Psychiatric health would be hardest of all to assess in such contexts.

Subsequent developments may all have brought impacts on human adaptation and health. These include (1) the emergence of taro agriculture and banana cultivation from at least 6,500 to 7,000 years ago (Denham *et al.* 2003); (2) the arrival of outsiders, including Austronesian-speakers in many coastal areas from the fourth millennium before the present (Lilley 1992; Spriggs 1997); (3) faunal introductions such as the pig at a much debated date not more recent than 1,900 years ago (Hide 2003); and (4) the introduction of the sweet potato from the Americas probably only a few hundred years ago, displacing taro in many places (Lilley 1992; Yen 2005). There is clearly great potential for health problems related to heavy reliance on staple crops, where these do not provide adequate protein and micronutrients in proportion to energy, are too bulky to be satisfactory foods for younger children, and/or bring disaster if they fail; and related also to village living conditions, somewhat denser and more sedentary. Sweet potato, given its productivity and quantitative dependence of many populations upon it, has major dietary implications and was no doubt involved in the build-up of denser, more settled populations in the highlands, certainly relative to hunter-gatherer subsistence and probably also relative to taro agriculture. The dietary role of pork even nowadays is much debated and regionally variable: probably its contribution was modest but real – smaller than an outsider might expect, on account of the social and symbolic importance and only occasional slaughter of domestic pigs (Hide 2003). Both sweet potato and pork are involved in the complex of factors that can precipitate pigbel amongst PNG highlanders (Lawrence 1992), although the antiquity of this disease is unclear. Pigs are potential, and in some cases actual, sources of parasitic zoonoses in New Guinea; but on the whole, relatively few zoonoses are recorded as causing human disease in PNG even today, a fact for which it can thank in large measure its relative isolation historically and the absence until recently of large placental mammals other than pig and dog (Owen 2005).

Even after the establishment of agriculture, population densities must be presumed to have remained moderate at most, as they were historically (only exceeding 30 persons per km^2 in parts of the highlands, the Sepik River basin and the Gazelle peninsula of East New Britain: Ward and Lea 1970). Settlements would have been relatively small: Forge (1972), in his review, found settlements to have fewer than 300 residents each historically in most areas, and somewhere over 1,000 in a few areas including the Sepik. Mobility between settlements increases their effective overall size from the viewpoint of sustaining infection, but even the largest language group in PNG would not have been large enough in earlier times to sustain acute viral infections with properties like those of measles, for example (Attenborough and Alpers 1995). So, even within a health profile dominated by infection, pre-colonial New Guinea would have been free of certain infections which have become a problem more recently, as measles now is. Furthermore, biogeographic isolation would have protected New Guinea for a long time, perhaps up to the colonial period, from the arrival of certain other pathogenic microorganisms, including a number of them causing sexually transmitted infections (Attenborough and Alpers 1995; Hughes 1997). Finally, the presence in many lowland New Guinea populations, both north and south of the main cordillera, of genetic polymorphisms believed to confer some malaria-protective benefit on carriers suggests the presence of malaria in New Guinea on a millennial time-scale (Attenborough and Alpers 1995). These polymorphisms number at least six – five of them affecting *P. falciparum* (α-thalassaemia, β-thalassaemia, glucose-6-phosphate dehydrogenase deficiency, ovalocytosis, and Gerbich negativity) and a sixth affecting *P. vivax* (Duffy antigen variation) (Serjeantson *et al.* 1992; Genton *et al.* 1995a; Mgone *et al.* 1996; Allen *et al.* 1997; Zimmerman *et al.* 1999, 2003; Cortés *et al.* 2004). The presence of so many such polymorphisms, in some cases clearly the product of multiple mutations, is hard to interpret in other than as adaptive to the presence of malaria in the ancestral populations for an evolutionarily significant period of time.

Putting together hints such as these, then, we have a few pencil marks towards a sketch of pre-colonial health in New Guinea: a set of populations with long exposure and adaptation to most facets of their complex tropical environment, but certainly not leading idyllically healthy lives, subject to many infections including malaria and yaws, though protected from some others by isolation and small social scale, and no doubt living with frequent trauma and arthritis. It is difficult to go beyond these broad qualitative indications.

Health impacts of colonization

How much had things changed by the time of the first systematic health surveys conducted under colonial régimes, which must form the baseline for the examination of quantitative trends to which I shall shortly turn? The earliest (late-nineteenth-century) written records, such as the diaries of Mikloucho-Maclay (1975) and the first Annual Reports of the Territories of Papua and New Guinea (discussed by Attenborough and Alpers 1995), do not capture the earliest contacts; but even they tend, in the earlier years at least, to be anecdotal, at best semi-quantitative. Early contacts with outsiders from colonizing powers, whether their formal representatives or not, had a potential to change health, deliberately and otherwise, in many ways. They not only introduced novel infections: they also introduced new foods and religions; forbade warfare; directed employment and labour migration; undertook vaccination; regulated disposal of the dead; implemented sanitation measures; conducted mass treatment campaigns; made dietary recommendations; instituted constraints on sexual behaviour; inspected premises for slaughtering and food preparation; and undertook environmental measures against malaria (Attenborough and Alpers 1995).

Public health in the initially distinct colonies of Papua and New Guinea has attracted a number of detailed historical accounts (Denoon *et al.* 1989; Burton-Bradley 1990; Spencer 1999; Davies 2002). In one famous instance, colonial prohibitions on consumption of human flesh unexpectedly interrupted the transmission of kuru in the Eastern Highlands during the 1950s (Alpers 1992). The overall impact of colonial activities on indigenous health was clearly mixed and is hard to assess. Some activities, including the introduction of novel infections, clearly had profound population health impacts, sometimes running ahead of the frontier of direct colonial contact. Other activities seem likely to have been quite local and/or temporary in any effects they may have had on health, bringing little lasting alteration of the disease environment of the country overall. In the gap, then – longest in the areas contacted earliest – between the pre-colonial era and the baseline studies for quantitative population health research, health patterns would have changed significantly in some respects, certainly including novel infections; probably not greatly in many others.

Mortality trends

In reviewing more recent trends, I start with the overview of health offered by mortality and life expectancy data. At a national level, the principal data sources are the two stratified sampling censuses of 1967 and 1971 and the three full censuses of 1980, 1990 and 2000. These can be supplemented with survey data which lack national coverage but are indicative and sometimes offer greater detail. There is not space here to discuss data problems, estimation methods and other complexities in depth as Riley and Lehmann (1992) do, but it is now possible to update the picture they drew.

Table 12.3 amplifies the findings collated by Riley and Lehmann (1992: Table 3.2), with other findings, mostly later ones reported by the PNG National Statistics Office (National Statistical Office 2003b: Tables III-1–III-4 and text). Mortality clearly has been and remains high by modern international standards. Nonetheless, although aspects of these findings are open to debate, it is clear that, on any indicator, a substantial mortality decline did take place in the second half of the twentieth century. Whatever the negative health impacts of modernization and other contemporary changes, they have yet to outweigh its benefits, at least at the level of mortality (unlike certain African countries since the advent of HIV/ AIDS). Infant mortality, as the most sensitive of these indicators to socio-economic and public health changes, reflects most strongly PNG's last

Table 12.3. *Estimates of mortality indicators in Papua New Guinea, 1946–2000*

	Crude mortality/1000 population	Infant mortality/1000 live births		Mortality 1–4 years/1000 live births		Life expectancy at birth		Life expectancy at age 25	
		Males	Females	Males	Females	Males (years)	Females (years)	Males (years)	Females (years)
1946	30–35	252	251			31	32		
1966	21	161	157			44	43		
1960s[a]						53	51	36	36
1971	20	130	120			49	50		
1971		142	125	92	90	40	41	31	32
1980	13	78	66	47	43	49	51	34	35
1991		88	76	57	55	52	51	39	37
1996	12	73		24		55	54		
2000	11	67	61	27	23	54	55	38	38

Note:

[a] Combining four studies of small groups from highlands, lowland foothills and islands, published 1967–69: van de Kaa (1971).

Source: Data mostly collated from various sources by Riley and Lehmann (1992) and National Statistical Office (2003b).

half-century of such changes, and they do cover the greater part of that period. Mortality improvements are inherently hard to sustain, however, and these figures, taken at face value, appear to show that in recent times improvements have become less consistent as well as more slight.

A similar picture, of improvement slowing at some point in the 1990s, appears to emerge from the other indicators also; and, of most concern, it is a slowing at a level short of the improvements one might hope for on the development experience of, for example, Kerala, Sri Lanka, Costa Rica and China (Caldwell 1986). Riley and Lehmann (1992) attribute the driving roles in the earlier improvement to health service provision, especially the availability of penicillin for pneumonia and chloroquine for malaria, together with other anti-malaria measures, and pertussis and tetanus immunization. Immunization campaigns are likely to have played a particularly large part in accounting for such improvements as have been seen in both infant and one to four years mortality. As Ulijaszek (1995) has pointed out, however, while immunization has brought demonstrable improvements which will have contributed to the alleviation of mortality (e.g., for measles; but see also Duke 2003), there are signs of trends in the opposite direction, even before 1990, in conditions not targeted by immunization (e.g., diarrhoea and malnutrition). This among other factors may have contributed to the unevenness of mortality trends.

Consistently over time (and consistently across all provinces in the year 2000 except Manus), infant and child mortality rates are higher for males than females. In keeping with so many dimensions of PNG life, however, mortality and mortality trends are highly variable regionally, as shown in Table 12.4. Here, for the 1980 and 2000 censuses only, key mortality indicators are shown province by province (updating Allen 1992: Table 2.2; on the basis of National Statistical Office 2003b: Tables III-1–III-4). The provinces are arranged not in the conventional order but so as to highlight trends: on the left, in ascending order of the 1980 to 2000 change in life expectancy (in years), and on the right, in descending order of the change in infant mortality across the same period (in deaths/1,000; the third numerical column in each case). These orders are highly correlated though not identical. Mortality at one to four years follows a very similar ordering to infant mortality, suggesting similarity in the underlying factors. Life expectancy at 25 years of age (e_{25}) also follows a similar ordering to life expectancy at birth (LEB; e_0), but the similarity is less close. In all cases, there is a weak negative correlation, never accounting for more than half the variation, between the 1980 value and the subsequent change: thus there is an unsurprising tendency, most noticeable for life expectancy at 25 years, for initially better results to show less

Table 12.4. *Mortality indicator changes between 1980 and 2000 by province: life expectancy at birth (e_0) and at 25 years (e_{25}); infant mortality rate (IMR) and 1–4 years mortality rate (1–4MR). Under each heading the third column is the difference between the first two. Females and males are combined throughout.*

	e_0			e_{25}				IMR			1–4MR		
	1980	2000	'00–'80	1980	2000	'00–'80		1980	2000	'00–'80	1980	2000	'00–'80
PNG	49.6	54.2	4.6	34.6	37.9	3.3	PNG	72	64	−8	45	25	−20
Milne Bay	57.1	54.1	−3.0	39.2	38.3	−0.9	Gulf	71	103	32	48	57	9
Gulf	47.3	46.4	−0.9	32.3	34.5	2.2	Milne Bay	50	69	19	27	28	1
N. Solomons	59.6	59.6	0.0	39.6	41.1	1.5	Morobe	62	80	18	39	38	−1
Madang	50.7	51.1	0.4	34.6	36.5	1.9	Madang	62	78	16	39	35	−4
Morobe	50.9	51.7	0.8	34.8	37.3	2.5	N. Solomons	33	47	14	17	15	−2
E. Highlands	53.1	55.4	2.3	36.1	38.0	1.9	W. Sepik	104	105	1	74	58	−16
NCD	56.7	59.2	2.5	37.2	37.9	0.7	E. Highlands	55	54	−1	33	19	−14
E. Sepik	49.3	52.2	2.9	36.4	37.4	1.0	E. New Britain	57	54	−3	34	19	−15
W. Sepik	42.1	46.0	3.9	30.6	34.4	3.8	W. New Britain	60	55	−5	37	19	−18
E. New Britain	52.8	57.1	4.3	36.0	39.5	3.5	Northern	67	59	−8	43	22	−21
W. Highlands	51.9	56.2	4.3	37.6	38.0	0.4	New Ireland	62	52	−10	36	17	−19
Central	51.3	56.4	5.1	34.8	38.1	3.3	Manus	55	45	−10	34	14	−20
New Ireland	52.7	57.9	5.2	36.4	40.0	3.6	Central	59	47	−12	36	15	−21
Northern	49.2	54.5	5.3	33.7	37.6	3.9	NCD	35	22	−13	20	5	−15
Enga	47.1	52.5	5.4	34.1	36.8	2.7	E. Sepik	94	79	−15	56	36	−20
W. New Britain	51.3	56.7	5.4	34.9	39.2	4.3	Western	83	66	−17	53	26	−27
Chimbu	50.2	56.8	6.6	36.6	39.2	2.6	Enga	91	69	−22	58	28	−30
Western	47.7	54.3	6.6	33.8	38.2	4.4	Chimbu	87	54	−33	51	19	−32
Manus	51.8	58.6	6.8	34.8	39.9	5.1	W. Highlands	81	48	−33	46	15	−31
S. Highlands	43.8	55.2	11.4	33.2	38.4	5.2	S. Highlands	116	61	−55	75	23	−52

Note:

To the left, provinces are arranged in ascending order of change in life expectancy at birth; to the right, they are arranged in descending order of change in infant mortality rate.

Source: Data from National Statistical Office (2003b).

improvement than initially poorer ones. The weakness of this trend, however, and counter-examples visible in the table confirm that in no sense have inherent obstacles to improvement been encountered yet. The absolute mortality levels observed convey the same message.

In Gulf, Milne Bay, Morobe, Madang and North Solomons Provinces, there have been large rises in infant mortality (whether viewed absolutely or proportionally), and these provinces have also seen rises or very small declines in one to four years mortality. Accordingly, LEB has declined in Milne Bay and Gulf, and remained essentially static in the other three. All five provinces are below the national figure for improvement in life expectancy at age 25 years also, but only in Milne Bay has this indicator actually declined: in Gulf and Morobe especially, adults appear relatively protected from the adverse changes affecting children. With the civil conflicts there during this time, it is not surprising to see North Solomons among these five provinces, as a result of rising infant mortality. What is perhaps surprising is that overall life expectancy there has remained static: North Solomons had the highest life expectancy of all provinces in 1980, and perhaps the conditions which permitted that outcome, whatever they may have been, still had sufficient effect in the year 2000 to protect its life expectancy from actually declining. There is no space here to discuss possible method problems or data errors, but demographic data collected in a conflict zone should clearly be interpreted cautiously.

At the other end of the spectrum, Southern Highlands Province has the largest improvement recorded in PNG on all these indicators. With the worst infant mortality of all in 1980, Southern Highlands had the most scope for improvements, and the data suggest that these have ensued. But with civil strife there too, this province's demographic results also require particularly cautious interpretation. Other highlands provinces, Western Highlands, Chimbu and Enga, also registered large improvements over initially high levels of child mortality; but Eastern Highlands levels were already lower, and its mortality indicators have improved less, in the case of infant mortality hardly at all. Western Province and East Sepik have also shown marked alleviations of initially high levels of infant and child mortality. Manus and West New Britain have shown conspicuous rises in life expectancy despite only moderate falls in infant and child mortality, bringing Manus to the third position on both life expectancy measures, behind only North Solomons and National Capital District on e_0, North Solomons and New Ireland on e_{25}, and almost to the point where a 25-year-old can 'expect' to live to 65 years of age.

While the right-hand half of Table 12.4 explicitly focuses on measures of pre-adult death, and LEB is also most substantially affected by that, life expectancy at 25 years focuses attention on premature adult death. Not surprisingly, this parameter is least closely tied to the other indicators listed here. Thus Gulf, West Sepik, Morobe and East New Britain Provinces all have more favourable trends in e_{25} than one might expect from their trends in e_0; while in East Sepik, Enga, Chimbu and most dramatically Western Highlands, significant benefits in child survival seem not to have flowed through to adult survival as well as one might have hoped. On the figures, National Capital District also belongs in this latter category, but conditions are probably unique there – on the one hand, relatively good and improving child survival patterns set a high benchmark; and on the other, an e_{25} no higher than the national average, exhibiting only marginal net improvement in 20 years despite the capital's facilities, probably provides the clearest signal apparent in this table of encroaching health problems (infectious and non-infectious) of modernization.

Table 12.4 arranges provinces according to absolute mortality trends rather than relative trends or absolute outcomes. Some features of mortality variation not highlighted by this procedure nonetheless deserve attention. For example, West Sepik (Sandaun) Province sits inconspicuously near the middle of both listings, being at neither extreme in its mortality trends; but its absolute mortality levels are extreme. Already one of the more disadvantaged provinces in 1971, by 1980 it had clearly the shortest life expectancies of any province, and the highest pre-adult mortality rates except for the Southern Highlands Province. By the year 2000, strikingly unlike Southern Highlands, infant mortality had not improved at all; and although other measures had improved modestly, West Sepik was the worst placed province on all measures, with Gulf now not far behind. The deterioration in Gulf Province has been more dramatic but has not resulted in a worse outcome in absolute terms. Worse figures than this are reported from specific localized populations in West and East Sepik Provinces; for example, IMRs among the Abelam (cited above), the Saniyo-Hiowe (400 per 1000: Townsend 1985) and the Mianmin (200 per 1000: Attenborough *et al.* in prep.).

It would be very interesting to renew the question (cf. Caldwell 1986; Inaoka n.d.) where PNG fits in terms of the contemporary relationship between economic variables and mortality improvement. Has PNG been able, as Sri Lanka was, for example (Caldwell 1986; Caldwell *et al.* 1989), to achieve good health and mortality outcomes relative to other nations at a similar economic level; or similar, or worse ones? Unfortunately there are data obstacles to answering this question rigorously. Diversity of income is as striking as many of PNG's other dimensions of diversity (Hanson *et al.*

Table 12.5. *Life expectancies at birth and trends in life expectancy at birth (*e_0*) for countries with gross national income (GNI) between US$ 400 and US$ 600 per caput, 2003*

	GNI/caput US$ 2003	e_0 2003	Years added to e_0 since 1970
Zimbabwe	480	33	− 22
Lesotho	590	35	− 14
Moldova	590	69	4
Uzbekistan	420	70	7
Benin	440	51	9
Mauritania	530	53	11
Mongolia	480	64	11
Timor-Leste	430	50	11
Guinea	430	49	12
Senegal	550	53	12
Comoros	450	61	13
Pakistan	470	61	13
Sudan	460	56	13
India	530	64	15
PNG	**510**	**58**	**15**
Solomon Islands	600	69	15
Bangladesh	400	62	18
Vietnam	480	69	20
Yemen	520	60	22

Note:
Countries are arranged in ascending order of life expectancy trend.
Source: Data from UNICEF (2004a).

2001), so much so that an average should not attract too much attention. Furthermore, with 87% of the population of PNG still living in rural areas and predominantly still covered by traditional means of access to land, food, shelter and community support, resource measures based only on the formal money economy are far from satisfactory for this purpose. Nevertheless, for what they may be worth, some selected international comparison figures are presented in Table 12.5, where absolute and trend life expectancy figures are shown for countries (many with significant traditional resource access also) with annual per caput incomes of US$400 to US$600. On the estimates of the United Nations Children's Fund (UNICEF), PNG emerges neither as well placed as Bangladesh, Vietnam and Yemen, which have both higher life expectancies and larger gains in life expectancy, nor as unfortunate as those southern African countries worst afflicted by AIDS, or even, in terms of the trend rather than the absolute figures, the former Soviet republics included in the table. Rather, PNG is near the upper end of an intermediate group, with gains in life expectancy similar to India and Solomon Islands, but absolute life expectancy for 2003 (though higher than the NSO figure for 2000 cited above) below theirs. Within this income range, no clear relationship between income and life expectancy emerges.

Leading causes of death are hard to identify in the absence of comprehensive vital registration systems. Figures at a national level tend to come from hospitals and health centres, and since many deaths take place away from such centres, there is an obvious potential for distortion in the data available. Local surveys, such as those collated by Riley and Lehmann (1992) for the 1960s to 1980s, confirm, however, that the pattern of the past has been one dominated by infectious causes of death: particularly respiratory infections and, especially (but not exclusively) in the lowlands, malaria. Diarrhoeal disease is important but less so than in many developing countries. Neonatal and congenital causes are important in early childhood, as are accidents and violence in adults. On World Health Organization summary data (World Health Organization Regional Office for the Western Pacific 2005), the leading causes of mortality in the year 2000 were, in descending order, pneumonia, perinatal conditions, malaria, tuberculosis and meningitis; and of morbidity, obstetric causes, pneumonia, malaria and perinatal conditions. Thus the recent picture has much in common with the earlier one. This chapter cannot be comprehensive in examining trends in these and other conditions; but trends in selected causes of mortality and morbidity will be examined more closely in what follows.

Health problems of modernization

First, I shall consider some evidence on trends in health problems associated with modernization. There can be no doubt but that significant health and nutritional benefits have accompanied modernization in PNG. At Ok Tedi, where mine employment and elements of a 'modern' lifestyle were introduced over very few years in the 1980s to a previously very remote region in Western Province, and also in Simbu Province, where data extend back more than four decades documenting somewhat less dramatic socioeconomic changes associated with cash-cropping for example, improvements in nutritional health were evident for at least a large part of that time: see review by Ulijaszek (1995). Whether trends in single locations or contemporaneous contrasts between more and less economically developed locations are considered, modernization appears linked, usually, with faster physical growth of children (Ulijaszek 1995; Müller *et al.* 2001) and higher birth weights (Ulijaszek 2001; Müller *et al.* 2002a; but see also Lourie 1986; Tracer *et al.* 1998).

A review of the impact of cash-cropping on nutritional health (Heywood and Hide 1994) found cash cropping to be associated with increased child growth, especially in the highlands from where the best data are available, but probably also in the lowlands where infant mortality declined over the same period. The evidence did not bear out pessimistic anticipations of rising malnutrition. As one might expect, there was a concomitant secular increase not only in adult weight and height, but also in the prevalence of various syndromes and diseases of modernization. While it is difficult to establish thresholds for individuals beyond which growth and anthropometric increases bring more cost than benefit, two broad conclusions seem clear: (a) that modernization has brought substantial nutritional health improvements to many PNG communities; and (b) that there have also been health costs at a population level to the nutritional changes that accompany modernization.

No cases of myocardial infarction were found in a large series of autopsies in Rabaul, New Britain, in the 1920s and 1930s (T. C. Backhouse, reported Riley and Lehmann 1992). In 1966 a survey of the Enga community of Murapin in the PNG highlands found no evidence of coronary artery disease, cerebrovascular disease, hypertensive disease or diabetes (Sinnett 1977). Serum cholesterol, serum triglycerides and blood pressure were all low, and showed no increase with advancing age, or even fell. While few studies from that period are so comprehensive, it seems reasonable to accept that these results are quite representative of PNG communities under conditions little influenced by modernization. Those cardiovascular diseases which are associated with modernization, and many of their biological risk factors, are essentially recent phenomena in PNG.

It has been proposed, on the basis of African experience, that in general hypertension is the first disorder of modernization to emerge as the process gets under way, followed by obesity, type 2 diabetes mellitus (or non-insulin-dependent diabetes, NIDDM, usually mature-onset, of which impaired glucose tolerance (IGT) is a precursor), cerebrovascular disease and finally ischaemic heart disease (Trowell 1981). By the 1960s certain more urbanized communities, for example in the Port Moresby region, were providing evidence of hypertension, with blood pressures that had begun to rise with age, weight increases and higher, though still not high, prevalence of type 2 diabetes (Sinnett *et al.* 1992). At the Ok Tedi mine site, blood pressure rose significantly within three years in the mid 1980s, among a range of other human biological changes (Lourie 1987; Ulijaszek 1995). Hypertension has now been described in a rural Purari delta population (Ulijaszek 1997). The first description of a myocardial infarction in an indigenous Papua New Guinean was that of Conyers (1971). More cases were identified through the 1970s, and by the 1980s death was attributable to ischaemic heart disease in 11% of autopsy samples in Port Moresby, but only in 2% of similar samples in the smaller town of Goroka (Misch 1988). A further series of cases with high case fatality in Port Moresby hospital confirmed the trend (Kevau 1990).

Type 2 diabetes, similarly, was a very rare diagnosis in the 1960s (Ogle 2001). A study of five PNG communities in 1991, urban, peri-urban and rural, lowland and highland, used an existing 'modernity score' to track the development of diabetes and other heart disease risk factors (Hodge *et al.* 1996). This score was based on education, employment, access to and experience of urban centres, type of housing and so on (King and Collins 1989). A subsequent study has found distance from urban centre to be a useful prime index of modernization (Ulijaszek *et al.* 2005). Hodge *et al.* (1996) found 'modernity score' to be positively associated with overweight and type 2 diabetes. Furthermore, they found cholesterol concentrations to be higher than previously recorded in PNG and to show age-related increases; the same was true of triglyceride concentrations in women. In an urban community, total cholesterol was similar to Australian levels while HDL cholesterol was lower. The authors point out that PNG's earlier freedom from heart disease by no means signals that Papua New Guineans are protected from developing blood lipid levels that bring risk in other populations – no evidence of special adaptation here – and that without preventive strategies the incidence of coronary heart disease can be expected to increase further.

For the three coastal communities included in the study of Hodge *et al.* (1996), there are also diabetes data from earlier studies (Dowse *et al.* 1994). The majority of the results on both NIDDM IGT exhibit an upward trend (Table 12.6), especially those from Koki; but these results are also diverse. Koki is a squatter settlement near the centre of Port Moresby. Most residents of Koki originate from Wanigela, about 200 km southeast of Port Moresby. As we might expect, it has both the highest modernity score and the highest IGT and NIDDM prevalences, very high on a global scale. Wanigela has medium modernity scores, very similar to Kalo, which geographically is situated between the other two. But in frank diabetes mellitus particularly, Kalo is much less severely affected. The authors could not explain this last contrast in lifestyle terms, and proposed that the two

Table 12.6. *Trends in age-standardized prevalence of impaired glucose tolerance (IGT) and non-insulin-dependent diabetes (NIDDM) in three coastal communities in Papua New Guinea*

	1977/1983	1991
IGT (%)		
Koki	18.0	34.2
Wanigela	6.7	14.1
Kalo	14.1	10.3
NIDDM (%)		
Koki	16.7	27.5
Wanigela	9.8	11.1
Kalo	1.2	1.5

Source: Dowse *et al.* (1994), where sample sizes, confidence intervals, age standardization method and diagnostic criteria are detailed.

(largely endogamous) Austronesian-speaking groups of Wanigela origin may have a high frequency of a specific genotype, perhaps more susceptible to the diabetogenic effects of a modernizing life style than any other known. Neither this lifestyle nor the hypothesized genotype currently affect the majority of the populations of PNG to the extent seen in Wanigelans: on the other hand, it has become clear that any genetic protection non-Austronesian-speakers may have is only relative (King *et al.* 1989). The contrast between Koki and Wanigela underlines the potency of lifestyle factors, without denying that genetic factors may be important too. Prevalence in the country as a whole is rising but not yet high, and above all diverse (Ogle 2001).

Modernization does not only take place *in situ*: people migrate towards it, on a large scale in the case of PNG's larger towns. Studies among migrants to Port Moresby have drawn attention to the emergence of cardiovascular and diabetes risk factors in these groups also, and to how this varies markedly according to their geographical origin within PNG, length of stay in Port Moresby, occupation and sex (Natsuhara *et al.* 2000).

The role of a modern lifestyle in increasing the prevalence of cardio-vascular disease, in which NIDDM is a major risk factor, has recei-ved enormous attention and other widely accepted risk factors include (1) high dietary intake of fats, especially saturated fats; (2) high dietary intake of sodium, in connexion with hypertension and haemorrhagic stroke; (3) excessive alcohol consumption; (4) low levels of physical activity; (5) obesity, especially with centripetal fat distribution; and (6) psychosocial stress. These classical risk factors account for some but not all of the observed variation in cardiovascular disease within- and between-populations around the world. Newer additional hypotheses implicate early-life undernutrition, marked catch-up growth and/or infectious agents such as the common bacteria *Chlamydia pneumoniae* and *Helicobacter pylori* in the disease processes leading to elevated risk of cardiovascular disease (Pollard 1997). These agents, perhaps potentiated only in the presence of the classical risk factors already discussed, could, if either or both of these hypotheses are correct, be particularly devastating for populations currently experiencing a rapid transition from under-development to modernization. PNG in the near future may perhaps represent just such a case.

Discussions of modernization and health frequently select, as I have here, hypertension, ischaemic heart disease, type 2 diabetes and their risk factors for special attention; and there are good reasons for this. But these are by no means the only non-infectious diseases that show a rise associated with modernization: dozens have been listed (Trowell and Burkitt 1981), including, for example, asthma (Anderson and Woolcock 1992).

Infections, old and new

Is the rise of diseases of modernization the main feature of population health in PNG to emerge as the nation modernizes? If we extrapolate existing trends, can we foresee an automatic dwindling of the diseases of underdevelopment, as the classic diseases of modernization rise?

Other processes, especially affecting communicable disease, ensure that the prospects are not so simple. Despite some declines in particular infections in conformity with the epidemiological transition model, it is infectious diseases collectively that still represent the country's main health problem, causing large numbers of deaths, especially deaths of children and younger adults, as well as many non-fatal episodes of ill-health (Carrad 1987; Riley and Lehmann 1992). Pneumonia, due to a variety of bacterial and viral pathogens, and malaria, due to four *Plasmodium* species as discussed above, predominate among these (Cattani 1992; Riley *et al.* 1992).

There is not enough space here to explore all the important themes, so despite their prominence in the statistics, I shall comment only briefly on respiratory infections. The continuing importance of pneumonia (acute lower respiratory infections, ALRI), and of specific acute (e.g., influenza) and chronic (e.g., tuberculosis) respiratory infections is clear. Pneumonia is especially salient in the highlands, where malaria is less prevalent, but important everywhere. At Tari in the Southern Highlands Province, in the period 1977 to 1983, ALRI accounted for 53% of infant deaths, 34% of child deaths (one to four years), 18% of deaths in people over 60 years of age and 23% of deaths overall (Lehmann 2002). By comparison, neonatal/congenital abnormalities accounted for 21% of infant deaths, other febrile illness (presumably including malaria in lower-lying parts of the study area) for 8% of them, gastroenteritis for 5% and measles for 4%. Mortality due to ALRI had actually risen since 1972 to 1974, for which there are several possible reasons including penicillin resistance. However, after a pneumococcal vaccine trial in the area in 1981 it fell significantly, even though half the participants had received only a placebo vaccine. The vaccine has been shown to be effective (Riley 2002), and this development appears to offer the best way forward for public health approaches to ALRI. Unfortunately the benefits it offers have yet to be translated into lasting benefits for large numbers of Papua New Guineans, because the vaccine has not become available at a price that PNG can afford.

Since the Second World War and into the period of national independence, PNG has developed ways of bringing the benefits of its health care system beyond the cities and towns to its predominantly rural population, notably a network of aid posts and health centres linked to the urban hospitals. At a minimum, government and non-government (e.g., mission) facilities alike have been able to offer first aid, antibiotics (e.g., procaine-penicillin injections) and malaria medications (e.g., chloroquine/amodiaquine), and have run maternal and child health clinics. Immunization programmes have been a particularly important part of PNG's public health system in both rural and urban areas. Indeed, as already discussed, PNG's successes in countering infectious disease at a national level have essentially been with those infections against which an immunization programme has been widely implemented – and not, to date, with pneumonia, influenza, diarrhoea or malaria.

A further concern arose during the 1990s and remains current, to the effect that even those public health gains which had been made by then are not secure; that health systems are experiencing some deterioration owing to funding and organizational problems; and that population health improvements are stalling as a result. For example, real per caput expenditure on health is declining, geographical coverage is uneven, many aid posts have closed, and harassment of staff, theft of equipment and vandalism of facilities are reported to be frequent (Ministry of Health 2001). In these circumstances it is not surprising that existing immunization programmes only reach half of all children, or that, as discussed elsewhere, more than half of women giving birth do so without medical or para-medical assistance, and overall mortality and morbidity outcomes in the past 15 years have tended to be disappointing. Connell (1997) argues strongly that both health systems and population health status have deteriorated substantially since the 1980s and that the prospects for improvement are poor; see also Edwards (1994).

Malaria

Among PNG's myriad health problems, only malaria – the most frequent outpatient diagnosis and the second most frequent diagnosis amongst health facility admissions in many endemic areas – rivals pneumonia in importance at the present time (Ministry of Health 2001). In some provinces and age categories, it equals pneumonia as a primary cause of death. At its maximum, malaria transmission in PNG occurs on a holoendemic scale comparable with equatorial Africa.

Malaria is an environmental disease among other things, dependent on warmth (minimum ambient temperatures of 18 °C for *Plasmodium falciparum* and 16 °C for *P. vivax* for completion of the parasite's life cycle in the mosquito) and aquatic breeding places for *Anopheles* mosquitos. It is thus not surprising that its transmission in humid equatorial PNG is difficult to disrupt (Cattani 1992). In PNG, environmental temperature depends mainly on altitude, and so malaria is highly endemic in the lowlands (except during the marked dry season found in some southern lowland areas). Transmission decreases with elevation above sea level, becoming unstable and more epidemic in character around 1,300 to 1,600 m, and ceasing above 1,700 to 1,800 m (Müller *et al.* 2003). With that said, malaria prevalence and transmission are highly variable geographically, even on a local (village to village) scale, with mosquito biting rates of up to 500 bites/person/night in some places, much less in others a few kilometres or less away (Cattani *et al.* 1986a; Burkot *et al.* 1988; Attenborough *et al.* 1997).

Plasmodium infections do not always result in clinical illness, but most frequently do so in those with the least effective immune protection, especially children under ten years of age, pregnant women, and residents of non-endemic areas. While its seriousness is not in question, severe and lethal malaria is less frequent in PNG than in areas of comparable endemicity in Africa, perhaps because of the multiple genetic protections against malaria already discussed (more numerous and varied than

Africa's principal genetic protection, the sickle-cell gene), and/or because of the higher prevalence than in most parts of Africa of vivax malaria, which may confer some cross-protection to the more dangerous falciparum malaria (Maitland *et al.* 1996; Smith *et al.* 2001; Bruce and Day 2003; Müller *et al.* 2003; Mayxay *et al.* 2004). High rates of spleen enlargement in adults as well as children are another distinctive feature of malaria in PNG, especially in the highland fringes and associated with high risk of premature death from indeterminate causes (Crane 1986).

Malaria's prehistory in New Guinea has been discussed above. The beginning of systematic data collection there came with the work of Robert Koch and his colleague Otto Dempwolff in Kaiser-Wilhelmsland (German New Guinea), just over a century ago (Ewers 1972). Malaria was a major cause of mortality among Europeans, Javanese and Chinese in the newly founded German colony and the high death rate caused survivors to abandon some of their settlements. Thirteen out of 45 Europeans died of malaria during one three-month period of 1891 at Finschhafen (now in Madang Province). Of 273 Chinese at Stephansort, near the present town of Madang, 125 died during 1898, mostly of malaria.

Koch and Dempwolff also studied malaria among the indigenous people between 1899 and 1903 and found it to be highly endemic through much of the area, though they also considered some villages and some off-shore islands to be malaria-free or to be experiencing malaria as a new disease. Koch was impressed with the importance of malaria as a cause of illness and death; and, we may now wryly read in the light of the subsequent retreat from such ambitions, optimistically instructed Dempwolff not only to study malaria amongst village people but also to attempt its eradication. Both around the Madang coast and on the Bismarck Islands, they found the prevalence of malaria (as indicated by parasitaemia at all ages, with all species) to be generally between 15% and 25%. These are low figures by comparison with more recent ones (e.g., approximately 35% to 45% for Madang area communities in 1981 to 1982, reported by Cattani *et al.* 1986b). With geographical variations, seasonal fluctuation and the possibility of different or less sensitive methods all confounding the comparison, however, we cannot be sure that malaria was less prevalent in 1900 around Madang than it is now. At any rate, despite all efforts made in terms of malaria treatment and control campaigns, there is not even a hint here of any downward trend over the twentieth century.

The locations which now have New Guinea's highest recorded endemicities of malaria, in the Sepik basin, were not covered in this early German colonial research. But one such location, Maprik District, provides one of the best data sets in New Guinea for malaria, from the 1950s, during the

Table 12.7. *Malaria prevalence (%) in Maprik, East Sepik*
Province: proportion of positive blood slides

	Date	P.f.	P.m.	P.v.	c.r.
All ages	1957	16.1	17.0	29.5	62.6
	1963	13.6	21.2	9.7	44.5
	1984	22.6	19.5	7.7	49.8
1–6 years	1957	24.1	25.6	38.5	90.3[a]
	1963	19.8	19.8	16.0	55.6
	1984	50.0	37.5	0.0	87.5
7–15 years	1957	11.7	13.0	28.2	62.3[a]
	1963	13.2	25.3	8.4	46.9
	1984	24.3	21.6	21.6	67.5
>15 years	1957	7.1	7.3	18.7	38.2[a]
	1963	10.4	20.4	7.0	37.8
	1984	14.0	11.3	4.7	30.0

Notes:
P.f. = *Plasmodium falciparum*; P.m. = *Plasmodium malariae*; P.v. =
Plasmodium vivax; c.r. = crude parasitaemia rate, i.e., % of blood slides
positive for any *Plasmodium* species.
[a] Usually, crude rates may be lower than the sum of the species-specific rates,
owing to infections with multiple species (not apparent here), but are not higher.
Desowitz and Spark do not comment on why they are 2–10% higher for the 1957
age-specific data from Peters (1960). Explanations invoking the possibility that in
some cases the species was indeterminate or was the rare *P. ovale* would have to
cover the fact that the all-ages data do not have a higher total crude rate.
Source: Desowitz and Spark (1987).

Australian colonial period, to quite recent times. The available data
(Table 12.7) allow us to look at trends from the 1950s to 1980s, in the
main *Plasmodium* species as well as in the crude rate of infection (Desowitz
and Spark 1987). The crude rate went down and up again, but what should
perhaps strike and concern us most is the changing species balance. While
P. malariae remained more or less constant (and higher at the location
studied than at most others), *P. vivax* declined; and *P. falciparum*, the most
dangerous of the species, increased in prevalence, both relatively and
absolutely. In 1990 to 1992, surveys in another part of the Maprik
District, Wosera, undertaken in preparation for vaccine trials, again
found high prevalence of malaria at 60% (at all ages, with all species,
ranging from 49% to 70% between villages: Genton *et al.* 1995b), a similar
level overall to those of the 1950s shown in Table 12.7, and up from those
of the 1960s and 1980s. *Plasmodium falciparum* was the predominant
species (observed in 55% of all positive blood slides: Genton *et al.*

Table 12.8. *Malaria prevalence (%) in Karimui, South Simbu: proportion of positive blood slides*

	P.f.	P.m.	P.v.	c.r.
1965	7.1	5.6	6.7	19.4
1971	4.0	0.7	2.4	7.1
1981	12.3	3.9	14.1	30.3
2001/02	21.3	3.3	6.6	31.2

Notes:
P.f. = *Plasmodium falciparum*; P.m. = *Plasmodium malariae*; P.v. = *Plasmodium vivax*; c.r. = crude parasitaemia rate, i.e., % of blood slides positive for any *Plasmodium* species.
Source: Müller *et al.* (2005).

1995b), a position it had approached in the 1980s (45%) but not in the 1960s (31%) or the 1950s (25%) (Desowitz and Spark 1987). There is little seasonality to confound this picture, although local differences within Maprik may do so to some extent. Most recently, a significant drop in the prevalence of *P. falciparum* but not *P. vivax* has been observed in the Maprik area (Kasehagen *et al.* 2006). Still, the overall impression remains that the problem is not declining.

The shift in species balance is not a trend purely local to Maprik or the Sepik basin. The 1980s Madang figures show a predominance of *P. falciparum*, although we do not know the earlier species balance there. A local survey in the highlands fringes of West Sepik Province (Attenborough *et al.* 1996) shows a predominance of *P. falciparum* also, though again we have no time series data. On the Karimui plateau, Simbu province, at 900 to 1,200 m, malaria is more prevalent than at similar altitudes in some other locations, but here we do have another good time series (Table 12.8; Müller *et al.* 2005). The crude parasitaemia rate rose, later than in the Sepik lowlands, and most strikingly *P. falciparum* prevalence rose both absolutely and relatively.

Are the shifts in species balance that we have seen in several locations linked to economic modernization and globalization; or do they merely coincide with these developments? The changes matter either way: perhaps it is more important to identify specific factors that might be responsible than attempt to answer that question in broad terms. In reality, both overall prevalence and species balance have changed in ways too complex and too frequently reversed to be plausibly tied to a single process, whether we label it modernization or not. No single factor explains all aspects of the pattern, but ones which may be relevant to variations in malaria epidemiology both in time and space include (Müller *et al.* 2003): (1) vector control programmes,

for example, DDT and dieldrin residual spraying; (2) large-scale drug administration (in the past mainly with chloroquine); (3) use of bednets, especially
when insecticide-impregnated; (4) drug resistance evolved by *Plasmodium*,
especially *P. falciparum*; (5) variations in the ecology, behaviour and vectorial
capacity of the mosquitos which act as vectors in PNG (prinicipally the
Anopheles punctulatus group, now known to number at least 10 species in
PNG, rather than the three recognized before 1993: Beebe and Cooper 2002);
and (6) relevant variations in human numbers, behaviour and genetics.

In Maprik, in 1957, no mosquito-spraying operations had yet taken place
and anti-malarial drugs were available only for the treatment of acute cases.
By 1963, there was still no spraying locally (it started shortly afterwards);
but anti-malarial drugs, especially chloroquine, were more readily available,
and this may have led to a reduction in the prevalence of *P. vivax*. In 1984,
spraying had virtually ceased, but chloroquine was still more accessible and
this appears to account for a further reduction in *P. vivax* rates. Mechanisms
for these trends and especially for the rise of *P. falciparum* remain to be
clarified. The clinical drug-resistance of some *P. falciparum* strains, especially to chloroquine, was not in evidence until the 1970s; but by the 1980s it
was highly prevalent (Müller *et al.* 2003), so clearly it could be important in
the rise of falciparum apparent in the 1984 and 1990 to 1992 Maprik data.
The recent drop in *P. falciparum* prevalence in Maprik may be related to the
introduction in the year 2000 of a new standard first-line treatment (chlor-
oquine/amodiaquine plus sulphadoxine-pyrimethamine).

In Karimui, the 1965 survey preceded control measures. The 1971 survey,
immediately following a mass drug (chloroquine and pyrimethamine) administration campaign, and early in the period of indoor residual DDT spraying to
control vectors, found substantially reduced overall malaria rates but also a
species balance apparently tipped towards *P. falciparum*. Vector control measures became less effective from 1978 onwards, and were abandoned in 1984, in
line with similar trends elsewhere as it became apparent that eradication would
not be achieved; thus the 1981 survey took place at a time of failing control.
This is no doubt linked to the large increase in both *P. vivax* and *P. falciparum*
prevalence found in this survey, relative to both 1971 and 1965 levels. In the
following 20 years from 1981, not only did control measures cease but also
chloroquine-resistant *P. falciparum* became abundant, which may play a key
part in explaining why by 2001 *P. falciparum* had become much more prevalent while vivax had declined. Population growth, a shift from dispersed long-
houses to more centralized villages, and correspondingly larger, more clustered
land cultivation patterns may also have contributed (Müller *et al.* 2005).

The rise of chloroquine-resistant falciparum malaria seems to have a
clear role in the shifting species balance and the absolute rise of

P. falciparum. The timing of the DDT indoor residual house spraying programme, relative to the rise of *P. falciparum* and the decline of *P. vivax*, suggests that it may have a role, though again the mechanism is unclear. One suggestion is that it led to a shift from indoor late-night biting to outdoor dawn and dusk biting (Spencer *et al.* 1974), perhaps by younger mosquitoes or different species, with different propensities to transmit *P. falciparum* and *P. vivax* (Bockarie *et al.* 1996). However, this remains to be resolved. Also some evidence (discussed above) might support hypotheses that reduction in vivax prevalence is conducive to an absolute as well as relative rise of *P. falciparum.* Among possible candidate factors suggested as having favoured *P. vivax* initially, related to its long-lasting liver stages and other biological characteristics, are insecticide spraying (irregular or otherwise) and mass drug administration. Vines (1970) found *P. vivax* to be more prevalent in sprayed areas, and more recent results elsewhere have also shown a shift to *P. vivax* in the course of effective vector-control programmes. Indiscriminate anti-malarial drug use patterns and the rise of drug resistance in *P. falciparum* particularly in the post-vector-control period appear to have reversed any such initial advantage to *P. vivax*, however.

It can plausibly be argued that before the Australian colonial intrusions that began in 1930, the central highlands of PNG were generally free of endemic malaria, except in low-lying and swampy areas. Before mid-century, legislation regulated recruitment of plantation labourers from above 1,200 m, which delayed the spread of malaria into the highlands (Radford *et al.* 1976). Eventually, however, under the influence of post-war economic development, and with relaxation of controls on recruitment and of compulsory malaria prophylaxis for highlanders working at the coast, epidemic malaria outbreaks in the highlands became a more significant phenomenon than before, with a more fluctuating and epidemic pattern than in the lowlands, and with marked seasonality. Between the late 1940s and the early 1960s, these outbreaks were – successfully, at the time – counteracted by means of a DDT-spraying programme in the most densely populated areas, including Eastern Highlands, Western Highlands and Simbu Provinces. By 1974, the health authorities were no longer attempting to control the spread of infection through controls on the mass movement of people, or through vector control. Sharp (1982) reported a parasite rate in non-epidemic situations in western Enga Province of 12% in 1979, all due to *P. vivax*; but a parasite rate of 75% (in surveys angled towards febrile individuals) below 1,600 m during epidemics there in 1980, nearly all due to *P. falciparum.* Recent surveys have shown increased malaria in outlying highlands areas over 1,200 m, though not in all areas of this altitude (I. Müller, personal communication). Malaria is now a troublesome health problem throughout the highlands provinces (Müller *et al.* 2002b).

Two further factors not mentioned so far promise, or threaten, to exert impacts on trends in malaria prevalence. On the one hand, we still have the hope of the vaccine, on which much has been achieved in research terms, although there are still obstacles to its general implementation to bring about a major reduction in the population health burden imposed by malaria (Brown and Reeder 2002; Genton *et al.* 2003). Even when it is available, the malaria vaccine is expected to bring a reduction in frequency and severity rather than a prospect of eradication. On the other hand, there is at least one further reason to predict an upward trend in malaria, unless the vaccine or other control measures can offset it – climate change. While the smaller Pacific nations have other reasons to be concerned about global climate change, a major consideration for PNG is the possibility of its effects on vector-borne disease including malaria – even the possibility that such effects are already under way, contributing to the trend in highlands malaria just discussed. Available data for New Guinea are insufficient to say at present, but projected climate change would be sufficient to raise the altitude cut-offs for the transmission of malaria by an amount that could be significant in terms of the location and movements of human populations. In the Ethiopian highlands, monthly night-time temperatures are reported to have risen by about 1.5 °C from 1980 to 1993, and with them – though there can be debate about what this means – has risen the incidence of falciparum malaria and the altitude cut-off for malaria there (Tulu 1996; Kovats *et al.* 2001).

One of the main points of the chapter is then, using malaria as a prime example, to underline the continuing salience of communicable disease in PNG's overall health profile. It is not that some primaeval pattern simply continues unchanged: but rather that malaria is still there, continuing unabated, but shifting in its expression as its prevalence rises, its species balance changes and its geographical territory expands, under the influence of evolving drug resistance, changing strategies of treatment and prevention and their implementation, modernizing ways of life and patterns of movement, perhaps dietary trends, possibly climate change and no doubt other factors yet to be definitely identified.

HIV/AIDS

While malaria is my prime example of the continuing prominence of communicable disease, some counterbalance is required to illustrate the variety of those communicable disease problems which show little sign of spontaneous decline as modernization proceeds. Other vector-borne diseases could be chosen, especially since many of them might be expected to spread or intensify under climate change: for example, a number of arboviruses; or Bancroftian filariasis (Kazura and Bockarie 2003). Given the practical reality in rural PNG that acute febrile illnesses are often treated presumptively as malaria until there are indications otherwise (Müller *et al.* 2003), it is likely that the incidence of such illnesses tends to be underestimated. Similarly, tuberculosis remains an important problem, undefeated and likely to be exacerbated as AIDS increasingly takes hold. But it is AIDS itself, strongly contrasting with malaria in several respects, including the historical recency of its presence in PNG, which will constitute my main counterbalancing example.

HIV/AIDS was first diagnosed in PNG in 1987 (Centre for International Economics 2002; National AIDS Council Secretariat and Department of Health 2002). Initially thereafter, numbers of cases reported remained small; but from the mid 1990s the upward trend became steeper, more than trebling between 1996 and 1998, and continuing to rise thereafter without evidence of plateauing in the data available so far. By March 2002, 5,239 people (2,665 male, 2,391 female, 183 sex not recorded) had been diagnosed as infected, more of them through presenting clinically with an AIDS-defining illness than through a positive HIV test; and 273 deaths had been reported (National AIDS Council Secretariat and Department of Health 2002). Seventy-one percent of these cases were detected in the National Capital District, 12% in Western Highlands Province, and 5% each in Eastern Highlands and Morobe Provinces, although no province was case-free. The manner of transmission was unrecorded for most cases, but where recorded, exposure was predominantly by heterosexual sex (72% for males, 75% for females), with vertical transmission the only other frequently recorded exposure category. Age at diagnosis, where recorded, was most commonly in the 20 to 34 years age range for both sexes: males outnumbered females in all adult age categories except those of 15 to 29 years.

These figures, based as they are on reported cases, need to be interpreted cautiously. Numbers of AIDS diagnoses and HIV-positive test results naturally depend on the availability and uptake of diagnostic and testing facilities. The higher female numbers during childbearing years, for example, probably reflect routine testing at the Port Moresby General Hospital antenatal clinic. Therefore we cannot be confident that the available data mirror the historical course of the epidemic in PNG to date or give an accurate measure of current incidence or prevalence, especially beyond the major urban centres where facilities are concentrated. In fact, we can be confident that the available data seriously underestimate the true extent of

HIV infection, although the degree of underestimation is hard to gauge: the reported cases may represent 5% to 25% of the true numbers according to some estimates made (Centre for International Economics 2002). It is unclear whether the true trend since 1987 or since 1996 is more steep, or less so, than appears from the available figures; whether the age/sex and transmission category figures of the recorded cases are representative of all cases; and whether, when the urban bias due to the availability of facilities is allowed for, HIV/AIDS is still a problem much more of Port Moresby and other urban centres than of the rural population.

Nonetheless, it is difficult to doubt that HIV infection is rising in prevalence, spreading geographically, and likely to continue both trends. The National AIDS Council has run a vigorous public campaign, including promotion of condom use; but a number of factors militate against its message. These include (Lemeki 2003): (1) sexual mores and attitudes to fertility such that one or both sexes are often reluctant to use condoms; (2) contexts for sexual interaction which can include, on a significant scale, multiple sexual partnerships, sexual violence and exchange of sex for cash, whether commercial or simply occasional; (3) much poverty, economic inequality and urban-rural migration; (4) high levels of other sexually transmitted infections in some areas and of tuberculosis; and (5) impoverished facilities for HIV/AIDS treatment as well as diagnosis. Comparisons with the situation in the worst affected southern African and South East Asian countries are being made.

The Centre for International Economics (2002), in a report for the Australian government aid agency AusAID, reviewed many of the points just summarized, and concluded, while acknowledging many uncertainties, that PNG appears poised on the brink of a major epidemic. It raised the prospects that by the year 2020 there could be (1) a decline in PNG's already low life expectancy, by up to 20% to 25%; (2) a national population between 13% and 38% smaller than it would have been without AIDS; and (3) negative economic growth. Concern as to the likely future impact of HIV/AIDS on PNG's population health is supported by subsequent reports that HIV/AIDS has become the second most common admission diagnosis at Port Moresby General Hospital after tuberculosis, and its most common cause of death (G. Tau, reported McBride 2005); and that the prevalence of HIV antibody in 300 patients (both sexes, ages where known, ranging from 10 to 69 years) presenting to the emergency department at Port Moresby General Hospital was 18% (Curry *et al.* 2005). HIV-positive individuals in the study of Curry *et al.* were of both sexes, and ranged in age from 18 to 60 years: their presenting illnesses were predominantly respiratory tract infections (most often pulmonary tuberculosis) or gastrointestinal tract infections.

Conclusion: double jeopardy?

In this chapter, with a view to exploring the health patterns of PNG as it makes its complex transitions from tradition to modernity, I have traversed a very wide range of topics, from prehistoric genetic adaptations to the nutritional and lifestyle diseases and emergent infections associated with modernization. How do the various themes link with each other?

There is some validity in evolutionary arguments (e.g., Boyden 2004) according to which a high level of adaptedness can be expected to build up in humans (or any other organisms) exposed for many millennia to a constant or only slowly changing environment. Rapid and radical change in environment or way of life may cause adaptedness to break down in some respects, and with this breakdown in adaptedness, quite plausibly, comes the emergence of new health problems. Thus adaptation may help explain the relative success with which Papua New Guinean populations have accommodated to malaria: its breakdown may help explain the rise of circulatory disease, diabetes and asthma there (and elsewhere). But this approach can only take us so far. An adaptive equilibrium does not guarantee good health and long life. As we have seen, modernization has been accompanied by rising life expectancy and the alleviation of some old health problems, even while new ones have emerged.

One intellectual move which may be appropriate here is to expand the concept of adaptation beyond the strictly evolutionary model which I have followed thus far. Perhaps vaccination programmes, for example, can be viewed as cultural adaptations – so long as this does not lead us to count on them as automatic. A 'people will adapt' attitude, sometimes encountered in the climate change literature for example, risks complacency. If we are to think in terms of successful biocultural adaptation, then surely thought-through responses must be part of that. It will take more than the slow ticking of genetic evolution, the exercise of physiological homeostasis and the unplanned accretion of historical and cultural change to meet the complex challenges of current health trends in PNG.

There is also, unquestionably, considerable validity in health transition models. It is clear that in some of PNG's highly diverse communities, more of them than in the past, and with more reach from the urban into the rural areas, the syndromes and diseases associated with modernization are becoming epidemiologically important. That is a trend that can be expected to continue. Some communities may be genetically predisposed, and/or predisposed by the coexistence of early undernutrition and highly prevalent bacterial infection, to the development of type 2 diabetes and ischaemic heart disease, although the role of modern diet and activity patterns is better established. Non-infectious disease is now a matter for real and rising concern (Temu 1991; Tefuarani *et al.* 2002). But, here too, in the PNG context, reality is more complex than many of the models, and it would be dangerously misleading to ascribe a kind of automatism to the health transition process, important as it is. It is by no means clear yet that time and economic development will be sufficient to ensure that lifestyle diseases eventually overtake infections, or even that gains so far in the control of infection and increase of life expectancy are secure for the future.

By reviewing trends in selected aspects of PNG's population health with special focus – mortality, type 2 diabetes, malaria, HIV/AIDS – and touching briefly on many others – including pneumonia, influenza, tuberculosis, measles, smallpox, sexually transmitted infections, yaws, diarrhoea, dysentery, pigbel, arboviruses, filariasis, maternal mortality, heart disease, asthma, arthritis, trauma – I have aimed to illustrate empirically some of PNG's many simultaneous and successive health trends, too complex to be drawn into a single overarching model unless that itself is very complex. Some observed trends bear out the expected declines of early-acting infections, increases in life expectancy and rises of late-acting non-infectious disease. Others, however, follow their own dynamic: malnutrition, diarrhoea and even measles continue to cause uncontrolled problems; tuberculosis is increasing; and most saliently, the emergence of HIV/AIDS threatens a large increase in the infectious death toll and even a reduction in life expectancy; while malaria shows no sign of conforming to health transition expectations, rather developing in increasingly dangerous ways, with a rise in the proportion due to *P. falciparum*, increasing drug resistance in *P. falciparum*, territorial expansion across the highlands, and the possibility of aggravation under climate change.

The PNG Ministry of Health (2001: 3) itself acknowledges in its current National Health Plan that 'The poor status of the health of the people, in particular rural women and children, is not improving.' The Ministry has numerous proposals to address this situation, though as it documents, there are funding and organizational hurdles to overcome in implementing them. It is beyond the scope of this chapter to review such proposals. Perhaps the strongest avenue in conventional public health terms would lie in redoubled immunization programmes – the improved coverage of vaccines already offered, the implementation of vaccines already feasible and tested, and the development and implementation of ones still at the research stage; of course all this too would present funding problems.

The search for low-cost alternative ways to improve population health is not easy. Maybe one possibility lies in regulation: PNG has a history of innovation in public health regulation, as witnessed by its ban on the advertisement of baby feeding bottles and breast-milk substitutes, and requirement of a prescription for the sale of feeding bottles (Biddulph 1983). Perhaps it is time for another imaginative step in this direction – if the right step can be found. Any suggestion to follow Bhutan's lead by banning the sale of tobacco products and public smoking (BBC News 2004) may attract scepticism in PNG for some time to come, despite the near universal prevalence of cough.

Since some of the problems are social – for example in the integration of health services into communities – maybe some social solutions, beyond purely technological and public health ones, are needed too. Perhaps a hint in this direction is given by Macintyre's (2004) case study from the Lihir islands, New Ireland Province. Contrary to the argument sometimes made that traditional ideas and cultural practices tend to be at odds with modern medical services and impede their uptake, she found, encouragingly, that women there have embraced new practices in relation to medical treatment and hospital birth where they saw benefits in them, and have experienced dramatic health benefits in consequence.

To conclude, then, despite such encouraging signs as those provided by Macintyre's study, there remain major population health problems for PNG. On the one hand, for a complex of environmental, biological, economic and sociocultural reasons, communicable diseases still play an extremely strong part in PNG's health profile, and the immediate outlook for most Papua New Guineans is not currently one dominated by the expected 'diseases of modernization'. On the other hand, the rise of those mainly non-infectious diseases and their risk factors, especially in the urban centres and the areas most influenced by them, is well attested and substantial. There is thus, as my title suggests, a kind of double jeopardy now at work – as for indigenous minorities in developed countries (Kunitz 1994), though with differences of scale and pattern – whereby both old causes and new, infectious and non-infectious, diminish the prospects of Papua New Guineans for long and healthy lives.

Acknowledgments

I thank Professor Ryutaro Ohtsuka for his invitation to participate in the Tokyo International Union of Anthropological and Ethnographic Sciences meeting on which this volume is based; Professor Tai-Ichiro Takemoto for his invitation to give a subsequent presentation at Nagasaki University Medical School; audiences in Tokyo and Nagasaki for their comments and questions; Drs Michael Alpers, Robin Hide, Ivo Müller and Stanley Ulijaszek for their critical readings of part or all of this chapter in draft; Drs Nigel Beebe, Colin Filer, Deirdre Joy, Judith Littleton, Christine McMurray, Kim Streatfield and Zhongwei Zhao for drawing my attention to particular points and sources; and the editors of this volume for their assistance and forbearance. All remaining errors and omissions are my own responsibility.

References

Adels, B. R. and Gajdusek, D. C. (1963). Survey of measles patterns in New Guinea, Micronesia and Australia. *American Journal of Hygiene* **77**, 317–43.

Allen, B. J. (1983). A bomb or a bullet or the bloody flux? Population change in the Aitape inland, Papua New Guinea, 1941–1945. *Journal of Pacific History* **18**, 218–35.

   (1989). Infection, innovation and residence: illness and misfortune in the Torricelli foothills from 1800. In *A Continuing Trial of Treatment: Medical Pluralism in Papua New Guinea*, ed. S. Frankel and G. Lewis. Dordrecht: Kluwer, pp. 35–68.

   (1992). The geography of Papua New Guinea. In *Human Biology in Papua New Guinea: The Small Cosmos*, ed. R. D. Attenborough and M. P. Alpers. Oxford: Clarendon Press, pp. 36–66.

Allen, S. J., O'Donnell, A., Alexander, N. D. E., *et al.* (1997). α + -thalassemia protects children against disease caused by other infections as well as malaria. *Proceedings of the National Academy of Sciences of the USA* **94**, 14736–41.

Alpers, M. P. (1992). Kuru. In *Human Biology in Papua New Guinea: The Small Cosmos*, ed. R. D. Attenborough and M. P. Alpers. Oxford: Clarendon Press, pp. 313–34.

Alpers, M. P. and Attenborough, R. D. (1992). Human biology in a small cosmos. In *Human Biology in Papua New Guinea: The Small Cosmos*, ed. R. D. Attenborough and M. P. Alpers. Oxford: Clarendon Press, pp. 1–35.

Anderson, H. R. and Woolcock, A. J. (1992). Chronic lung disease and asthma in Papua New Guinea. In *Human Biology in Papua New Guinea: The Small Cosmos*, ed. R. D. Attenborough and M. P. Alpers. Oxford: Clarendon Press, pp. 289–301.

Attenborough, R. D. (2002). Ecology, homeostasis and survival in human population dynamics. In *Human Population Dynamics: Cross-Disciplinary Perspectives*, ed. H. Macbeth and P. Collinson. Cambridge: Cambridge University Press, pp. 186–208.

Attenborough, R. D. and Alpers, M. P. (1995). Change and variability in Papua New Guinea's patterns of disease. In *Human Populations: Diversity and Adaptation*, ed. A. J. Boyce and V. Reynolds. Oxford: Oxford University Press, pp. 189–216.

Attenborough, R. D., Gardner, D. S. and Gibson, F. D. (1996). *Malaria and Filariasis amongst the Mianmin of the Highland Fringes of Sandaun Province, Papua New Guinea: A Report.* Canberra: Department of Archaeology and Anthropology, Australian National University.

Attenborough, R. D., Burkot, T. R. and Gardner, D. S. (1997). Altitude and the risk of bites from mosquitoes infected with malaria and filariasis among the Mianmin people of Papua New Guinea. *Transactions of the Royal Society of Tropical Medicine and Hygiene* **91**, 8–10.

Ballard, C., Brown, P., Bourke, R. M. and Harwood, T. (eds.) (2005). *The Sweet Potato in Oceania: A Reappraisal.* Pittsburgh & Sydney: Ethnology Monographs and Oceania Monographs.

Barnett, J. (2002). Resilence and adaptation. In *Asia-Pacific Network Workshop on Ethnographic Perspectives on Resilience to Climate Variability in Pacific Island Countries*, ed. J. Barnett and M. Busse. Christchurch, New Zealand: Macmillan Brown Centre for Pacific Studies, University of Canterbury, pp. 5–11.

BBC News (2004). Bhutan forbids all tobacco sales. *BBC.* http://news.bbc.co.uk/2/hi/south_asia/4012639.stm (accessed December 2005).

Beebe, N. W. and Cooper, R. D. (2002). Distribution and evolution of the *Anopheles punctulatus* group (Diptera: Culicidae) in Australia and Papua New Guinea. *International Journal for Parasitology* **32**, 563–74.

Benjamin, A. L. and Dramoi, V. (2002). Outbreak of measles in the National Capital District, Papua New Guinea in 2001. *Papua New Guinea Medical Journal* **45**, 178–84.

Biddulph, J. (1983). Legislation to protect breastfeeding. *Papua New Guinea Medical Journal* **26**, 9–12.

Bockarie, M. J., Alexander, N., Bockarie, F., *et al.* (1996). The late biting habit of parous *Anopheles* mosquitoes and pre-bedtime exposure of humans to infective female mosquitoes. *Transactions of the Royal Society of Tropical Medicine and Hygiene* **90**, 23–5.

Boyden, S. (2004). *The Biology of Civilisation: Understanding Human Culture as a Force in Nature.* Sydney: University of New South Wales Press.

Brinkhoff, T. (2003). *City population.* http://www.citypopulation.de (accessed 25 November 2005).

Brown, G. V. and Reeder, J. C. (2002). Malaria vaccines. *Medical Journal of Australia* **177**, 230–1.

Bruce, M. C. and Day, K. P. (2003). Cross-species regulation of *Plasmodium* parasitemia in semi-immune children from Papua New Guinea. *Trends in Parasitology* **19**, 271–7.

Burkot, T. R., Graves, P. M., Paru, R., Wirtz, R. A. and Heywood, P. F. (1988). Human malaria transmission studies in the *Anopheles punctulatus* complex: sporozoite rates, inoculation rates and sporozoite densities. *American Journal of Tropical Medicine and Hygiene* **39**, 135–44.

Burton, J. (1983). A dysentery epidemic in New Guinea and its mortality. *Journal of Pacific History* **18**, 236–61.

Burton-Bradley, B. G. (ed.) (1990). *A History of Medicine in Papua New Guinea: Vignettes of an Earlier Period.* Kingsgrove, New South Wales: Australasian Medical Publishing Company.

Caldwell, J. C. (1986). Routes to low mortality in poor countries. *Population and Development Review* **12**, 171–220.

Caldwell, J. C., Gajanayake, I., Caldwell, P. and Peiris, I. (1989). Sensitization to illness and the risk of death: an explanation for Sri Lanka's approach to good health for all. *Social Science and Medicine* **28**, 365–79.

Caldwell, J. C., Findley, S., Caldwell, P., *et al.* (eds.) (1990). *What We Know about Health Transition: The Cultural, Social and Behavioural Determinants of Health.* Canberra: Health Transition Centre, Australian National University.

Carrad, E. V. (1987). *Review of Disease Patterns in Papua New Guinea.* Port Moresby: University of Papua New Guinea Press.

Cattani, J. A. (1992). The epidemiology of malaria in Papua New Guinea. In *Human Biology in Papua New Guinea: The Small Cosmos*, ed. R. D. Attenborough and M. P. Alpers. Oxford: Clarendon Press, pp. 302–12.

Cattani, J. A., Moir, J. S., Gibson, F. D., *et al.* (1986a). Small-area variations in the epidemiology of malaria in Madang Province. *Papua New Guinea Medical Journal* **29**, 11–17.

Cattani, J. A., Tulloch, J. L., Vrbova, H., *et al.* (1986b). The epidemiology of malaria in a population surrounding Madang, Papua New Guinea. *American Journal of Tropical Medicine and Hygiene* **35**, 3–15.

Centre for International Economics (2002). *Potential Economic Impacts of an HIV/AIDS Epidemic in Papua New Guinea.* Canberra: AusAID.

Clark, J. T. and Kelly, K. M. (1993). Human genetics, paleoenvironments and malaria: relationships and implications for the settlement of Oceania. *American Anthropologist* **95**, 612–30.

Connell, J. (1997). Health in Papua New Guinea: a decline in development. *Australian Geographical Studies* **35**, 271–93.

Conyers, R. A. J. (1971). Myocardial infarction in a New Guinean. *Medical Journal of Australia* **ii**, 412–17.

Cortés, A., Benet, A., Cooke, B. M., Barnwell, J. W. and Reeder, J. C. (2004). Ability of *Plasmodium falciparum* to invade Southeast Asian ovalocytes varies between parasite lines. *Blood* **104**, 2961–6.

Crane, G. G. (1986). Recent studies of hyperreactive malarious splenomegaly (tropical splenomegaly syndrome) in Papua New Guinea. *Papua New Guinea Medical Journal* **29**, 35–40.

Curry, C., Bunungam, P., Annerud, C. and Babona, D. (2005). HIV antibody seroprevalence in the emergency department at Port Moresby General Hospital, Papua New Guinea. *Emergency Medicine Australasia* **17**, 359–62.

Davies, M. (2002). *Public Health and Colonialism: The Case of German New Guinea 1884–1914.* Wiesbaden: Harrassowitz Verlag.

Dengler, L. and Preuss, J. (2003). Mitigation lessons from the July 17, 1998 Papua New Guinea tsunami. *Pure and Applied Geophysics* **160**, 2001–31.

Denham, T. P., Haberle, S. G., Lentfer, C., *et al.* (2003). Origins of agriculture at Kuk swamp in the highlands of New Guinea. *Science* **301**, 189–93.

Denoon, D., Dugan, K. and Marshall, L. (1989). *Public Health in Papua New Guinea: Medical Possibility and Social Constraint, 1884–1984*. Cambridge: Cambridge University Press.

Desowitz, R. S. and Spark, R. A. (1987). Malaria in the Maprik area of the Sepik region, Papua New Guinea: 1957–1984. *Transactions of the Royal Society of Tropical Medicine and Hygiene* **81**, 175–6.

Dowse, G. K., Spark, R. A., Mavo, B., *et al.* (1994). Extraordinary prevalence of non-insulin-dependent diabetes mellitus and bimodal plasma glucose distribution in the Wanigela people of Papua New Guinea. *Medical Journal of Australia* **160**, 767–74.

Duke, T. (2003). The crisis of measles and the need to expand the ways of delivering vaccines in Papua New Guinea. *Papua New Guinea Medical Journal* **46**, 1–7.

Edwards, K. N. (1994). Rural health service crisis in Papua New Guinea: causes, implications and possible solutions. *Papua New Guinea Medical Journal* **37**, 145–51.

Ewers, W. H. (1972). Malaria in the early years of German New Guinea. *Journal of the Papua and New Guinea Society* **6**, 3–30.

Fenner, F., Henderson, D. A., Arita, I., Jezek, Z. and Ladnyi, I. D. (1988). *Smallpox and Its Eradication*. Geneva: World Health Organization.

Fix, A. G. (2003). Simulating hemoglobin history. *Human Biology* **75**, 607–18.

Foley, W. A. (1992). Language and identity in Papua New Guinea. In *Human Biology in Papua New Guinea: The Small Cosmos*, ed. R. D. Attenborough and M. P. Alpers. Oxford: Clarendon Press, pp. 136–49.

Forge, A. (1972). Normative factors in the settlement size of neolithic cultivators (New Guinea). In *Man, Settlement and Urbanism*, ed. P. J. Ucko, R. Tringham and G. W. Dimbleby. London: Duckworth, pp. 363–76.

(1990). The power of culture and the culture of power. In *Sepik Heritage: Tradition and Change in Papua New Guinea*, ed. N. Lutkehaus, C. Kaufmann, W. E. Mitchell, L. Osmundsen and M. Schister. Durham, North Carolina: Carolina Academic Press, pp. 160–70.

Froment, A. (2001). Evolutionary biology and health of hunter-gatherer populations. In *Hunter-Gatherers: An Interdisciplinary Perspective*, ed. C. Panter-Brick, R. H. Layton and P. Rowley-Conwy. Cambridge: Cambridge University Press, pp. 239–66.

Gajdusek, D. C. (1990). Raymond Pearl Memorial Lecture, 1989: Cultural practices as determinants of clinical pathology and epidemiology of venereal infections: implications for predictions about the AIDS epidemic. *American Journal of Human Biology* **2**, 347–51.

Garner, P. A., Talwat, E. N., Hill, G., Reid, M. S. and Garner, M. F. (1986). Yaws reappears. *Papua New Guinea Medical Journal* **29**, 247–52.

Genton, B., Al-Yaman, F., Mgone, C. S., *et al.* (1995a). Ovalocytosis and cerebral malaria. *Nature* **378**, 564–5.

Genton, B., Al-Yaman, F., Beck, H.-P., *et al.* (1995b). The epidemiology of malaria in the Wosera area, East Sepik Province, Papua New Guinea, in preparation for vaccine trials. I. Malariometric indices and immunity. *Annals of Tropical Medicine and Parasitology* **89**, 359–76.

Genton, B., Anders, R., Alpers, M. P. and Reeder, J. C. (2003). The malaria vaccine development program in Papua New Guinea. *Trends in Parasitology* **19**, 264–70.

Greenburg-Hudd, S. (1998). *Dental palaeopathology and subsistence in protohistoric and historic skeletal remains from Papua New Guinea.* M.Litt. thesis, Archaeology & Anthropology, Australian National University, Canberra.

Groube, L. M. (1993). Contradictions and malaria in Melanesian and Australian prehistory. In *A Community of Culture: The People and Prehistory of the Pacific*, ed. M. Spriggs, D. E. Yen, W. Ambrose, R. Jones, A. Thorne and A. Andrews. Canberra: Department of Prehistory, Research School of Pacific Studies, Australian National University, pp. 164–86.

(2000). *Plasmodium falciparum*: the African genesis. In *Australian Archaeologist: Collected Papers in Honour of Jim Allen*, ed. A. Anderson and T. Murray. Canberra: Coombs Academic Publishing, Australian National University, pp. 131–44.

Groube, L. M., Chappell, J., Muke, J. and Price, D. (1986). A 40 000-year-old occupation site at Huon Peninsula, Papua New Guinea. *Nature* **324**, 453–5.

Hanson, L. W., Allen, B. J., Bourke, R. M. and McCarthy, T. J. (2001). *Papua New Guinea Rural Development Handbook*. Canberra: The Australian National University.

Heywood, P. F. and Hide, R. L. (1994). Nutritional effects of export-crop production in Papua New Guinea: a review of the evidence. *Food and Nutrition Bulletin* **15**, 233–49.

Hide, R. (2003). *Pig Husbandry in New Guinea: A Literature Review and Bibliography*. Canberra: Australian Centre for International Agricultural Research.

Hodge, A. M., Dowse, G. K., Erasmus, R. T., *et al.* (1996). Serum lipids and modernization in coastal and highland Papua New Guinea. *American Journal of Epidemiology* **144**, 1129–42.

Hope, G. S. and Haberle, S. (2005). The history of the human landscapes of New Guinea. In *Papuan Pasts: Cultural, Linguistic and Biological Histories of Papuan-Speaking Peoples*, ed. A. K. Pawley, R. D. Attenborough, J. Golson and R. L. Hide. Canberra: Pacific Linguistics, The Australian National University, pp. 541–54.

Hope, G. S., Golson, J. and Allen, J. (1983). Palaeoecology and prehistory in New Guinea. *Journal of Human Evolution* **12**, 37–60.

Hughes, J. (1997). A history of sexually transmitted diseases in Papua New Guinea. In *Sex, Disease and Society: A Comparative History of Sexually Transmitted Diseases and HIV/AIDS in Asia and the Pacific*, ed. M. Lewis, S. Bamber and M. Waugh. Westport, CT: Greenwood Press, pp. 231–47.

Hume, J. C. C., Lyons, E. J. and Day, K. P. (2003). Malaria in antiquity: a genetics perspective. *World Archaeology* **35**, 180–92.

Inaoka, T. (1995). Health and survival in modernizing Papua New Guinea societies. *Anthropological Science* **103**, 339–47.

(n.d.). The modernization of Papua New Guinea and the changes in disease patterns – recent health transition and emerging/re-emerging infectious diseases. *JCAS Area Studies Research Reports* **2**, 201–17 [In Japanese: transl. A. Callaway].

Joy, D. A., Feng, X., Mu, J., *et al.* (2003). Early origin and recent expansion of *Plasmodium falciparum. Science* **300**, 318–21.

Kasehagen, L., Müller, I., McNamara, D. T. *et al.* (2006). Changing patterns of *Plasmodium* blood stage infections in the Wosera region of Papua New Guinea as monitored by light microscopy and high through-put PCR, *American Journal of Tropical Medicine & Hygiene* **67**, 588–96.

Kazura, J. W., and Bockarie, M. J. (2003). Lymphatic filariasis in Papua New Guinea: interdisciplinary research on a national health problem. *Trends in Parasitology* **19**, 260–3.

Kevau, I. H. (1990). Clinical documentation of twenty cases of acute myocardial infarction in Papua New Guineans. *Papua New Guinea Medical Journal* **33**, 275–80.

King, H. and Collins, A. M. (1989). A modernity score for individuals in Melanesian society. *Papua New Guinea Medical Journal* **32**, 11–22.

King, H., Finch, C., Collins, A. M., *et al.* (1989). Glucose tolerance in Papua New Guinea: ethnic differences, association with environmental and behavioural factors and the possible emergence of glucose intolerance in a highland community. *Medical Journal of Australia* **151**, 204–10.

Kirk, R. L. (1992). Population origins in Papua New Guinea – a human biological overview. In *Human Biology in Papua New Guinea: The Small Cosmos*, ed. R. D. Attenborough and M. P. Alpers. Oxford: Clarendon Press, pp. 172–97.

Kovats, R. S., Campbell-Lendrum, D. H., McMichael, A. J., Woodward, A. and Cox, J. S. (2001). Early effects of climate change: do they include changes in vector-borne disease? *Philosophical Transactions of the Royal Society of London, Series B: Biological Sciences* **356**, 1057–68.

Kunitz, S. J. (1994). *Disease and Social Diversity: The European Impact on the Health of Non-Europeans.* New York: Oxford University Press.

Lavu, E. K., Oswyn, G. and Vince, J. D. (2002). Sickle-cell/β+-thalassaemia in a Papua New Guinean: the first reported case of the sickle gene in Papua New Guinea. *Medical Journal of Australia* **176**, 70–1.

Lawrence, G. (1992). Pigbel. In *Human Biology in Papua New Guinea: The Small Cosmos*, ed. R. D. Attenborough and M. P. Alpers. Oxford: Clarendon Press, pp. 335–44.

Leclerc, M. C., Durand, P., Gauthier, C., *et al.* (2004). Meager genetic variability of the human malaria agent *Plasmodium vivax. Proceedings of the National Academy of Sciences of the USA* **101**, 14455–60.

Lehmann, D. (2002). Demography and causes of death among the Huli in the Tari basin. *Papua New Guinea Medical Journal* **45**, 51–62.

Lemeki, M. (2003). *Sexually transmitted diseases and HIV/AIDS: knowledge and sexual behaviour among Eastern Highlands youth, Papua New Guinea.* MA thesis, Australian National University, Canberra.

Lilley, I. (1992). Papua New Guinea's human past: the evidence of archaeology. In *Human Biology in Papua New Guinea: The Small Cosmos*, ed. R. D. Attenborough and M. P. Alpers. Oxford: Clarendon Press, pp. 150–71.

Livingstone, F. B. (1989). Simulation of the diffusion of the β-globin variants in the Old World. *Human Biology* **61**, 297–309.

Lourie, J. A. (1986). Trends in birthweight over 43 years at Kwato, Milne Bay Province. *Papua New Guinea Medical Journal* **29**, 337–43.

(ed.) (1987). *Ok Tedi Health and Nutrition Project Papua New Guinea 1982–1986: Final Report.* Port Moresby and Tabubil: University of Papua New Guinea and Ok Tedi Mining Limited.

Lourie, J. A., Budd, G. and Anderson, H. R. (1992). Physiological adaptability in Papua New Guinea. In *Human Biology in Papua New Guinea: The Small Cosmos*, ed. R. D. Attenborough and M. P. Alpers. Oxford: Clarendon Press, pp. 268–80.

Macintyre, M. (2004). 'Thoroughly modern mothers': maternal aspirations and declining mortality on the Lihir islands, Papua. *Health Sociology Review* **13**, 43–53.

Maitland, K., Williams, A. I., Bennett, N. M., *et al.* (1996). The interaction *Plasmodium falciparum* and *P. vivax* in children on Espiritu Santo island, Vanuatu. *Transactions of the Royal Society of Tropical Medicine and Hygiene* **90**, 614–20.

Manning, L. A. and Ogle, G. D. (2002). Yaws in the periurban settlements of Port Moresby, Papua New Guinea. *Papua New Guinea Medical Journal* **45**, 206–12.

Mayxay, M., Pukrittayakamee, S., Newton, P. N. and White, N. J. (2004). Mixed-species malaria infections in humans. *Trends in Parasitology* **20**, 233–40.

McBride, W. J. (2005). HIV/AIDS in Papua New Guinea: an unfolding disaster? *Emergency Medicine Australasia* **17**, 304–6.

McMurray, C. (2004). *Morbidity and Mortality Patterns in the Pacific.* United Nations Economic and Social Commission for Asia and the Pacific. http://www.unescap.org/esid/psis/population/popseries/apss163/index.asp (accessed 12 September 2005).

Mgone, C. S., Koki, G., Paniu, M. M., *et al.* (1996).Occurrence of the erythrocyte band 3 (*AEI*) gene deletion in relation to malaria endemicity in Papua New Guinea. *Transactions of the Royal Society of Tropical Medicine and Hygiene* **90**, 228–31.

Mikloucho-Maclay, N. N. (1975). *New Guinea Diaries, 1871–1883.* Madang (PNG): Kristen Pres.

Ministry of Health (2001). *National Health Plan 2001–2010: Health Vision 2010.* Port Moresby: Ministry of Health, Independent State of Papua New Guinea.

Misch, K. A. (1988). Ischaemic heart disease in urbanized Papua New Guinea: an autopsy study. *Cardiology* **75**, 71–5.

Müller, I., Vounatsu, P., Allen, B. J. and Smith, T. (2001). Spatial patterns of child growth in Papua New Guinea and their relation to environment, diet, socio-economic status and subsistence activities. *Annals of Human Biology* **28**, 263–80.

Müller, I., Betuela, I. and Hide, R. L. (2002a). Regional patterns of birthweights in Papua New Guinea in relation to diet, environment and socio-economic factors. *Annals of Human Biology* **29**, 74–88.

Müller, I., Taime, J., Ibam, E., *et al.* (2002b). Complex patterns of malaria epidemiology in the highlands region of Papua New Guinea. *Papua New Guinea Medical Journal* **45**, 200–5.

Müller, I., Bockarie, M., Alpers, M. P. and Smith, T. (2003). The epidemiology of malaria in Papua New Guinea. *Trends in Parasitology* **19**, 253–9.

Müller, I., Tulloch, J., Marfurt, J., Hide, R. and Reeder, J. C. (2005). Malaria control in Papua New Guinea results in complex epidemiological changes. *Papua New Guinea Medical Journal*, **48**, 151–7.

Naraqi, S., Feling, B. and Leeder, S. R. (2003). Disease and death in Papua New Guinea. *Medical Journal of Australia* **178**, 7–8.

National AIDS Council Secretariat and Department of Health (2002). *HIV/AIDS Quarterly Report: March 2002*. Port Moresby: National AIDS Council Secretariat.

National Statistical Office (2002). *Papua New Guinea 2000 Census: Final Figures*. Port Moresby: National Statistical Office of Papua New Guinea.

 (2003a). *Papua New Guinea 2000 Census: National Report*. Port Moresby: National Statistical Office of Papua New Guinea.

 (2003b). *Recent Fertility and Mortality Indices and Trends in Papua New Guinea: A Report Based on the Analysis of 2000 Census Data*. Port Moresby: National Statistical Office of Papua New Guinea.

Natsuhara, K., Inaoka, T., Umezaki, M., *et al.* (2000). Cardiovascular risk factors of migrants in Port Moresby from the highlands and island villages, Papua New Guinea. *American Journal of Human Biology* **12**, 655–64.

Norgan, N. G. (1997). Human adaptability studies in Papua New Guinea: original aims, successes and failures. In *Human Adaptability: Past, Present and Future*, ed. S. J. Ulijaszek and R. Huss-Ashmore. Oxford: Oxford University Press, pp. 102–25.

Ogle, G. D. (2001). Type 2 diabetes mellitus in Papua New Guinea – an historical perspective. *Papua New Guinea Medical Journal* **44**, 81–7.

Ohtsuka, R. (1986). Low rate of population increase of the Gidra Papuans in the past: a genealogical-demographic analysis. *American Journal of Physical Anthropology* **71**, 13–23.

Owen, I. L. (2005). Parasitic zoonoses in Papua New Guinea. *Journal of Helminthology* **79**, 1–14.

Pawley, A. K., Attenborough, R. D., Golson, J. and Hide, R. L. (eds.) (2005). *Papuan Pasts: Cultural, Linguistic and Biological Histories of Papuan-Speaking Peoples*. Canberra: Pacific Linguistics, The Australian National University.

Peters, W. (1960). Studies on the epidemiology of malaria in New Guinea. *Transactions of the Royal Society of Tropical Medicine and Hygiene* **54**, 242–60.

Poka, H. and Duke, T. (2003). In search of pigbel: gone or just forgotten in the highlands of Papua New Guinea? *Papua New Guinea Medical Journal* **46**, 135–42.

Pollard, T. M. (1997). Environmental change and cardiovascular disease: a new complexity. *Yearbook of Physical Anthropology* **40**, 1–24.

Radford, A. J., van Leeuwen, H. and Christian, S. H. (1976). Social aspects in the changing epidemiology of malaria in the highlands of New Guinea. *Annals of Tropical Medicine and Parasitology* **70**, 11–23.

Riley, I. D. (1983). Population change and distribution in Papua New Guinea: an epidemiological approach. *Journal of Human Evolution* **12**, 125–32.

(2002). Pneumonia vaccine trials at Tari. *Papua New Guinea Medical Journal* **45**, 44–50.

Riley, I. D. and Lehmann, D. (1992). The demography of Papua New Guinea: migration, fertility and mortality patterns. In *Human Biology in Papua New Guinea: The Small Cosmos*, ed. R. D. Attenborough and M. P. Alpers. Oxford: Clarendon Press, pp. 67–92.

Riley, I. D., Lehmann, D. and Alpers, M. P. (1992). Acute respiratory infections. In *Human Biology in Papua New Guinea: The Small Cosmos*, ed. R. D. Attenborough and M. P. Alpers. Oxford: Clarendon Press, pp. 281–8.

Riley, J. C. (2001). *Rising Life Expectancy: A Global History*. Cambridge: Cambridge University Press.

Secretariat of the Pacific Community (2004). *Pacific Island Populations 2004*. Secretariat of the Pacific Community (Demography/Population Programme). http://www.spc.int/demog/ (accessed 12 September 2005).

Serjeantson, S. W., Board, P. G. and Bhatia, K. K. (1992). Population genetics in Papua New Guinea: a perspective on human evolution. In *Human Biology in Papua New Guinea: The Small Cosmos*, ed. R. D. Attenborough and M. P. Alpers. Oxford: Clarendon Press, pp. 198–233.

Sharp, P. T. (1982). Highlands malaria: malaria in Enga Province of Papua New Guinea. *Papua New Guinea Medical Journal* **25**, 253–60.

Sinnett, P. F. (1977). *The People of Murapin*. Faringdon: E. W. Classey.

Sinnett, P. F., Kevau, I. H. and Tyson, D. (1992). Social change and the emergence of degenerative cardiovascular disease in Papua New Guinea. In *Human Biology in Papua New Guinea: The Small Cosmos*, ed. R. D. Attenborough and M. P. Alpers. Oxford: Clarendon Press, pp. 373–86.

Smith, T., Genton, B., Baea, K., *et al.* (2001). Prospective risk of morbidity in relation to malaria infection in an area of high endemicity of multiple species of *Plasmodium*. *American Journal of Tropical Medicine and Hygiene* **64**, 262–7.

Specht, J. (2005). Revisiting the Bismarcks: some alternative views. In *Papuan Pasts: Cultural, Linguistic and Biological Histories of Papuan-Speaking Peoples*, ed. A. K. Pawley, R. D. Attenborough, J. Golson and R. L. Hide. Canberra: Pacific Linguistics, The Australian National University, pp. 235–88.

Spencer, M. (1999). *Public Health in Papua New Guinea 1870–1939*. Brisbane: Australian Centre for International and Tropical Health and Nutrition.

Spencer, T. E. T., Spencer, M. and Venters, D. (1974). Malaria vectors in Papua New Guinea. *Papua New Guinea Medical Journal* **17**, 22–30.

Spriggs, M. (1997). *The Island Melanesians*. Oxford: Blackwell.

Swadling, P. and Hide, R. L. (2005). Changing landscape and social interaction: looking at agricultural history from a Sepik-Ramu perspective. In *Papuan Pasts: Cultural, Linguistic and Biological Histories of Papuan-Speaking peoples*, ed. A. K. Pawley, R. D. Attenborough, J. Golson and R. L. Hide. Canberra: Pacific Linguistics, The Australian National University, pp. 289–327.

Swadling, P. and Hope, G. (1992). Environmental change in New Guinea since human settlement. In *The Naive Lands: Prehistory and Environmental Change in Australia and the Southwest Pacific*, ed. J. Dodson. Melbourne: Longman Cheshire, pp. 13–42.

Tefuarani, N., Sleigh, A. and Hawker, R. (2002). Congenital heart diseases: a future burden for Papua New Guinea. *Papua New Guinea Medical Journal* **45**, 175–7.

Teitelbaum, M. S. (1975). Relevance of demographic transition theory for developing countries. *Science* **188**, 420–5.

Temu, P. I. (1991). Adult medicine and the 'new killer diseases' in Papua New Guinea: an urgent need for prevention. *Papua New Guinea Medical Journal* **34**, 1–5.

Townsend, P. K. (1985). Infant mortality in the Saniyo-Hiowe population, Ambunti District, East Sepik Province. *Papua New Guinea Medical Journal* **28**, 177–82.

Tracer, D. P., Sturt, R. J., Sturt, A. and Braithwaite, L. M. (1998). Two decade trends in birth weight and early childhood growth in Papua New Guinea. *American Journal of Human Biology* **10**, 483–93.

Trowell, H. C. (1981). Hypertension, obesity, diabetes mellitus and coronary heart disease. In *Western Diseases: Their Emergence and Prevention*, ed. H. C. Trowell and D. P. Burkitt. London: Edward Arnold, pp. 3–32.

Trowell, H. C. and Burkitt, D. P. (eds.) (1981). *Western Diseases: Their Emergence and Prevention*. London: Edward Arnold.

Tulu, A. N. (1996). *Determinants of malaria transmission in the highlands of Ethiopia: the impacts of global warming on morbidity and mortality ascribed to malaria*. Ph.D. thesis, University of London.

Tyson, D. C. (1987). *An ecological analysis of child malnutrition in an Abelam community, Papua New Guinea*. Ph.D. thesis, Australian National University, Canberra.

Ulijaszek, S. J. (1995). Development, modernisation and health intervention. In *Health Intervention in Less Developed Nations*, ed. S. J. Ulijaszek. Oxford: Oxford University Press, pp. 82–136.

(1997). Modernization, nutritional status and hypertension in a rural Papua New Guinea population (abstract). *Annals of Human Biology* **24**, 281.

(2001). Secular trend in birthweight among the Purari delta population, Papua New Guinea. *Annals of Human Biology* **28**, 246–55.

Ulijaszek, S. J., Koziel, S. and Hermanussen, M. (2005). Village distance from urban centre as the prime modernization variable in differences in blood pressure and body mass index of adults of the Purari delta of the Gulf province, Papua New Guinea. *Annals of Human Biology* **32**, 326–38.

United Nations Children's Fund (2004a). *The State of the World's Children 2005: Childhood Under Threat*. New York: UNICEF.

(2004b). *Progress for Children*. New York: UNICEF.

van de Kaa, D. J. (1971). *The demography of Papua and New Guinea's indigenous population*. PhD thesis, Australian National University, Canberra.

Vines, A. P. (1970). *An Epidemiological Sample Survey of the Highlands, Mainland and Islands Regions of the Territory of Papua and New Guinea*. Port Moresby: Department of Public Health, Territory of Papua and New Guinea.

Ward, R. G. and Lea, D. A. M. (eds.) (1970). *An Atlas of Papua and New Guinea.* Port Moresby and Glasgow: University of Papua and New Guinea and Collins – Longman.

Webb, S. (1995). *Palaeopathology of Aboriginal Australians: Health and Disease across a Hunter-Gatherer Continent.* Cambridge: Cambridge University Press.

Wiesenfeld, S. L. (1967). Sickle-cell trait in human biological and cultural evolution. *Science* **157**, 1134–40.

Wigley, S. C. (1977). The first hundred years of tuberculosis in New Guinea. In *The Melanesian Environment*, ed. J. H. Winslow. Canberra: Australian National University Press, pp. 471–84.

   (1991). Tuberculosis and Papua New Guinea: the Australian connection. In *History of Tuberculosis in Australia, New Zealand and Papua New Guinea*, ed. A. J. Proust. Canberra: Brolga Press, pp. 103–17.

World Health Organization (2000). *The World Health Report 2000 – Health Systems: Improving Performance.* Geneva: World Health Organization.

World Health Organization Regional Office for the Western Pacific (2005). *Countries and Areas: Papua New Guinea.* World Health Organization. http://www.wpro.who.int/countries/png/ (accessed 12 September 2005).

Yen, D. E. (2005). Reflection, refraction and recombination. In *The Sweet Potato in Oceania: A Reappraisal*, ed. C. Ballard, P. Brown, R. M. Bourke and T. Harwood. Pittsburgh and Sydney: Ethnology Monographs and Oceania Monographs, pp. 181–7.

Zimmerman, P. A., Woolley, I., Masinde, G. L. *et al.* (1999). Emergence of FY*A[null] in a *Plasmodium vivax* – endemic region of Papua New Guinea. *Proceedings of the National Academy of Sciences of the USA* **96**, 13973–7.

Zimmerman, P. A., Patel, S. S., Maier, A. G., Bockarie, M. J. and Kazura, J. W. (2003). Erythrocyte polymorphisms and malaria parasite invasion in Papua New Guinea. *Trends in Parasitology* **19**, 250–2.

Index